BRAIN OSCILLATIONS, SYNCHRONY, AND PLASTICITY

BRAIN OSCILLATIONS, SYNCHRONY, AND PLASTICITY

Basic Principles and Application to Auditory-Related Disorders

Jos J. Eggermont
Departments of Physiology and Pharmacology, and Psychology
University of Calgary, Canada

ELSEVIER

ACADEMIC PRESS
An imprint of Elsevier

Academic Press is an imprint of Elsevier
125 London Wall, London EC2Y 5AS, United Kingdom
525 B Street, Suite 1650, San Diego, CA 92101, United States
50 Hampshire Street, 5th Floor, Cambridge, MA 02139, United States
The Boulevard, Langford Lane, Kidlington, Oxford OX5 1GB, United Kingdom

Library of Congress Cataloging-in-Publication Data
A catalog record for this book is available from the Library of Congress

British Library Cataloguing-in-Publication Data
A catalogue record for this book is available from the British Library

ISBN 978-0-12-819818-6

For information on all Academic Press publications
visit our website at https://www.elsevier.com/books-and-journals

Publisher: Nikki Levy
Acquisitions Editor: Melanie Tucker
Editorial Project Manager: Hilary Carr
Production Project Manager: Kiruthika Govindaraju
Cover Designer: Mark Rogers

Typeset by SPi Global, India

Contents

Preface

One may take two approaches to the topic of oscillations, synchrony and networks in the brain especially in the study of neurological and psychiatric disorders. These are the molecular approach and the systems-imaging approach. The first is reductionist in kind, and the second is synthetic. I have chosen to emphasize the latter approach, partially because I am more familiar with it but mostly because it is noninvasive and widely applied to humans. Some of the imaging findings are fortunately highly correlated with molecular results, especially in Schizophrenia and Alzheimer's disease, so I included these molecular data. I also touch upon the role of various neurotransmitters in the generation of, for example gamma oscillations.

Oscillations in the brain are organized hierarchically, with frequencies from less than 1 Hz to more than 100 Hz. These are named according to common frequency bands 1–4 Hz (delta waves), 4–8 Hz (theta waves), 8–12 Hz (alpha waves), 13–30 Hz (beta waves), and >40 Hz (gamma waves). The approximately 0.1 Hz very slow waves feature not only in the electroencephalography (EEG) but also as the dominant fluctuation is the functional magnetic resonance imaging (fMRI). The higher frequency rhythms can only be seen in the EEG and magnetoencephalography (MEG).

These synchronized local rhythms are either local and in that case with frequencies in the beta and gamma range, but they are more long ranging for the lower frequencies, particularly the delta and theta frequency range and to a lesser extent the alpha range. The low frequencies are capable of organizing higher frequency oscillations whose amplitude or phase is locked to the phase of the lower frequency ones. In this way, higher frequency ones (e.g., gamma) may be synchronized between far away regions in the brain. The alpha rhythm, which is the dominant oscillatory activity in awake adults, probably originates from an interaction between local negative feedback circuits in the thalamus and the cortex. When the feedback circuits in the thalamus work in isolation, they can produce the faster sleep spindles. Faster EEG rhythms such as beta and gamma are produced by feedback circuits between cortical pyramidal neurons and local inhibitory inter neurons. Some types of theta may originate in medial temporal circuits and are projected to other parts of the cortex through the hippocampus. Physiological EEG rhythms, thus, arise in different neural circuits and probably have different functions. The alpha rhythm may play a role in attention and semantic memory but may also have a modulatory effect on other rhythms. Beta rhythms are often related to motor processes and gamma rhythms to perception and consciousness. The theta rhythm has been associated with working memory.

The basic question is how the low frequency—including the very slow—waves become synchronized or transmitted over large distances? There are two possibilities: through the nerve tracts as trains of rhythmic action potentials and/or as volume-conducted activity. Notably, the very slow fluctuations in the fMRI, basically reflecting blood oxygen level-dependent (BOLD) fluctuations, are correlated or anticorrelated across large brain areas. Their electrophysiological equivalents are very slow changes in membrane potentials, caused by synchronous synaptic potentials and measurable as local field potentials (LFPs) and very slow waves in the EEG.

How can all this be measured? Connections between parts of the brain can be measured as existing nerve tracts using diffusion tensor imaging with tractography. These are called structural connections and serve the basic action potential-based transmission between neurons and networks. Connections can also be estimated from correlated changes in the various networks, such as the very slow BOLD responses. These correlations reflect functional connectivity that is nondirectional and may be more far ranging as structural connectivity by way of common inputs. A structural MRI technique, voxel-based morphology, allows the measurement of gray matter density and volume, has been applied in relation to disorders such as tinnitus and hearing loss, and may diagnose atrophy of specific brain regions.

At the scalp level, correlations in voltage or current time series between scalp electrodes or magnetic sensors can be used to infer functional networks at the sensor level. A better way is to first estimate intracranial sources (sometimes called equivalent dipoles) from the scalp activity and then estimate correlations between the local source voltage- or current-time series. Given the large number of frequency bands detectable in the EEG or MEG, correlations at the sensor or source level are often conducted after filtering the time series into applicable frequency bands, from delta to gamma. This may result in different connectivities between identified networks, the latter often based on fMRI. Especially, but not exclusively, the gamma band activity correlates strongly with the networks identified based on the correlations between the very slow BOLD responses.

The brain is a complex system, built up from modules (e.g., auditory cortex) into local networks (i.e., the auditory system network) that connect with among others salience-, attention-, and default-mode networks. Less than two decades ago, this network approach to the human brain was pioneered using fMRI and based on long-distance correlation of the very low frequency fluctuations in the BOLD response. Networks are made from structural connections and from synchrony of brain oscillations, both local and global.

I will focus my attention, but not exclusively, to networks involving the auditory system or will interact therewith during perception and cognition. In this book, I will also discuss various forms of noninvasive neuromodulation, such as transcranial magnetic, direct current, alternating current stimulation, and focused ultrasound, that have been successfully used to modulate endogenous cortical rhythms in normal participants. Promising results were reported in neurological diseases, but so far negative ones in tinnitus.

Do the various neurological disorders have substrates in the neural networks and their connectivity? As we have seen, distributed neural systems may communicate through synchronization of their oscillatory activity. Synchronization here refers to the existence of a consistent relationship between activity patterns of two or more spatially separated neuronal groups. It often implies that there is a consistent relation between the phases of the oscillatory activity of two brain regions. Synchronization in different frequency bands has been associated with various cognitive functions and the integration of information in the healthy brain. In neurological disorders, this process of functional integration can become disrupted and gives rise to various symptoms of cognitive dysfunction. The fact that structural hubs, particularly in the default mode network, play such a striking role in brain network controllability may further help to explain the growing body of evidence indicating that diseases can preferentially target hub areas.

The roadmap for this book involves exploring and expanding the questions and statements outlined above. The book consists of four divisions: I Oscillations (Chapters 1–4), II Synchrony (Chapters 5 and 6), III Plasticity (Chapters 7–9), and IV Disorders (Chapters 10–12).

In Chapters 1–3, I emphasize the different roles of the various brain oscillations in perception, cognition, and execution or action in the auditory domain when compared with the visual and somatosensory ones. In contrast, cognitive and action aspects are typically less sense dependent and are discussed independent of modality (Chapter 4). Are long-distance functional connections exclusively formed by the synchrony of low-frequency oscillations or do they need action potentials traveling along the long nerve tracts? (Chapter 5). How does the combination of low- and high-frequency rhythms allow perceptual and cognitive networks to form? (Chapter 6). Can brain rhythms, neural synchrony, and networks be modulated and changed by external stimuli? (Chapter 8). How are brain networks formed and change across the lifespan? (Chapter 9). Network hubs appear to be energy demanding and are the most vulnerable part of the networks and may become damaged in hearing loss (Chapter 7), neurological, and psychiatric disorders. Finally, I will describe the communalities in the vulnerability of hubs connecting networks in tinnitus, dyslexia, autism, schizophrenia, and Alzheimer's disease, all of which show in addition disorders in auditory perception (Chapters 10–12).

Thanks to Hilary Carr, Elsevier's content manager, for timely and consistent helping with problems in the mastering of their new content and manuscript-managing system. Thanks also to my wife Mary for listening, reading, and commenting on some parts of the book. Having a University office—courtesy of the Psychology Department—with a view of the Rocky Mountains, although occasionally distracting, provided an inspirational and much appreciated working environment.

CHAPTER

1

Brain rhythms, neural synchrony, and networks in health and disease

The brain is a complex system, built up from modules (e.g., auditory cortex) forming local networks (e.g., the auditory network) that connect with, among others, salience-, attention- and default mode-networks. Less than two decades ago, this network view was pioneered using functional magnetic resonance imaging (fMRI) and based on long-distance voxel-by-voxel correlation of the ultra-low frequency (~ 0.1 Hz) fluctuations in the blood oxygen level dependent (BOLD) response. Networks are based on structural connections, but can also be inferred from local and global synchrony of brain oscillations—defining functional connections. In this book, I emphasize on networks involving the auditory, speech and language systems or those that interact therewith during perception and cognition. I will discuss neural synchrony at various levels from the action potential level, through local field potentials (LFPs) giving rise to oscillations recordable from the scalp. I will conclude with changes in oscillations and long-range connections that are vulnerable and are affected in a large number of psychiatric and neurological disorders, exemplified in this chapter by Parkinson's disease. Changes found in tinnitus, dyslexia, autism, schizophrenia, and Alzheimer's disease—disorders with a strong auditory component—are described in detail in Chapters 10, 11, and 12.

1.1 Brain rhythms

1.1.1 Introduction

Brain rhythms are semiperiodic changes in the amplitude and frequency of the electroencephalogram (EEG) and magnetoencephalogram (MEG). The nomenclature for classifying neural oscillations in the EEG

https://doi.org/10.1016/B978-0-12-819818-6.00009-1

originated in a study by Berger (1929). He first identified and named the dominant rhythm; alpha (8–12 Hz). It has over the years been expanded to include delta (1–4 Hz), theta (4–8 Hz), beta (15–30 Hz), gamma (30–80 Hz), and the high-frequency gamma range (> 80 Hz). In addition, an ultraslow (~ 0.1 Hz) rhythm involving large areas of the neocortex is manifested in the blood oxygen level dependent (BOLD) response. Correlations in this rhythm across voxels initially resulted in the identification of the default mode network (DMN) observed in the absence of specific sensory inputs or tasks (Fox and Raichle, 2007). The DMN comprises several subnetworks that span the cortex from frontal to parietal. The DMN and the dorsal attention network, as well as the DMN and the task-positive network show anticorrelated fluctuations (Hu et al., 2017; Fig. 1.1).

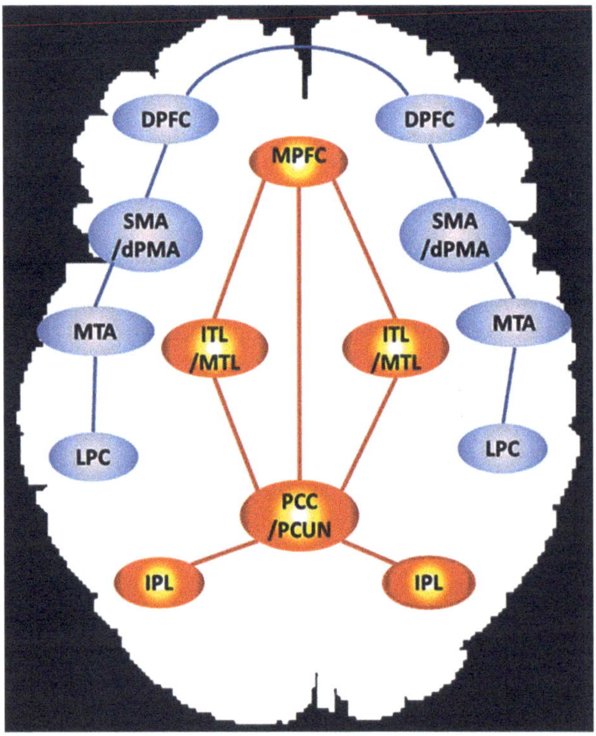

FIG. 1.1 Simplified illustration of the principal regions for the two anticorrelated networks, the DMN *(red)* and TPN *(blue)*. *DMN*, default mode network; *DPFC*, dorsolateral PFC; *dPMA*, dorsal premotor areas; *IPL*, inferior parietal lobe; *ITL/MTL*, inferior/mesial temporal lobe; *LPC*, lateral parietal cortex; *MPFC*, medial prefrontal cortex; *MTAs*, middle temporal area; *POC/PCUN*, posterior cingulate cortex and adjacent precuneus; *SMA*, supplementary motor area; *TPN*, task positive network. *Reprinted with permission from Hu, M.-L., Zong, X.-F., Mann, J.J., Zheng, Liao, Y-H., Li, Z.-C., et al., 2017. A review of the functional and anatomical default mode network in schizophrenia. Neurosci. Bull. 33 (1), 73–84. Copyright 2017. Springer.*

1.1.2 Properties of the dominant electroencephalogram rhythms

1.1.2.1 Ultraslow fluctuations

Ultraslow (< 0.1 Hz) voltage fluctuations in EEG recordings (Steriade et al., 1993) are a likely electrophysiological correlate of the spontaneous BOLD response fluctuations in fMRI. These slow EEG fluctuations are themselves correlated with changes in the power of higher frequency rhythms in the EEG (Buzsáki and Draguhn, 2004). The finding that slow fluctuations may modulate the power of higher-frequency activity has been referred to as hierarchical nesting of frequencies and is an important principle in brain organization. More rapid fluctuations reflected in beta or gamma activity coordinate neuronal activity across small spatial scales, whereas slower fluctuations in EEG power, for example, delta and theta bands, are required for long-range coordination. There is evidence for such a frequency-space trade-off in the brain (Buzsáki et al., 2012; Steriade, 2006). Organized by increasing oscillation frequency content, one recognizes (Wang, 2010) first of all the very slow (< 1 Hz) oscillations of UP and DOWN states, discovered by Steriade et al. (1993). During these large-amplitude slow EEG oscillations, virtually all cell types in the cerebral cortex repetitively switch between the UP state (where neurons are depolarized and fire at a low rate) and the DOWN state (where neurons are hyperpolarized to below − 80 mV and do not fire). Luczak et al. (2007) simultaneously recorded populations of 50–200 cortical neurons in layer V of anesthetized and awake rats. Each neuron displayed a bursting spike pattern during the spontaneous UP states. In auditory cortex, these bursts had a stereotyped make up, that was similar between spontaneous UP state onset periods and stimulus onset periods, but the rate and also precise relative timing of spikes varied between stimuli (Luczak et al., 2013), suggesting a modulatory function of stimulation on the intrinsic bursting pattern.

1.1.2.2 Delta

The delta rhythm has the lowest frequency (1–4 Hz) and highest amplitude of the classical brainwaves. Delta waves begin to appear in stage 3: sleep state, and by stage 4 nearly all EEG activity is dominated by delta waves. In the awake state, delta waves may be involved in integration of cerebral activity with homeostatic processes, and are prominent in several disorders such as tinnitus (Chapter 10). Considerable evidence also suggests that delta oscillations are implicated in attention, salience detection, motivation, and subliminal perception (Knyazev, 2012; Chapter 4). Delta oscillations also appear to drive temporal prediction accuracy (Arnal et al., 2015) and successful speech understanding (Steinmetzger and Rosen, 2017; Chapter 3).

1.1.2.3 Theta

Two types of theta rhythm have been described. The hippocampal theta rhythm (Bland, 1986) and cortical theta rhythms have no clear relationship. In human EEG studies, the term theta refers to frequency components in the 4–8 Hz range, regardless of their source. Prestimulus theta-band power in the hippocampus and neocortex predicts whether a memory for a subsequent event will be formed. Theta band activities over the midfrontal cortex are involved in cognitive control (Cavanagh and Frank, 2014).

1.1.2.4 Alpha

The alpha rhythm (8–12 Hz) represents large-amplitude EEG oscillations that spontaneously occur in awake and relaxed states, particularly with eyes closed. Thalamocortical alpha oscillations are characteristic in sensory and motor systems, which often but incorrectly are attributed to an "idling" brain state. Alpha band oscillations are increasingly viewed as an active mechanism that controls sensory information processing as a function of cognitive relevance (Sadaghiani et al., 2012) by two mechanisms—inhibition and gating of sensory information (Klimesch, 2012). In humans, the specific members of the 8–12 Hz family rhythms are known as alpha oscillations of the visual system, mu rhythms of the sensorimotor system, and tau rhythms of the auditory system (Lehtelä et al., 1987), but generally I will all label them as alpha.

1.1.2.5 Beta

The second identified brain rhythm, called beta, was initially observed in the primary motor cortex by Hans Berger in 1931 (Miller et al., 2012). Recent studies indicate that beta oscillations (15–30 Hz) are not limited to the motor system, but more generally are involved in sensorimotor integration and top-down signaling (Alavash et al., 2017) as well as in valance judgments (Lipsman et al., 2014).

1.1.2.6 Gamma

The third identified brain rhythm, gamma, was prominently evoked by sensory stimuli in the olfactory bulb (Adrian, 1950), and later studied extensively in that structure by Freeman (1978). Gamma band (30–80 Hz) synchronization is found in many cortical areas, such as the visual system in the cat, from retina to cortex (Gray et al., 1989) and the auditory system in humans (Pantev et al., 1991). It is induced by different stimuli or tasks, and is related to several cognitive processes (Fries, 2009). Periodic events in the frequency range above the gamma band are referred to as high-frequency oscillations.

1.1.2.7 Occurrence in cortical areas and layers

In summary, high-frequency beta and gamma oscillations are related to local processing, while the rhythm of low-frequency oscillations at delta, theta, and alpha frequencies provides the timing structure for long-range communication (Buzsáki and Watson, 2012). Scheeringa et al. (2016) showed that alpha, beta, and gamma oscillations represent activity in different cortical layers in the human brain. Gamma-band EEG power was positively correlated with the BOLD signal in the superficial layers, beta-power was negatively correlated to deep layer BOLD, and alpha-power was correlated with both deep and superficial layer BOLD responses. Groppe et al. (2013) characterized intracortically recorded—electrocorticography (ECoG)—oscillations commonly observed over particular cortical areas when an individual is awake and immobile with eyes closed. They found substantial variability in the dominant frequencies and substantial overlap thereof across lateral central, temporal, and frontal areas. Surprisingly, theta was the most dominant type of oscillation found, whereas alpha was largely limited to parietal and occipital regions. Beta was prominent over the middle frontal gyrus, and the pars opercularis. Gamma/high gamma activity was too infrequent and variable across individuals to be reliably characterized. These findings support the importance of theta oscillations in general cortical function, suggest that alpha activity is primarily related to sensory processing and attention, and demonstrate that beta networks extend far beyond primary sensorimotor regions (Groppe et al., 2013).

1.1.3 Interaction of brain rhythms

The human cortex consists of approximately 100 billion neurons (Herculano-Houzel, 2009), quite a bit more than Adrian (1936) guessed as a basis for the following percipient statement: "Fortunately, however, the cortex need not be regarded as a system of several million independent units. The cells act in groups, small or large, and sometimes there is a fairly simple coordinated wave spreading over a wide area. Activity of this kind occurs only in special conditions, but it gives rise to relatively simple potential changes, and if very many units are involved the potentials are large–large enough to record in man through the unopened skull." Buzsáki et al. (2013) noted that the study of brain rhythms is among the fastest growing fields in human systems neuroscience. This may be because these brain rhythms form a hierarchical system that guides spiking-neural activity within and across circuits at multiple time scales. It is tempting to indicate the parallel with the hypothesized quantum-mechanical pilot waves that guide the propagation of elementary particles (Farmelo, 2019). Slow oscillations are particularly suited to set a functional bias throughout a large population of neurons, whereas oscillations at intermediary frequencies, such as in the theta and alpha ranges, are optimal to modulate, or to gate, the transfer of information across specific populations (Lopes da Silva, 2013). Examples are the role of theta in connecting the hippocampal formation and associated cortical areas (Mizuseki et al., 2009) and alpha in the transfer of information in thalamocortical systems (Amzica and Lopes da Silva, 2011). Oscillations at the higher frequencies, in the beta and gamma range, often engage relatively discrete populations in the transfer of specific information (Freeman, 2003) within and between neural networks (Chapter 4).

1.2 Neural synchrony

1.2.1 A note on correlation, synchronization, and resonance

In the broadest sense, *correlation* is any statistical association between two variables, though in common usage it most often refers to how these variables (e.g., neural-firing rate, spike-firing times, or oscillation frequency amplitude or phase) have a linear relationship with each other (Chapter 5). *Synchronization* occurs between neural populations and ensures that data streams (e.g., spike trains and oscillations) are received and transmitted correctly. At the level of such neural assemblies, synchronized activity of large numbers of neurons (network nodes) can give rise to macroscopic oscillations (e.g., alpha), which can be observed in the EEG. Oscillatory activity in groups of neurons generally arises from feedback connections between the neurons that result in the synchronization of their firing patterns. The interaction between neurons can give rise to oscillations at a different frequency than the firing frequency of individual neurons. In neuroscience, synchronization and correlation are often used in interchanging fashion. In physics, synchrony is only used when the oscillating objects are weakly coupled. This will nearly always be the case for cortical neural networks (Abeles, 1982). A *resonant* frequency is a natural frequency of vibration (oscillation) determined by the parameters of the oscillating object. Some of the implications of resonant frequencies are as follows: (1) It is easy to get an object to oscillate (synchronized with) at its resonant frequencies and difficult at other frequencies. (2) An oscillating object will respond at its resonant frequencies contained in a complex excitation (e.g., speech) and oscillate at some of those frequencies (reflecting e.g., syllabic, phonemic, and prosodic rates) essentially "filtering out" other frequencies present in the exciting signal (Doelling et al., 2019).

1.2.2 Synchrony at the action potential level

Action potential-firing synchrony forms the basis for neural assembly formation. Hebb (1949) defined a neural assembly as "a diffuse structure comprising cells in the cortex and diencephalon, capable of acting briefly as a closed system, delivering facilitation to other such systems." Hebb suggested that, depending on functional requirements, individual brain cells could participate in different cell assemblies and be involved in multiple computations. In other words (Varela et al., 2001), neural assemblies may be defined as distributed local networks of neurons transiently linked by reciprocal dynamic connections.

The quantification of the correlation in spiking activity occurring between pairs of such neurons is crucial in defining membership of these neurons in clusters or assemblies. Clusters typically refer to locally connected groups of neurons, whereby assemblies typically comprise spatially diffuse groups consisting of many local clusters. Neuron clusters as probed by microelectrode arrays are local and typically do not cross cortical-area boundaries. As a consequence, the spatial size of the clusters rarely exceeds a few mm and on average is ~ 1 mm, that is, they are roughly the size of fMRI voxels. Clusters are largely confined by cortical area boundaries and are affected by stimulus type (Eggermont, 2006). In the example shown here, common periodicities in spike firing, driven by spindling activity in the LFP, were removed by coherence-based correction. Uncorrected correlation gives rise to much larger clusters, which suggests that spike-spike correlations form subsets of oscillation-based clusters.

One could therefore ask: is spike-pair correlation sufficient or even meaningful as a basis for connectivity and clustering? Schneidman et al. (2006) showed for the vertebrate retina, that weak neuron-pair cross-correlations may coexist with strong population behavior in the responses of ten or more neurons. They found that this collective behavior could be predicted quantitatively by the observed pairwise correlations and did not need the assumption of higher-order interactions. Schneidman et al. (2006) noted that: "the success of a model that includes only pairwise interactions provides an enormous simplification in our description of the network. This may be important not only for our analysis, but also for the brain." Whether this also applies to the auditory (and visual) cortex remains to be investigated. Cross-correlation is sufficient for linear systems. It is also sufficient for nonlinear systems with only feed-forward connections, but it is not in the case of nonlinear systems with feedback connections. Because the thalamo-cortical system is obviously nonlinear and has strong feedback connections, one potentially needs to extent this analysis by incorporating nonlinear relations, for example, by using mutual information as a basis for clustering (Eggermont, 2006).

1.2.3 Synchrony at the local field potential level

Buzsáki and Draguhn (2004) viewed the synchronous activity of oscillating networks as the critical "middle ground" linking individual neuron activity to behavior. Coherent oscillations may play a role in coordination and communication between neural populations that are simultaneously engaged in a cognitive process. Oscillation-based synchrony is arguably an energy-efficient physical mechanism for temporal coordination. Buzsáki and Draguhn (2004) noted that network oscillations bias input selection, temporally link neurons into assemblies, and facilitate synaptic plasticity. These are mechanisms that cooperatively support temporal representation and long-term consolidation of information. Neural spike-pair correlation implies synchronization of action potentials on some time scale, which can occur with or without oscillations (Eggermont, 1990; Fig. 1.2A). The function of synchronous oscillations can be embedded in the more general perspective of correlation between spike times of neurons (Chapter 5).

Buschman et al. (2012) simultaneously recorded from multiple electrodes in dorsolateral prefrontal cortex (DLPFC) while monkeys switched between two behavioral tasks, i.e., responded to color versus orientation of

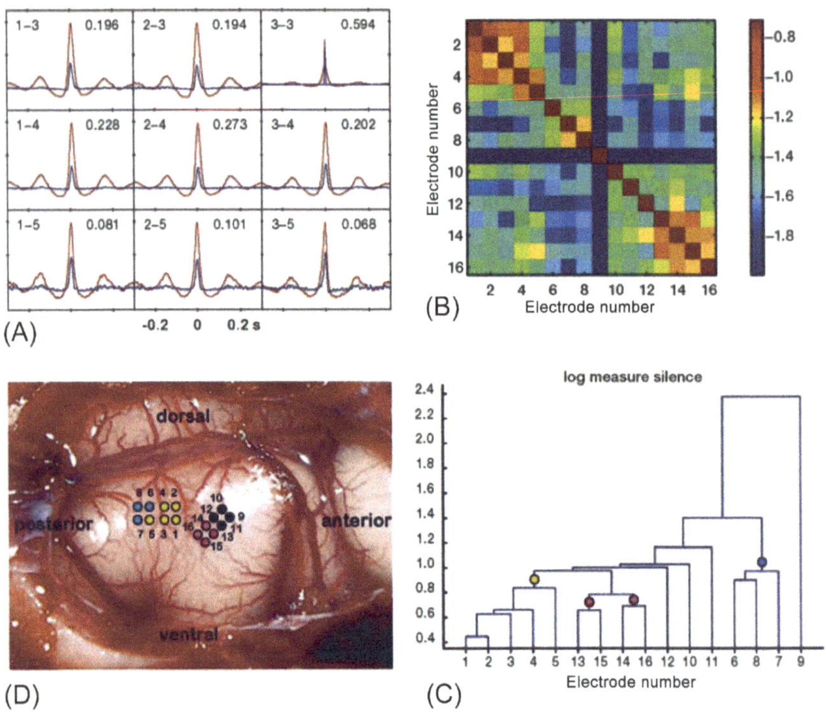

FIG. 1.2 Clustering procedure. The basis for the clustering procedure is the pairwise cross-correlation (A) shown here for a few pairs in their traditional form *(red curves)* and the corrected form *(blue curves)*. Each panel shows the correlograms scaled on their own extremes. The time base is from – 0.3 to 0.3 s. The full pairwise correlation matrix is shown for the spikes recorded on 16 electrodes in (B) where the base 10 logarithm of the corrected correlation is shown (*color bar* indicates these values). Based on the pairwise correlation matrix a hierarchical clustering procedure performed on the 16 electrodes (C) results in four multielectrode clusters indicated at the branching points with a colored dot. These *colors* are used to identify the positions of the electrodes belonging to the same cluster on the surface of the auditory cortex (D). Electrodes are numbered on the cortical surface. All electrodes are in primary auditory cortex. *From Eggermont, J.J., 2006. Properties of correlated neural activity clusters in cat auditory cortex resemble those of neural assemblies. J. Neurophysiol. 96, 746–764.*

objects. Oscillatory synchronization of LFPs formed neural ensembles that showed rule-specific increases in synchrony at beta (19–40 Hz) frequencies between electrodes. In addition, individual DLPFC neurons synchronized to the LFP ensemble corresponding to the current task (color versus orientation). Furthermore, the ensemble encoding of the behaviorally dominant orientation task showed increased alpha (6–16 Hz) synchrony when preparing to apply the alternative (weaker) color task. This suggested to Buschman et al. (2012) that: "beta-frequency synchrony selects the relevant rule ensemble, while alpha-frequency synchrony deselects (inhibits) a stronger, but currently irrelevant, ensemble." From simultaneous surface-EEG measurements and intracortical recordings of LFPs in the primary visual cortex of behaving monkeys, Musall et al. (2014) found that trial-by-trial fluctuations in EEG power could be explained by a linear combination of LFP power and inter-electrode temporal synchrony. This effect was observed in both stimulus and spontaneous conditions and was particularly strong in the gamma range (30–100 Hz).

In humans, synchrony between LFPs often in combination with individual neuron action potentials, may be measured from intracortical recordings, for example, ECoGs in patients with intractable epilepsy. Several of such studies have bridged the gap from single neurons to content-specific patterns in the LFP, showing both category and stimulus specific neuronal firing and LFP responses at low (< 10 Hz) and high (> 30 Hz) frequencies in the human temporal lobe. Thus, activity in the theta and gamma bands appears particularly relevant for neural representations (van Gerven et al., 2013).

1.2.4 Long-distance synchrony

Distant neural systems may communicate through synchronization of their activity (Varela et al., 2001). At the level of oscillatory brain activity, it often implies that there is a consistent relation between the phase of the oscillatory activity of two brain regions. Large-scale synchronization link neural sub-networks, which are relatively far apart in the brain (> 1 cm), and with transmission delays > 8–10 ms over polysynaptic pathways. In this case, phase-locking synchrony must reside in distant connections, maintained either by corticocortical fibers or thalamocortical reciprocal pathways. These pathways correspond to feed-forward and feedback connections (Varela et al., 2001). The transmission of information along nerve tracts relies on action potentials. The action potential input from one area to another area results in excitatory and inhibitory postsynaptic potentials, which, if temporally overlapping across a neuronal population, would give rise to LFPs that can be measured extracellularly. So, in addition to local processes that affect membrane potentials, long-range input from other brain areas also affects locally measured LFPs.

It appears that inhibitory interneurons provide an important mechanism for synchronization of the LFP oscillations of remote neural populations (Buzsáki and Wang, 2012).

The first studies showing long-range neural synchrony of gamma oscillations (60–120 Hz) were conducted in the animal visual system (Eckhorn et al., 1988; Gray et al., 1989). The source of these oscillations may in part be due to oscillations in the retina, which are strongly correlated with those in the visual thalamus and cortex (Castelo-Branco et al., 1998). This inter- and intra-cortical synchrony was considered as a mechanism for perceptual binding (Eckhorn et al., 1988; Gray et al., 1989) however, this idea remains controversial (Harris and Gordon, 2015; Merker, 2013). Harris and Gordon (2015) noted that although synchrony may not fully encode binding, visual tasks are indeed associated with local and long-range synchrony. If not for perceptual binding, what are the benefits of synchrony? For instance, exact spike timing controls how well information is transmitted from one brain region to another (Riehle et al., 1997). In addition, synchronous presynaptic inputs to a neuron may summate more effectively and lead to an action potential in a postsynaptic neuron (Bernander et al., 1994). Fries (2005) proposed that long-range coherence of oscillations ensures that a given region provides input in a temporal window in which the downstream target is receptive. This is incorporated in the 'communication-by-coherency' hypothesis (Fries, 2009), which states that communication between regions that have similar gamma-oscillation phase is more effective than communication between regions with gamma oscillations that are out of phase (Womelsdorf et al., 2007). Because the amplitude and frequency of gamma oscillations can change rapidly, both spontaneously and in response to changing stimulus variables such as visual contrast (Ray and Maunsell, 2010), the precise nature of the stimulus determines whether these gamma oscillations are coherent across different parts of the same brain region (Ahmed and Cash, 2013). However, see Chapter 4 for differences between the role of gamma in visual and auditory systems.

1.3 Brain networks

1.3.1 From oscillations and synchrony to brain networks

The combined structural, functional and effective connections between neural networks in the brain are commonly called the "connectome." Structural networks describe anatomical connectivity by nerve fiber tracts, which tends to be relatively stable but plastic over longer time scales. By contrast, functional networks are derived from statistical connections of time series data, such as trains of action-potentials, spontaneous EEG or MEG time series or slow fluctuations in the BOLD response, which are often represented as linear cross-correlations. Effective connectivity is estimated from connection strengths and the causal connection directionality between distinct macroscopic-scale functional regions from EEG/MEG and fMRI (Stam, 2010), or between single neurons or neuron clusters (Gourévitch and Eggermont, 2007). The amplitude of MEG/EEG signals depends on the size of the activated neuronal population, the degree of synchronization of the LFPs, the strengths of these LFPs and the spatial orientations of their source dipoles. In contrast to action potentials, LFPs and consequently MEG/EEG signals reflect both suprathreshold and subthreshold oscillations, the latter of which are not accompanied by action potentials (Schnitzler and Gross, 2005). A composite view of group-level networks organized into functional domains (Salman et al., 2019) is shown in Fig. 1.3. Note the locations of the anticorrelated default mode and attention networks.

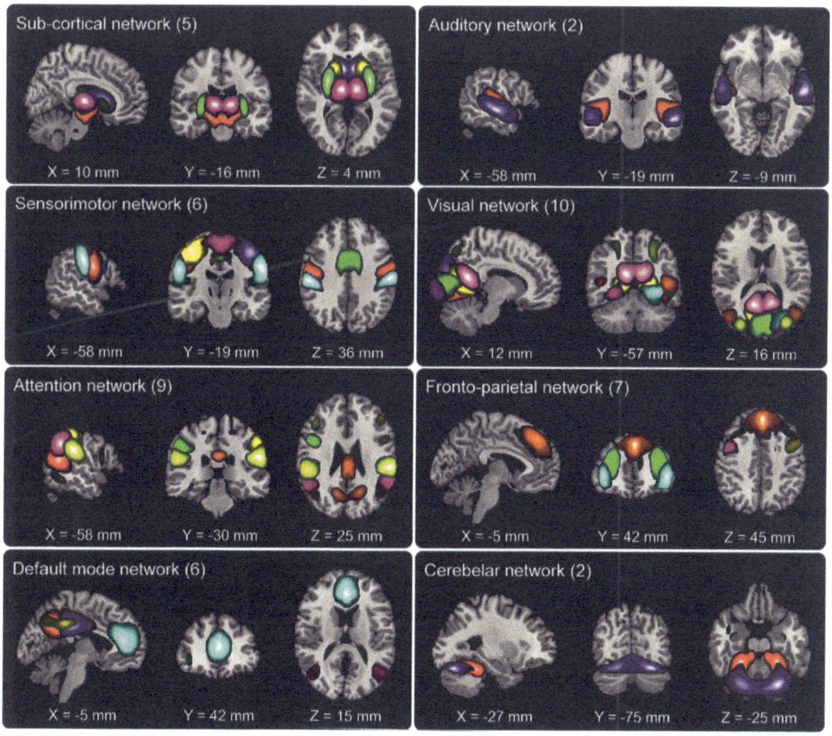

FIG. 1.3 Composite view of 47 group-level networks grouped into functional domains: 5 subcortical (SC), 2 auditory (AUD), 10 visual (VIS), 6 sensorimotor (SM), 9 attention (ATT), 7 frontoparietal (FRN), 6 default mode (DMN), and 2 cerebellar (CB) networks. Intensity of color represents *z*-scores. Component labels and peak activation coordinates can be found in previous work. *From Salman, M.S., Du, Y., Lin, D., Fu, Z., Fedorov, A., Damaraju, E., et al., 2019. Group ICA for identifying biomarkers in schizophrenia: 'adaptive' networks via spatially constrained ICA show more sensitivity to group differences than spatio-temporal regression. NeuroImage Clin. 22, 101747. Open access.*

Bressler and Richter (2015) hypothesized that top-down neocortical signals in the beta band convey behavioral context to low-level sensory neurons (Fig. 1.4). Top-down beta-frequency directed influences from extra-striate visual cortex (ESC) to primary visual cortex (V1), and from posterior parietal cortex (PPC) to ESC has been demonstrated. Top-down beta-frequency directed influences from PPC to V1 were shown in monkey visual cortex (Spaak et al., 2012). Beta-frequency activity also shows bidirectional influences between PPC and prefrontal cortex (PFC). Bressler and Richter (2015) predicted a beta-frequency directed influence from PFC to ESC, and speculated that large-scale distributed networks, self-organized at the highest hierarchical levels, are the source of top-down signals in the neocortex.

1.3.2 Graph theory to represent and quantify connectivity

A powerful technique to represent and analyze the connectivity- or correlation-matrix (cf. Fig. 1.2B) between brain areas is based on graph theory (Bullmore and Sporns, 2009; Stam, 2014). Graph theory represents networks by way of nodes and edges. In brain networks the nodes can be individual neurons, or they can be EEG or MEG activity at particular electrodes or field sensor locations, estimated intracranial sources, or voxels in fMRI studies of neural networks. The sensor and source locations and voxels represent large collections of neurons, depending on voxel size or spatial resolution as in EEG and MEG. A standard fMRI voxel has about the size of 10 mm^3 and contains on the order of 10^5 neurons and 10^9 synapses (Honey et al., 2009). The network edges in brain studies represent axonal connections (structural), or cross-correlation coefficients (functional) between fMRI, EEG, and MEG time series, or spike trains.

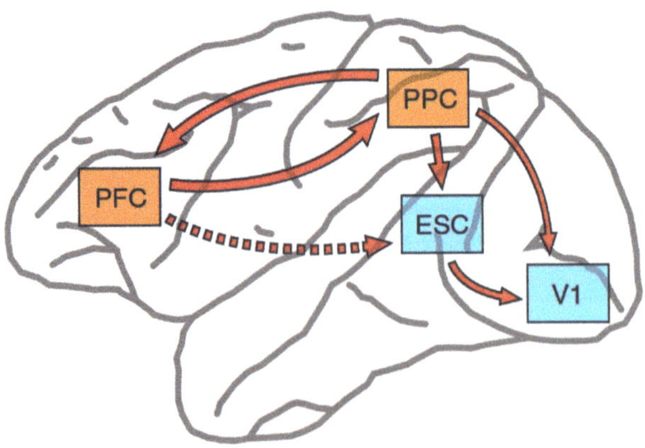

FIG. 1.4 Interareal beta-frequency directed functional connectivity in the neocortex. *ESC*, extra-striate cortex; *PFC*, prefrontal cortex; *PPC*, posterior parietal cortex; *V1*, primary visual cortex. *Reprinted with permission from Bressler, S.L., Richter, C.G., 2015. Interareal oscillatory synchronization in top-down neocortical processing. Curr. Opin. Neurobiol, 31, 62–66. Copyright 2015. Elsevier.*

Diffusion tensor imaging (DTI) was used to reconstruct the white matter connections of the brain network, with the strength of the connections defined as the level of myelination of the nerve tracts (Van den Heuvel and Fornito, 2014). They used graph measures to probe the organization of the human connectome, particularly path length, clustering, and the presence of hubs and modules (Fig. 1.5). A hub is a network node with a

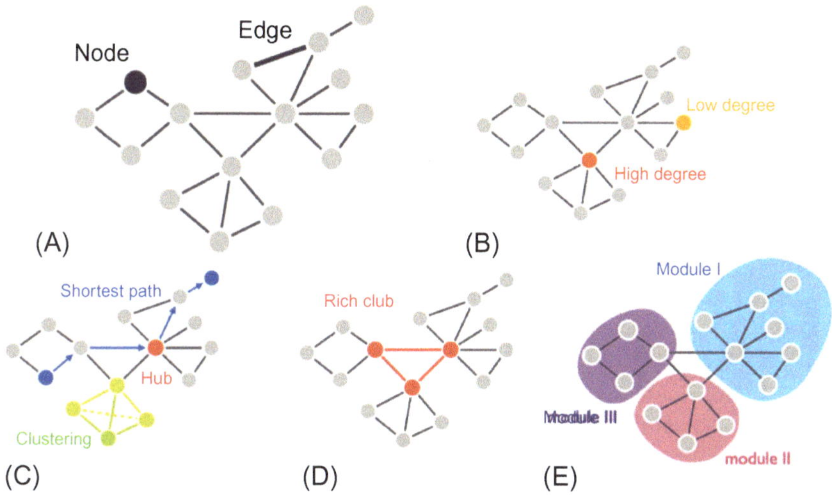

FIG. 1.5 A graph and examples of basic graph attributes. (Panel A) A network can be mathematically described as a graph, consisting of a collection of "nodes," and a collection of "edges" describing the collection of connections between the nodes of the network. When the number of network edges is sparse the layout of the edges describes the topological organization of the network. Basic network attributes include a network's degree distribution (Panel B), clustering, characteristic path length and modular structure. The degree of a node describes its number of connections to other nodes in the graph. Nodes with a high degree that take on a central position in the graph are often referred to as "hubs." (Panel C) The "clustering" of a graph provides insight into the level of local connectedness of the graph, describing how strong the connected neighbors of a node are connected themselves. In (C) the node depicted in *green* has three connected neighbors *(light green nodes)* together sharing two of the possible three edges between them (the dotted edge is missing), resulting in a clustering coefficient of two-thirds. The "characteristic path length" of a graph describes the average number of edges that have to be crossed to travel between any two nodes in the network. In (C) the shortest path between the *two blue nodes* is depicted by the *blue arrows*, which defines the shortest path traveling along four edges and three nodes of the network. The *red node* reflects an exemplary hub node, displaying a high degree, a short global path length and being involved in a large number of communication paths in the network. (D) Besides being individually rich in connectivity, high degree nodes can display an above chance level of interconnectivity, forming a densely connected "rich club." A closely related concept is the formation of a "core" within networks. (E) The modular structure of the network describes the tendency of nodes to form local connected clusters that share a relative high level of connectivity with each other than with other regions. Here, three modules can be distinguished, consisting of subsets of nodes predominantly connected to nods within their own module. *Reprinted with permission from Van den Heuvel, M.P., Fornito, A., 2014. Brain networks in schizophrenia. Neuropsychol. Rev. 24, 32–48. Copyright 2014. Elsevier.*

large number of connections (high degree of edges) to other nodes. The characteristic path length (degree) of a brain graph is defined as the average number of steps required to travel between nodes of the network (e.g., brain regions) and is often taken as a (inverse) metric of communication efficiency or information integration of a region (Stam and van Straaten, 2012). Modularity is calculated as the fraction of edges that fall within the given communities of a network and the expected fraction if the connections were distributed at random across the network. Hubs connect modules.

Networks in the brain are typically characterized by fMRI in the resting state (spontaneous activity; Fox and Raichle, 2007). A note of caution for the auditory system in the resting state is a difficult concept because the MRI environment is ubiquitous with loud scanner noise (Scarff et al., 2004). Langers and Melcher (2011), Yakunina et al. (2016), and Andoh et al. (2017) validated the use of continuous sampling, that is, with continuous scanner noise for resting-state studies of auditory networks that are of course activated by the scanner noise. Despite this, interleaved silent steady-state sampling resulted in more robust patterns of functional connectivity within the auditory network compared with continuous sampling during scanning (Andoh et al., 2017).

What constitutes the human auditory network, subcortical and cortical? Fig. 1.6 traces the human lemniscal auditory pathway, which is tonotopically organized, from cochlea to cortex (Saenz and Langers, 2014). The primary auditory cortex (A1) occupies the posterior part of Heschl's gyrus, whereas the secondary auditory areas (A2) are located on the planum temporale and the superior bank of the posterior superior temporal gyrus. The white-matter pathways, projecting from cortical auditory regions that are involved in higher-order auditory processing, are the arcuate fasciculus and the interhemispheric pathways in the posterior part of the corpus callosum. The arcuate fasciculus is involved in both sound and language processing, and connects caudal temporal cortex and inferior parietal cortex to locations in the frontal lobe (Curcic-Blake et al., 2017).

Hearing loss is associated with changes in the connectome (Chapter 7). The putative effects of hearing loss on attention, memory, and other cognitive functions are typically attributed to changes in prefrontal cortex (Eggermont, 2019). These changes may not be independent of changes in auditory cortex, especially when one realizes that many neurons in these brain areas are connected to the auditory cortex through the arcuate fasciculus, and also respond to sound. An fMRI study (Langers and Melcher, 2011) showed multiple networks of acoustically responsive brain centers outside the classic auditory cortex. These included four classically "nonauditory" areas, namely cinguloinsular cortex, mediotemporal limbic lobe, basal ganglia, and posterior orbitofrontal cortex (Fig. 1.7A).

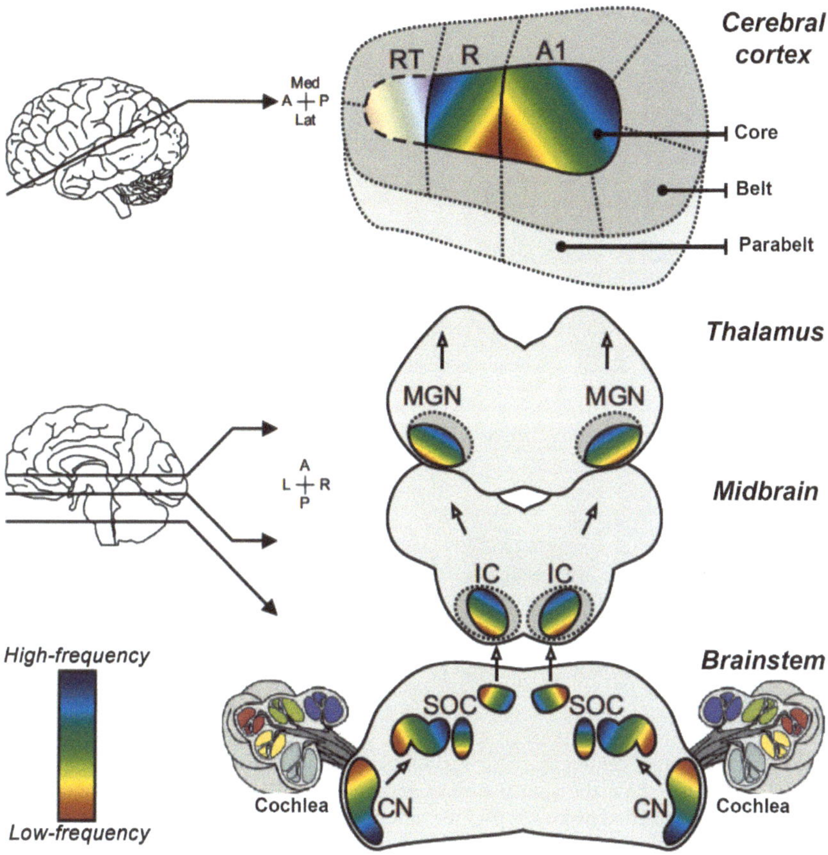

FIG. 1.6 The central auditory pathway. A. All nuclei that form part of the classical lemniscal auditory pathway are tonotopically organized. These include various subdivisions of the cochlear nucleus (CN), superior olivary complex (SOC), inferior colliculus (IC), and medial geniculate nucleus (MGN). In the auditory cerebral cortex in the superior part of the temporal lobe, expected divisions of core, belt, and parabelt are based on the nonhuman primate model of auditory cortical organization. Human neuroimaging consistently shows at least two primary tonotopic gradients (high-to-low-to-high) in the auditory cortex, homologous to primary fields A1 and rostral held R in the monkey cortex. In some primate studies, a third rostrotemporal held RT is delineated but neuroimaging evidence for a similar held in humans is sparse. *With permission from Saenz, M., Langers, D.R.M., 2014. Tonotopic mapping of human auditory cortex. Hear. Res. 307, 42–52. Copyright 2014. Elsevier.*

Functional connectivity analyses demonstrated coordinated activity between the involved brain structures. Resting-state fMRI revealed largely similar networks (Fig. 1.7B) meaning that although they are not auditory stimulus dependent, they may still be driven by the high spontaneous activity in the auditory system (Kennedy et al., 1978).

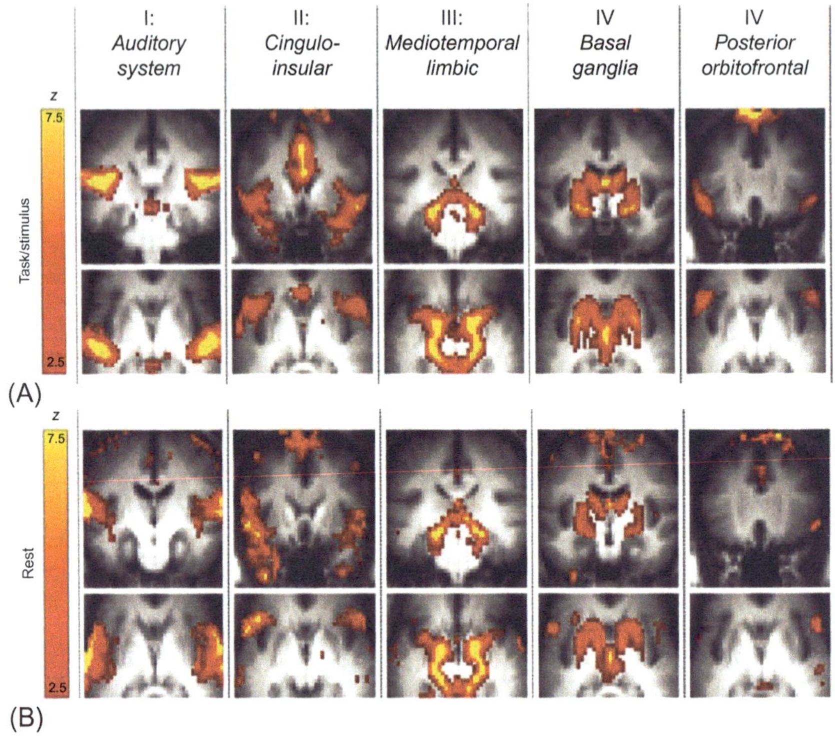

FIG. 1.7 Five brain centers showing sound-related responses (A) are also present at rest (B). (A) The maps show the spatial distribution of component amplitude, scaled to unit variance and thresholded at $z=2.5$. (B) Five independent components that closely matched those of components I–V in Panel A were independently extracted from the resting-state runs. *From Langers, D.R.M., Melcher, J.R., 2011. Hearing without listening: functional connectivity reveals the engagement of multiple nonauditory networks during basic sound processing. Brain Connect. 1 (3), 233–244. Open access.*

1.4 Network changes in neurological disorders

Synchronization in different EEG-frequency bands within and between brain areas has been associated with various cognitive functions and the integration of information in the healthy brain (Chapter 4). Abnormal synchronization has been associated with several neural disorders, including epilepsy (Truccolo et al., 2014) schizophrenia (Ford et al., 2007), dementia (Stam and van Straaten, 2012), and, in particular, basal ganglia disorders such as Parkinson's disease (PD). Here, we briefly illustrate some brain-rhythm and MRI-based network changes in PD.

1.4.1 Changes in brain oscillations

Whole head 151-channel MEG, eyes-closed, resting-state oscillatory activity was recorded in a group of PD patients with varying disease duration and severity (including early-stage patients) as well as in age-matched controls (Stoffers et al., 2007). Early-stage PD patients had widespread slowing in the frequencies of resting-state MEG activity relative to controls. Changes included a widespread increase in theta and low alpha power, as well as a loss of beta power over all but the frontal regions of interest (ROIs) and a loss of gamma power over all but the right occipital ROI.

1.4.2 Changes in network connectivity

One of the central questions in neuroscience is how brain networks are organized under normal conditions and how this connectivity breaks down in neurological disease (Stam and van Straaten, 2012). Graph-based network studies of the brain demonstrated that healthy brains self-organize towards so-called "small-world networks" characterized by a combination of dense local connectivity and a few critical long-distance connections (Chapter 6; Reijneveld et al., 2007).

Using structural and functional MRI, Chen et al. (2015) investigated the whole-brain resting-state functional connectivity patterns in PD. They found that: "The majority of the most discriminative functional connections were located within or across the DMN, cinguloopercular (CON) and frontal-parietal networks (FPN) and the cerebellum" (Fig. 1.8). These resting-state network alterations might play important roles in the pathophysiology of this disease.

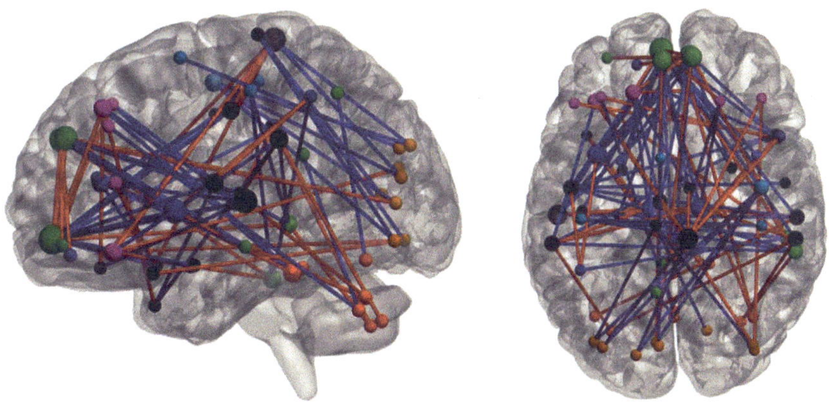

FIG. 1.8 Functional connections and change in PD. Regions (nodes) are color-coded by category [cinguloopercular network (CON), *blue*; default mode network (DMN), *green*; cerebellum, *red*; visual network, *orange*; sensorimotor network, *cyan*; frontal-parietal network, *purple*; and others, *black*] and size-coded by weight. *Red lines* (edges) represent increased functional connections, and *blue lines* represent decreased functional connections. *PD*, Parkinson's disease. *From Chen, Y., Yang, W., Long, J., Zhang, Y., Feng, J., Li, Y., et al., 2015. Discriminative analysis of Parkinson's disease based on whole-brain functional connectivity. PLoS One 10 (4), e0124153. https://doi.org/10.1371/journal.pone.0124153. Open access.*

1.5 Summary

Brain rhythms with oscillation frequencies from ~0.1 to >100Hz may be recorded from the scalp, the lowest are also found in fluctuations of the BOLD signal of the fMRI. Low frequency (1–15Hz) rhythms underlie long distance functional connections, whereas higher frequency oscillations reflect local connectivity. Because the higher frequency rhythms are nested with the lower rhythms, they may also be synchronized between distant brain regions. The so formed functional and structural connections within and between brain regions are commonly studied by graph-theoretical methods leading to concepts as small-world, hubs, and neural modules. Many neurological disorders are characterized by dys-synchrony within and between hubs and brain modules. Here, we illustrate this for Parkinson's disease.

References

Abeles, M., 1982. Local Cortical Circuits. An Electrophysiological Study. Springer Verlag, Berlin.

Adrian, E.D., 1936. The electrical activity of the cortex. Proc. R. Soc. Med. 29, 197–200.

Adrian, E., 1950. The electrical activity of the mammalian olfactory bulb. Electroencephalogr. Clin. Neurophysiol. 2, 377–388.

Ahmed, O.J., Cash, S.S., 2013. Finding synchrony in the desynchronized EEG: the history and interpretation of gamma rhythms. Front. Integr. Neurosci. 7, 58. https://doi.org/10.3389/fnint.2013.00058.

Alavash, M., Daube, C., Wöstmann, M., Brandmeyer, A., Obleser, J., 2017. Large-scale network dynamics of beta-band oscillations underlie auditory perceptual decision-making. Netw. Neurosci. 1 (2), 166–191. https://doi.org/10.1162/netn_a_00009.

Amzica, F., Lopes da Silva, F.H., 2011. Cellular substrates of brain rhythms. In: Schomer, D.L., Lopes da Silva, F.H. (Eds.), Niedermeyer's Electroencephalography: Basic Principles, Clinical Applications and Related Fields, sixth Edition. Philadelphia, Lippincott Williams & Wilkins, pp. 33–64.

Andoh, J., Ferreira, M., Leppert, I.R., Matsushit, R., Pike, B., Zatorre, R.J., 2017. How restful is it with all that noise? Comparison of interleaved silent steady state (ISSS) and conventional imaging in resting-state fMRI. NeuroImage 147, 726–735.

Arnal, L.H., Doelling, K.B., Poeppel, D., 2015. Delta-beta coupled oscillations underlie temporal prediction accuracy. Cereb. Cortex 25, 3077–3085.

Berger, H., 1929. Über das Elektroenkephalogramm des Menschen. Arch. Psychiatr. Nervenkr. 87, 527–570.

Bernander, O., Koch, C., Usher, M., 1994. The effect of synchronized inputs at the single neuron level. Neural Comput. 6, 622–641. https://doi.org/10.1162/neco.1994.6.4.622.

Bland, B.H., 1986. The physiology and pharmacology of hippocampal formation theta rhythms. Prog. Neurobiol. 26, 1–54.

Bressler, S.L., Richter, C.G., 2015. Interareal oscillatory synchronization in top-down neocortical processing. Curr. Opin. Neurobiol. 31, 62–66.

Bullmore, E., Sporns, O., 2009. Complex brain networks: graph theoretical analysis of structural and functional systems. Nat. Rev. Neurosci. 10, 186–198.

Buschman, T.J., Denovellis, E.L., Diogo, C., Bullock, D., Miller, E.K., 2012. Synchronous oscillatory neural ensembles for rules in the prefrontal cortex. Neuron 76, 838–846.

Buzsáki, G., Draguhn, A., 2004. Neuronal oscillation in cortical networks. Science 304, 1926–1929.

Buzsáki, G., Wang, X.-J., 2012. Mechanisms of gamma oscillations. Annu. Rev. Neurosci. 35, 203–225.

Buzsáki, G., Watson, B.O., 2012. Brain rhythms and neural syntax: implications for efficient coding of cognitive content and neuropsychiatric disease. Dialogues Clin. Neurosci. 14, 345–367.

Buzsáki, G., Anastassiou, C.A., Koch, C., 2012. The origin of extracellular fields and currents—EEG, ECoG, LFP and spikes. Nat. Rev. Neurosci. 13 (6), 407–420.

Buzsáki, G., Logothetis, N., Singer, W., 2013. Scaling brain size, keeping timing: evolutionary preservation of brain rhythms. Neuron 80, 751–764.

Castelo-Branco, M., Neuenschwander, S., Singer, W., 1998. Synchronization of visual responses between the cortex, lateral geniculate nucleus, and retina in the anesthetized cat. J. Neurosci. 18 (16), 6395–6410.

Cavanagh, J.F., Frank, M.J., 2014. Frontal theta as a mechanism for cognitive control. Trends Cogn. Sci. 18 (8), 414–421.

Chen, Y., Yang, W., Long, J., Zhang, Y., Feng, J., Li, Y., et al., 2015. Discriminative analysis of Parkinson's disease based on whole-brain functional connectivity. PLoS One 10 (4). https://doi.org/10.1371/journal.pone.0124153.

Curcic-Blake, B., Ford, J.M., Hubl, D., Orlov, N.D., Sommer, I.E., Waters, F., et al., 2017. Interaction of language, auditory and memory brain networks in auditory verbal hallucinations. Prog. Neurobiol. 148, 1–20.

Doelling, K.B., Assaneo, M.F., Bevilacqua, D., Pesaran, B., Poeppel, D., 2019. An oscillator model better predicts cortical entrainment to music. Proc. Natl. Acad. Sci. U. S. A. 116 (20), 10113–10121.

Eckhorn, R., Bauer, R., Jordan, W., Brosch, M., Kruse, W., Munk, M., et al., 1988. Coherent oscillations: a mechanism of feature linking in the visual cortex? Biol. Cybern. 60, 121–130.

Eggermont, J.J., 1990. The Correlative Brain. Theory and Experiment in Neural Interaction. Springer Verlag, Berlin.

Eggermont, J.J., 2006. Properties of correlated neural activity clusters in cat auditory cortex resemble those of neural assemblies. J. Neurophysiol. 96, 746–764.

Eggermont, J.J., 2019. The Auditory Brain and Age-related Hearing Impairment. Academic Press, London, UK.

Farmelo, G., 2019. The Universe Speaks in Numbers. Basic Books, New York City, NY.

Ford, J.M., Krystal, J.H., Mathalon, D.H., 2007. Neural synchrony in schizophrenia: from networks to new treatments. Schizophr. Bull. 33 (4), 848–852.

Fox, M.D., Raichle, M.E., 2007. Spontaneous fluctuations in brain activity observed with functional magnetic resonance imaging. Nat. Rev. Neurosci. 8, 700–711.

Freeman, W.J., 1978. Spatial properties of an EEG event in the olfactory bulb and cortex. Electroencephalogr. Clin. Neurophysiol. 44, 586–605.

Freeman, W.J., 2003. The wave packet: an action potential for the 21st century. J. Integr. Neurosci. 2 (1), 3–30.

Fries, P., 2005. A mechanism for cognitive dynamics: neuronal communication through neuronal coherence. Trends Cogn. Sci. 9, 474–480.

Fries, P., 2009. Neuronal gamma-band synchronization as a fundamental process in cortical computation. Annu. Rev. Neurosci. 32, 209–224.

Gourévitch, B., Eggermont, J.J., 2007. Evaluating information transfer between auditory cortical neurons. J. Neurophysiol. 97, 2533–2543.

Gray, C.M., König, P., Engel, A.K., Singer, W., 1989. Oscillatory responses in cat visual cortex exhibit inter-columnar synchronization which reflects global stimulus properties. Nature 338, 334–337.

Groppe, D.M., Bickel, S., Keller, C.J., Jain, S.K., Hwang, S.T., Harden, C., et al., 2013. Dominant frequencies of resting human brain activity as measured by the electrocorticogram. NeuroImage 79, 223–233.

Harris, A.Z., Gordon, J.A., 2015. Long-range neural synchrony in behavior. Annu. Rev. Neurosci. 38, 171–194.

Hebb, D.O., 1949. The Organization of Behavior. John Wiley & Sons, Inc, New York.
Herculano-Houzel, S., 2009. The human brain in numbers: a linearly scaled-up primate brain. Front. Hum. Neurosci. 3, 31. https://doi.org/10.3389/neuro.09.031.2009.
Honey, C.J., Sporns, O., Cammoun, L., Gigandet, X., Thiran, J.P., Meuli, R., et al., 2009. Predicting human resting-state functional connectivity from structural connectivity. Proc. Natl. Acad. Sci. U. S. A. 106 (6), 2035–2040.
Hu, M.-L., Zong, X.-F., Mann, J.J., Zheng, J.-J., Liao, Y.-H., Li, Z.-C., et al., 2017. A review of the functional and anatomical default mode network in schizophrenia. Neurosci. Bull. 33 (1), 73–84.
Kennedy, C., Sakurada, O., Shinohara, M., Jehle, J., Sokoloff, L., 1978. Local cerebral glucose utilization in the normal conscious macaque monkey. Ann. Neurol. 4, 293–301.
Klimesch, W., 2012. Alpha-band oscillations, attention, and controlled access to stored information. Trends Cogn. Sci. 16 (12), 606–617.
Knyazev, G.G., 2012. EEG delta oscillations as a correlate of basic homeostatic and motivational processes. Neurosci. Biobehav. Rev. 36, 677–695.
Langers, D.R.M., Melcher, J.R., 2011. Hearing without listening: functional connectivity reveals the engagement of multiple nonauditory networks during basic sound processing. Brain Connect. 1 (3), 233–244.
Lehtelä, L., Salmelin, R., Hari, R., 1987. Evidence for reactive magnetic 10-Hz rhythm in the human auditory cortex. Neurosci. Lett. 222, 111–114.
Lipsman, N., Kaping, D., Westendorff, S., Sankar, T., Lozano, A.M., Womelsdorf, T., 2014. Beta coherence within human ventromedial prefrontal cortex precedes affective value choices. NeuroImage 85, 769–778.
Lopes da Silva, F., 2013. EEG and MEG: relevance to neuroscience. Neuron 80, 1112–1128.
Luczak, A., Barthó, P., Marguet, S.L., Buzsáki, G., Harris, K.D., 2007. Sequential structure of neocortical spontaneous activity in vivo. Proc. Natl. Acad. Sci. U. S. A. 104 (1), 347–352.
Luczak, A., Bartho, P., Harris, K.D., 2013. Gating of sensory input by spontaneous cortical activity. J. Neurosci. 33 (4), 1684–1695.
Merker, B., 2013. Cortical gamma oscillations: the functional key is activation, not cognition. Neurosci. Biobehav. Rev. 37, 401–417.
Miller, K.J., Hermes, D., Honey, C.J., Hebb, A.O., Ramsey, N.F., Knight, R.T., et al., 2012. Human motor cortical activity is selectively phase-entrained on underlying rhythms. PLoS Comput. Biol. 8 (9), e1002655. https://doi.org/10.1371/journal.pcbi.1002655.
Mizuseki, K., Sirota, A., Pastalkova, E., Buzsáki, G., 2009. Theta oscillations provide temporal windows for local circuit computation in the entorhinal-hippocampal loop. Neuron 64, 267–280.
Musall, S., von Pföstl, V., Rauch, A., Logothetis, N.K., Whittingstall, K., 2014. Effects of neural synchrony on surface EEG. Cereb. Cortex 24, 1045–1053.
Pantev, C., Makeig, S., Hoke, M., Galambos, R., Hampson, S., Gallen, C., 1991. Human auditory evoked gammaband magnetic fields. Proc. Natl. Acad. Sci. U. S. A. 88, 8996–9000.
Ray, S., Maunsell, J.H., 2010. Differences in gamma frequencies across visual cortex restrict their possible use in computation. Neuron 67, 885–896.
Reijneveld, J.C., Ponten, S.C., Berendse, H.W., Stam, C.J., 2007. The application of graph theoretical analysis to complex networks in the brain. Clin. Neurophysiol. 118, 2317–2331.
Riehle, A., Grün, S., Diesmann, M., Aertsen, A., 1997. Spike synchronization and rate modulation differentially involved in motor cortical function. Science 278, 1950–1953.
Sadaghiani, S., Scheeringa, S.R., Lehongre, K., Morillon, B., Giraud, A.-L., D'Esposito, M., et al., 2012. Alpha-band phase synchrony is related to activity in the fronto-parietal adaptive control network. J. Neurosci. 32 (41), 14305–14310.
Saenz, M., Langers, D.R.M., 2014. Tonotopic mapping of human auditory cortex. Hear. Res. 307, 42–52.
Salman, M.S., Du, Y., Lin, D., Fu, Z., Fedorov, A., Damaraju, E., et al., 2019. Group ICA for identifying biomarkers in schizophrenia: 'adaptive' networks via spatially constrained

ICA show more sensitivity to group differences than spatio-temporal regression. NeuroImage Clin. 22, 101747.

Scarff, C.J., Dort, J.C., Eggermont, J.J., Goodyear, B.G., 2004. The effect of MR scanner noise on auditory cortex activity using fMRI. Hum. Brain Mapp. 22, 341–349.

Scheeringa, R., Koopmans, P.J., van Mourik, T., Jensen, O., Norris, D.G., 2016. The relationship between oscillatory EEG activity and the laminar-specific BOLD signal. Proc. Natl. Acad. Sci. U. S. A. 113 (24), 6761–6766.

Schneidman, E., Berry II, M.F., Segev, R., Bialek, W., 2006. Weak pairwise correlations imply strongly correlated network states in a neural population. Nature 440, 1007–1012.

Schnitzler, A., Gross, J., 2005. Normal and pathological oscillatory communication in the brain. Nat. Rev. Neurosci. 6, 285–296.

Spaak, E., Bonnefond, M., Maier, A., Leopold, D.A., Jensen, O., 2012. Layer-specific entrainment of gamma-band neural activity by the alpha rhythm in monkey visual cortex. Curr. Biol. 22, 2313–2318.

Stam, C.J., 2010. Use of magnetoencephalography (MEG) to study functional brain networks in neurodegenerative disorders. J. Neurol. Sci. 289, 128–134.

Stam, C.J., 2014. Modern network science of neurological disorders. Nat. Rev. Neurosci. 15 (10), 683–695.

Stam, C.J., van Straaten, E.C.W., 2012. The organization of physiological brain networks. Clin. Neurophysiol. 123, 1067–1087.

Steinmetzger, K., Rosen, S., 2017. Effects of acoustic periodicity and intelligibility on the neural oscillations in response to speech. Neuropsychologia 95, 173–181.

Steriade, M., 2006. Grouping of brain rhythms in corticothalamic systems. Neuroscience 137, 1087–1106.

Steriade, M., Contreras, D., Curró Dossi, R., Nuñez, A., 1993. The slow (<1 Hz) oscillation in reticular thalamic and thalamocortical neurons: scenario of sleep rhythm generation in interacting thalamic and neocortical networks. J. Neurosci. 13 (6), 3284–3299.

Stoffers, D., Bosboom, J.L.W., Deijen, J.B., Wolters, E.C., Berendse, H.W., Stam, C.J., 2007. Slowing of oscillatory brain activity is a stable characteristic of Parkinson's disease without dementia. Brain 130, 1847–1860.

Truccolo, W., Ahmed, O.J., Harrison, M.T., Eskandar, E.N., Cosgrove, G.R., Madsen, J.R., et al., 2014. Neuronal ensemble synchrony during human focal seizures. J. Neurosci. 34 (30), 9927–9944.

Van den Heuvel, M.P., Fornito, A., 2014. Brain networks in schizophrenia. Neuropsychol. Rev. 24, 32–48.

van Gerven, M., Maris, E., Sperling, M., Sharan, A., Litt, B., Anderson, C., et al., 2013. Decoding the memorization of individual stimuli with direct human brain recordings. NeuroImage 70, 223–232.

Varela, F., Lachaux, J.P., Rodriguez, E., Martinerie, J., 2001. The brainweb: phase synchronization and large-scale integration. Nat. Rev. Neurosci. 2, 229–239.

Wang, X.-J., 2010. Neurophysiological and computational principles of cortical rhythms in cognition. Physiol. Rev. 90, 1195–1268.

Womelsdorf, T., Schoffelen, J.M., Oostenveld, R., Singer, W., Desimone, R., Engel, A.K., et al., 2007. Modulation of neuronal interactions through neuronal synchronization. Science 316, 1609–1612.

Yakunina, N., Kim, T.S., Tae, W.S., Kim, S.S., Nam, E.C., 2016. Applicability of the sparse temporal acquisition technique in resting-state brain network analysis. AJNR Am. J. Neuroradiol. 37, 515–520.

CHAPTER

2

The gamma, beta, and theta rhythms and nested oscillations

2.1 Introduction

Recently, evidence for plasticity in oscillatory coupling in theta, beta, and gamma bands has been obtained in memory circuits consisting of the hippocampus and its connected areas. This suggests a role in memory processing, through enhanced interregional neural communication, in the entorhinal-hippocampal and prefrontal-hippocampal memory circuits (Igarashi, 2015). Thus it makes sense to combine these three rhythms in this chapter.

Initially the gamma rhythm was extensively studied in the olfactory system (Freeman, 1978) and a decade later was taken up by Eckhorn et al. (1988) and the Singer group (Gray et al., 1989) for the visual cortex. They demonstrated gamma synchrony locally for multiunit activity in the primary visual cortex (V1), later between responses in V1 and secondary visual cortex (V2) and ultimately showing neural synchrony across visual cortices in both hemispheres (Engel et al., 1991a,b). Crucially, these synchronies were only manifested for stimulation with moving bars. This led to the idea of perceptual binding (Singer, 1993). We now know that increased gamma power also correlates with increased attention, faster reaction times, and that it precedes the initiation of movement.

Beta rhythms are predominantly found in the frontoparietal cortex, and have been identified with a sensorimotor function and are attenuated during voluntary movements. Beta correlates with decision speed and is crucial for perceptual decision-making.

Theta allows local communication across brain areas in the frontal areas. Theta may be used in cognitive control. Theta is involved in rhythmic auditory perception by active chunking of a continuous stream of speech into segments of ~ 200 ms. In the hippocampus the first strong evidence for theta-gamma coupling was found; this was limited to gamma frequencies below 90 Hz. I will first treat the roles of individual rhythms in some detail and then present the findings for theta-gamma coupling.

https://doi.org/10.1016/B978-0-12-819818-6.00014-5

2.2 The gamma rhythm

2.2.1 Gamma rhythms and neurotransmitters

2.2.1.1 Contributing neurons

Gamma rhythms (30–80 Hz) are ubiquitous in many brain regions, particularly in the hippocampus and neocortex. Cortical interneurons are classified into several subtypes that contribute to cortical oscillatory activity. Parvalbumin (PV)-expressing inhibitory interneurons are involved in oscillations of local field potentials (LFPs) in the gamma-frequency range. Kuki et al. (2015) found that PV cells play important roles in persistence of the UP state (Chapter 1) and in the balance between gamma and beta oscillations across cortical layers. Salkoff et al. (2015) found that both pyramidal and inhibitory neurons in entorhinal cortex receive large, coherent inhibitory synaptic inputs from spike firing in the gamma frequency range. These result from the recurrent and divergent/convergent nature of local pyramidal cell-fast-spiking interneuron anatomical connections in the entorhinal network. In addition, fast-spiking interneurons receive large, phasic excitatory synaptic inputs immediately before spike generation followed shortly by synaptic inhibition. This supports the principal-interneuron gamma generation model (Tiesinga and Sejnowski, 2009) and reveals how the synaptic connectivity between excitatory and inhibitory neurons supports the generation of LFP-gamma oscillations and spike-firing synchrony. To further explore the role of PV interneurons, Perrenoud et al. (2016) recorded the intracellular activity of pyramidal and PV-expressing neurons in the visual cortex of awake mice while acquiring LFPs to monitor gamma activity. They found that gamma activity arises when PV-expressing neurons synchronize their output in response to a correlated input, reflecting the general activation of the local cortical network. This did not require visual input. In contrast the output of pyramidal neurons only became entrained to gamma activity when the mice were visually stimulated. Thus gamma activity synchronizes pyramidal neurons specifically when the cortex is engaged in processing external inputs (Welle and Contreras, 2016).

2.2.1.2 Involvement of neurotransmitters in human cortex

Can excitatory and inhibitory activity be localized in human cortex and related to the co-occurring gamma activity? Measurements using magnetic resonance spectroscopy (MRS) in different cortical fields have been applied with substantially different results. Muthukumaraswamy et al. (2009) combined MRS to quantify the resting-state gamma-aminobutyric acid (GABA) concentration in single voxels of the visual cortex, with magnetoencephalography (MEG) to measure visual-stimulus-induced gamma oscillations, and functional magnetic resonance imaging (fMRI) to measure the blood oxygen level-dependent (BOLD) response to a simple visual grating stimulus. They demonstrated that across individuals the gamma frequency, but not amplitude, was positively correlated with the resting-state GABA concentration in visual cortex. In addition the magnitude of BOLD response was inversely correlated with resting-state GABA concentration and with gamma frequency.

Contrasting findings were reported by Cousijn et al. (2014). They also measured visually evoked gamma oscillations using MEG, and assessed GABA and glutamate concentrations with MRS. Gamma peak frequency and amplitude were measured at the location with maximal MEG activation and at the location of the selected MRS voxel in occipital cortex. Cousijn et al. (2014) did not find a correlation between gamma peak frequency and GABA concentration. They also did not find a correlation between evoked gamma activity and glutamate or between gamma and the ratio of GABA/glutamate. Cousijn et al. (2014) concluded "that cortical gamma oscillations do not have a consistent, demonstrable relationship to excitatory/inhibitory network activity as proxied by MRS measurements of GABA and glutamate."

Wyss et al. (2017) then investigated this mechanism in the primary auditory cortex (A1) where the highest densities of $GABA_A$ receptors in sensory areas have been found. They combined MRS with EEG to investigate the individual differences in GABA and glutamate concentration on the auditory-evoked gamma band response and found no significant relationship. In addition, they found no correlations between gamma peak frequency or power and concentrations of GABA, glutamate, and the GABA/glutamate ratio. These findings suggested to Wyss et al. (2017) that: "according to MRS measurements, and given the auditory stimuli used in this study, GABA and glutamate concentrations are unlikely to play a significant role in the inhibitory and excitatory drive in the generation of gamma band activity in the auditory cortex."

It is thus likely that not the concentration of GABA determines the occurrence of gamma activity but the synchronization of GABA transmitting neuronal activity as illustrated in the previous section.

2.2.2 Gamma generating circuits

Vinck et al. (2016) performed simultaneous LFP and single-unit recordings in four connected areas of the freely moving rat (primary somatosensory cortex barrel field, $S1_{BF}$, V1, perirhinal cortex, and hippocampal area CA1). $S1_{BF}$ spiking activity was strongly entrained by both beta and gamma $S1_{BF}$ oscillations, which were associated with deactivations and activations, respectively. Multiple classes of fast spiking and excitatory cells in $S1_{BF}$ showed prominent differences in rhythmic entrainment and in the extent to which phase locking was modulated by behavioral state (Fig. 2.1).

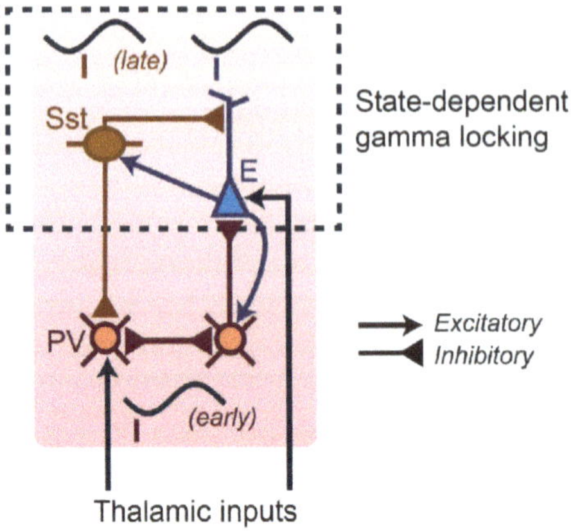

FIG. 2.1 Hybrid model of primary somatosensory barrel field cortex (S1BF) gamma generation. A class of fast-spiking (FS) interneurons (*red*: presumably Parvalbumin-positive, PV, cells) has strong mutual inhibitory connections, and does not have a substantial projection to other FS interneurons. These FS cells receive excitatory drive both from thalamic projections and from local recurrent excitatory inputs. Because of the mutual inhibitory connections, the FS cell network starts firing gamma synchronously and also entrains the excitatory cells. The excitatory (E) cells fire at a slightly later phase because the inhibitory synaptic potentials from I→E connections are slower than the I→I connections. Another group of FS cells (*yellow*: presumably Somatostatin-positive, Sst, cells) relies on the local excitatory cells to be entrained to the gamma rhythm does not receive a substantial thalamic drive and does not exhibit strong gamma phase locking. These cells fire late in the gamma cycle, few milliseconds after the excitatory cells, and might help to shut down the activity of both the early-firing FS cells and the excitatory cells. Early-firing FS cells show little state-dependence, whereas excitatory and late-firing FS cells show strong state dependence. *From Vinck, M., Bos, J.J., Van Mourik-Donga, L.A., Oplaat, K.T., Klein, G.A., Jackson, J.C., et al., 2016. Cell-type and state-dependent synchronization among rodent somatosensory, visual, perirhinal cortex, and hippocampus CA1. Front. Syst. Neurosci. 9, 187. doi:10.3389/fnsys.2015.00187. Open access.*

Ray and Maunsell (2015) discussed potential mechanisms that rely on rhythmic inhibition and underlie short- and long-range processing. One of these, the communication-through-coherence (CTC) hypothesis (Fries, 2009) proposes that when the activity of a neuronal assembly oscillates, the periodic fluctuations in excitability produce temporal windows for communication such that only in-phase oscillating assemblies can communicate effectively, thereby allowing flexible long-range communication between neuronal assemblies. Jia et al. (2013) had shown an increase in pairwise synchrony of neurons in visual cortical areas V1 and V2 due to gamma, but the conduction delay between V1 and V2 was on the order of ~ 3 ms, whereas gamma rhythms in the two areas differed by ~ 90 degrees, equating to ~ 5–8 ms delay for in-phase processing. Ray and Maunsell (2015) considered this inconsistent with CTC. Another mechanism focuses on coding of sensory information in the timing of the spikes relative to the phase of an ongoing oscillation [phase coding (PC)]. Fries et al. (2007) proposed that gamma oscillations might also be used for PC, "because the inhibition is strongest at the peak of gamma cycle (measured extracellularly), strong incoming excitation can overcome the inhibition earlier and fire a spike earlier in the gamma cycle, whereas weak excitation can only lead to a spike when the inhibition is weakest (at the trough of the gamma cycle). Thus, stimulus intensity can be encoded in the position of the spike with respect to the phase of gamma." Findings in humans appear to favor the phase-coding hypothesis. However, Ray and Maunsell (2015) see gamma oscillations just as a simple correlate of cognition, rather than a neural mechanism underlying it.

2.2.3 Gamma frequency and cognition

Gamma oscillations may relate to many cognitive processes (Fries, 2009; Jensen et al., 2007; Rodriguez et al., 1999; Tallon-Baudry and Bertrand, 1999; Chapter 4). The visual system is special as it presents evoked gamma band oscillations (60–120 Hz) not only in visual cortex but also in the retina and visual thalamus (Castelo-Branco et al., 1998). However, recordings in primate and cat visual thalamus and V1 suggest that the thalamus is not a major contributor to the gamma power in cortex (Koepsell et al., 2009). Gamma oscillations in the 20–80 Hz range in cortex are largely cortical in origin (Engel et al., 1991a; Tallon-Baudry et al., 2001). In an extensive review, Sedley and Cunningham (2013) described two classes of gamma oscillations: "broadband" and "narrowband" that have very different properties (and likely different roles). Broad-frequency range gamma (anywhere from 30 to 160 Hz) is predominantly transient following stimulus onset, and correlates positively with multiunit-spiking activity. Narrow-band gamma is centered ~ 40–50 Hz with a bandwidth of only 10–20 Hz and occurs in response to certain large size and high luminance visual stimuli presented at particular spatial frequencies. The magnitude of this type of gamma oscillation varies inversely with multiunit activity (Jia et al., 2011; Ray and Maunsell, 2011), which suggests an inhibitory function.

Human MEG and EEG studies using similarly large-sized visual stimuli with high luminance contrasts (e.g., gratings and checkerboards) also showed a narrow-band gamma oscillation, persisting for the duration of the stimulus (Adjamian et al., 2004; Muthukumaraswamy et al., 2009; Scheeringa et al., 2011). Studies in other sensory domains have not identified narrowband gamma oscillations. For example, auditory stimulus-induced gamma oscillations mainly occur at higher frequencies of ~ 80 Hz and occupy a broader frequency range and as in V1 generally occur predominantly transiently to stimulus transitions (Edwards et al., 2005; Griffiths et al., 2010; Sedley et al., 2012). While lower-frequency gamma oscillations are sometimes detected in response to auditory stimuli are less abundant and not occurring in electrocorticography (ECoG) studies in the absence of the ubiquitous higher frequency gamma (Nourski et al., 2019). Gamma in somatosensory cortex appears similar to that in auditory cortex (Bauer et al., 2006; Gross et al., 2007; Ray et al., 2008) having predominantly relatively high frequency, broadband, and transient for hundreds of milliseconds following stimulus onset.

Summarizing, narrow-band gamma oscillations may serve an inhibitory role, which could include mediating surround suppression in V1. The role of broad-band gamma seems to be largely excitatory.

2.3 The beta rhythm

2.3.1 Beta and the motor system

Beta-band oscillations (~ 13–30 Hz) have typically been associated with sensorimotor functions (Aumann and Prut, 2015; Brovelli et al., 2004) and are attenuated during voluntary movements or motor imagery (Pfurtscheller and Lopes da Silva, 1999; Turella et al., 2016). By using power- and coherence-spectral analysis, Brovelli et al. (2004) identified a beta-synchronized large-scale network linking precentral and postcentral areas in monkeys as they pressed a hand lever during the wait period of a visual discrimination task. Using Granger causality between LFP spectra, strong influences were observed from primary somatosensory cortex to both motor cortex and inferior posterior parietal cortex, with the latter area also exerting causal influences on motor cortex. Causal influences from motor cortex to postcentral sites, however, were weak in one monkey and not observed in the other.

Miller et al. (2012) studied the particular role of rhythms in human motor cortex both during rest and during engaging in different finger movements. Arrays of ECoG electrodes were placed directly on frontal, temporal, and parietal brain surfaces to measure LFPs. Movement of different fingers produces spatially focal and finger-specific changes in local neuronal population activity, along with spatially diffuse and finger-nonspecific decreases in 12–20 Hz beta-rhythm power. Apart from these power changes, there were independent interactions between the beta rhythm and local mean instantaneous firing rate. During rest, 5%–20% of the variation in local cortical activity was due to entrainment on the phase of the beta rhythm over the precentral and postcentral gyri, but the interaction was significantly diminished during movement. This shifting entrainment suggests that the beta rhythm is not simply a background process that is suppressed during movement but rather that the beta rhythm plays an active and important role in motor processing.

2.3.2 Audio-visual beta

Hipp et al. (2011) obtained EEG recordings in 24 participants with normal hearing, normal or corrected-to-normal vision, and no history of neurological or psychiatric illness. Participants were presented with an audiovisual stimulus (500 trials) and a subsequent visual-only control stimulus (100 trials). Visual stimulation—two moving bars approaching each other, overlapping, and diverging again—was identical under both conditions. In addition, during audiovisual trials a click-sound (20 ms duration, 60 dB sound pressure level) was played at the moment of bar overlap via a frontal loudspeaker. Participants reported their percept (bounce or pass of the moving bars) of the ambiguous stimulation through button-press (left and right thumb) after offset of a visual fixation-cross.

Based on corticocortical coherence between the stimulation and baseline intervals, Hipp et al. (2011) identified a widespread but highly structured cortical network (Fig. 2.2A) that showed enhanced beta-band coherence (15–23 Hz) during stimulation. Beta-band coherence in this network was enhanced for ~ 1 s around the time of bar overlap (Fig. 2.2B). They also quantified the detailed pattern of synchronization between the different nodes within the network (Fig. 2.2C). This revealed a hub-like structure in which the right posterior parietal cortex synchronized most prominently with other nodes of the network. Thus long-range beta-synchrony was enhanced in a highly structured network during stimulus presentation. Hipp et al. (2011) found that beta-synchrony was not only enhanced during stimulus processing but also predicted the subjects' percept of the stimulus. Comparing coherence within the identified network for trials in which the subjects perceived the stimulus as "bouncing" or "passing" yielded a significant difference (Fig. 2.2D) with enhanced beta-coherence for bounce trials. The strongest perception-related effects were found for synchronization with frontal regions (Fig. 2.2E). Thus beta-oscillatory neuronal synchronization mediates neuronal communication within frequency-specific, large-scale cortical networks (Hipp et al., 2011).

In the auditory modality, suppression of beta power is an obligatory response to sound (Mazaheri and Picton, 2005). Various studies observed a role of auditory beta oscillation in rhythm perception and predictive timing processing (Arnal et al., 2015; Bartolo and Merchant, 2015; Bartolo et al., 2014; Breska and Deouell, 2016; Fujioka et al., 2012; Ross et al., 2017). Beta suppression may thus be due to both exogenous and endogenous factors, the latter related to predictive sensory processing and motor planning. Florin et al. (2017) used MEG source analysis to reveal the pretarget oscillations of successful auditory detection of small pitch deviations from a sequence of standard tones. They found that "successful pitch change detection could be predicted from the amplitude of theta (4–8 Hz) oscillatory activity in the right inferior frontal gyrus and beta (12–30 Hz) oscillatory

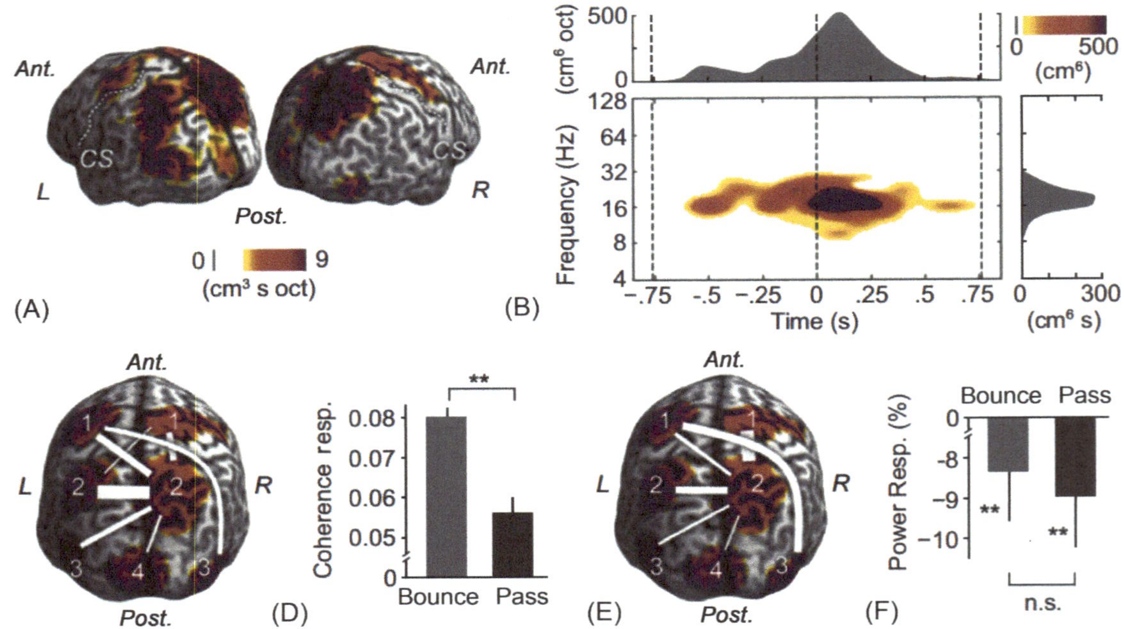

FIG. 2.2 Perception-related beta-synchrony network. (A) Spatial localization of cortical regions engaged in the network. For each cortical location, the *color* shows for how long, across which frequency range, and to how many other locations coherence was increased for bounce and pass trials relative to baseline (*CS*, central sulcus). (B) *Bottom left*: Spectrotemporal coherence profile of the network that displays between how many locations coherence was increased at a given time and frequency. *Top right*: Corresponding temporal and spectral coherence profiles. The spectral profile ranged from 15 to 23 Hz (full width at half maximum). (C) Coherence structure between the seven clusters forming the beta-network (stimulation versus baseline). Line-width represents relative coherence strength. 1, frontal cortex; 2, posterior parietal cortex; 3, lateral occipitotemporal cortex; 4, medial occipital cortex. (D) Coherence response relative to pre-stimulus baseline for bounce and pass percepts within the beta-network (mean ± SEM). (E) Structure of difference in coherence between bounce and pass percepts. Line-width represents relative coherence strength. (F) Power response relative to prestimulus baseline for bounce and pass percepts within the beta-network (mean ± SEM). *With permission from Hipp, J.F., Engel, A.K., Siegel, M., 2011. Oscillatory synchronization in large-scale cortical networks predicts perception. Neuron 69, 387–396. Copyright 2011. Elsevier.*

activity in the right auditory cortex." Alavash et al. (2017) recorded MEG in human listeners who judged acoustic stimuli composed of tone sweeps with respect to pitch or sweep direction. They found that faster decisions were negatively correlated with the power of neural oscillations within the alpha and mid-beta bands. Alavash et al. (2017) argued that this network effect is a manifestation of large-scale neural couplings underlying both auditory processing and manual responding.

2.4 The theta rhythm

"Theta oscillations represent the 'on-line' state of the hippocampus, and may be critical for temporal coding/decoding of active neuronal ensembles and the modification of synaptic weights" (Buzsáki, 2002). Local field oscillations at theta/alpha frequencies have also been observed in numerous cortical structures (Steriade, 2000). Intracranial recordings in humans indicate, however, that these patterns occur in different behavioral states and are not coherent with hippocampal theta waves (Raghavachari et al., 2001).

2.4.1 Emphasis on phase

The mutual information of two random variables quantifies the amount of information obtained about one random variable through observing the other random variable. By applying this to the encoding of natural sounds in the auditory cortex of alert animals, Kayser et al. (2009) found that temporal spike-train patterns and spatial population spiking were both highly informative. However, the relative phase of slow ongoing LFP theta rhythms at which the temporal or population spike-responses occurred provided much additional and complementary information. Overall, LFP phase regardless of frequency provided nearly three times more information than LFP energy, resulting in a highly significant difference. The information for the combined "Energy and Phase" signal provided more information than the summed information of energy and phase (Fig. 2.3).

2.4.2 Role of theta in speech

Several cortical areas are significantly associated with speech processing. The cortical responses reflect detailed information about the spectral and temporal content of speech, words, or speech-like stimuli—as a particular realization of the natural sounds presented in Fig. 2.3. Functional roles of cortical entrainment to speech may include encoding acoustic features, parsing syllabic boundaries, and selecting sensory information in

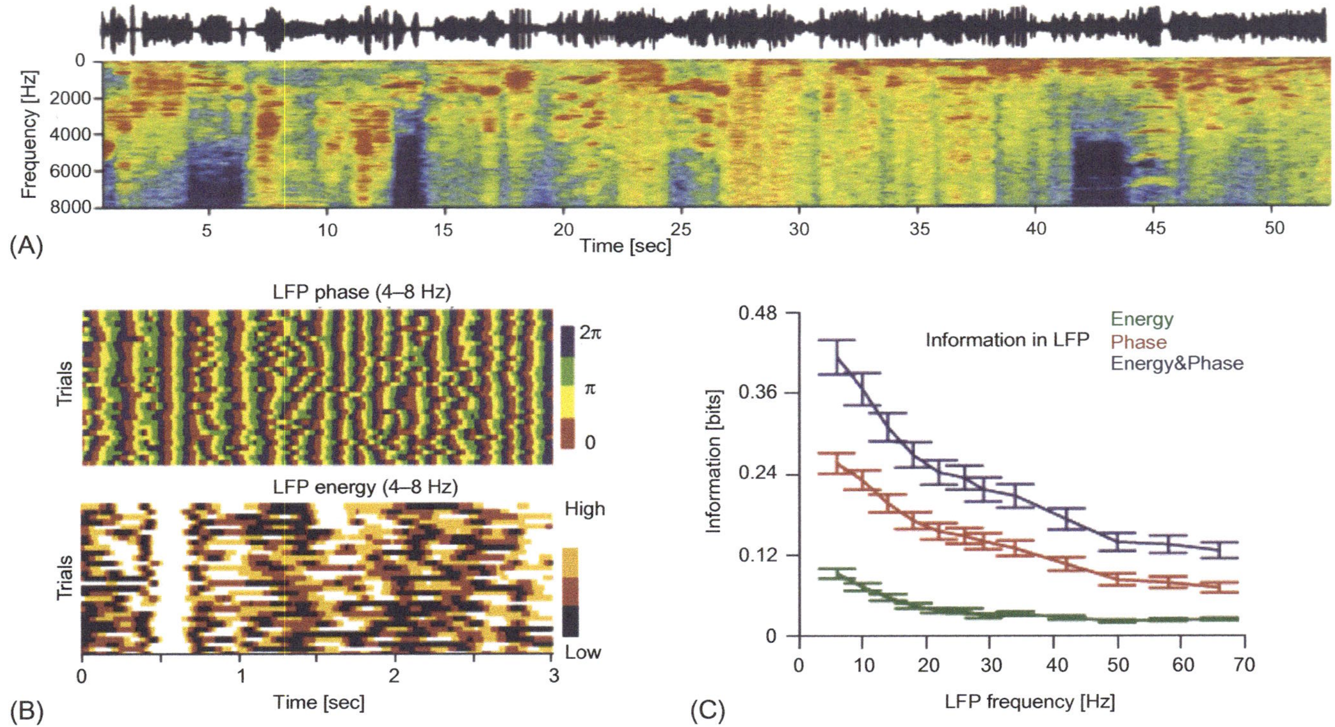

FIG. 2.3 Information in local field potentials (LFP). (A) Waveform and spectrogram (*red*, high power; *blue*, low power) of the acoustic stimulus sequence consisting of a 52s sequence of naturalistic sounds including environmental and animal sounds and segments of conspecific vocalizations. (B) Examples of low-frequency (4–8Hz) LFP phase and energy for 30 subsequent trials. Both signals are binned into four bins. (C) Mutual information between LFP (phase, energy, and the combined energy and phase signal) and stimulus. Error bars indicate mean and SEM across sites ($n=86$, $T=4$ms time window for the mutual information calculations). *With permission from Kayser, C., Montemurro, M.A., Logothetis, N., Panzeri, S., 2009. Spike-phase coding boosts and stabilizes the information earned by spatial and temporal spike patterns. Neuron 61, 597–608. Copyright 2009. Elsevier.*

complex listening environments. Theta-band entrainment (4–8 Hz) may encode speech features critical for intelligibility while delta-band entrainment (1–4 Hz) may be related to the perceived, non-speech-specific acoustic rhythm (Ding and Simon, 2014).

Luo and Poeppel (2007) showed that the phase pattern of theta-band responses recorded from human auditory cortex (ACx) with MEG reliably tracks and discriminates spoken sentences and that this discrimination ability is correlated with speech intelligibility. Their findings suggest that a ~ 200 ms temporal window (~ 1 period of theta oscillation) segments the incoming speech signal to track speech dynamics. Using time-inverted speech that is quite similar to normal spoken sentences in terms of low-frequency modulation characteristics but unintelligible, Howard and Poeppel (2010) found that discrimination based on theta-band phase patterns was equal for both types of stimuli. Thus theta-band phase patterns primarily reflect cortical processing of low-frequency modulations present in the acoustic signal that are required for intelligibility, rather than exclusively processing lexical, syntactic, or semantic aspects related to comprehension. They also observed that whereas theta-band phase patterns differ across such stimuli, power patterns do not. Speech elicits a phase-locked response in the ACx that is dominated by theta frequencies when observed by MEG. This phase-locked response could be new phase-locked activity superimposed on the ongoing theta oscillation or, alternatively, results from phase resetting of the ongoing oscillation. Howard and Poeppel (2012) found significant positive correlation not only during the onset response but also throughout the response period. In fact, transient increases in phase coherence were accompanied by transient increases in amplitude in accordance with a "signal plus background" model of the evoked response (see also Chapter 3, Section 3.3). The same group, Teng et al. (2017) showed "that the human auditory system tracks the temporal aspects of sound concurrently at 2 timescales corresponding to the EEG theta- and gamma-band ranges but not at the timescale corresponding to alpha, which has also been found to be related to auditory perception. They found a strong intertrial phase coherence in the theta band for stimuli of all modulation rates and in the gamma band for stimuli with corresponding modulation rates. The alpha band did not respond in a similar manner. Subsequently, Teng et al. (2018) found that cortical oscillations in the theta band even track sound with irregular temporal structure as found in sound with the modulation spectrum of speech. This suggests an auditory segmentation mechanism, complementary to entrainment, that operates on a timescale of ~ 200 ms. This chunking scheme is prevalent in auditory processing of sounds of various dynamics and may serve as a fundamental step for further perceptual analysis. Using MEG, Teng et al. (2019) showed that the temporal fine structure in speech entrains cortical responses between 3 and 6 Hz. They also showed that it is "insufficient to investigate solely the speech envelope to understand temporal speech segmentation, as the speech envelope and the temporal fine structure convey comparable, if not inseparable, temporal information."

2.4.3 Segregation of concurrent sounds

In a noisy room with multiple people talking, listeners with normal hearing can still recognize and understand attended speech and simultaneously ignore background noise and irrelevant speech. Timing synchrony of frequency components allows fusion into a more complex sound, and if the frequency components are harmonic, the fusion is more likely. In contrast, timing asynchrony is a major element to distinguish one stream from two streams (Eggermont, 2015a,b). The ability to segregate a single sound source from concurrent sources is crucial for auditory perception. Using high-density EEG, Kerlin et al. (2010) found that selective attention increases discrimination of the attended speech signal in auditory cortex in the range of 4–8 Hz. They demonstrated furthermore that this attentional gain is due to enhancement of the attended sentence and possibly suppression of the unattended stream (Wang et al., 2016; Clouter et al., 2017). Tóth et al. (2016) investigated the event-related oscillations evoked by complex tones, which could be perceived as a single sound, and tonal complexes with cues promoting the perception of two concurrent sounds by inharmonicity, onset asynchrony, and/or perceived source location difference of the component tones. They found that in two time-windows, the amplitude of theta oscillation was modulated by the presence vs. absence of these cues: 60–350 ms/6–8 Hz (early) and 350–450 ms/4–6 Hz (late). The early response appeared both in the passive and the active listening conditions; it did not closely match the task performance; and it had a frontocentral scalp distribution. The late response was only elicited in the active listening condition; it closely matched the task performance; and it had a centroparietal scalp distribution. The neural processes reflected by these responses are probably involved in the processing of concurrent sound segregation cues, in sound categorization, and response preparation and monitoring.

2.4.4 Hippocampal theta in human memory

The hippocampal theta rhythm is critical for learning and memory. Nyhus and Curran (2010) proposed that gamma and theta oscillations allow for the transient interaction between cortical structures and the hippocampus for the encoding and retrieval of episodic memories as described by the hippocampal memory indexing theory. Theta oscillations act to temporally order these individual episodic memory representations. Using seed-based fMRI, Robin et al. (2015) found evidence for a medial prefrontal-parietal network and a medial temporal lobe network, which were functionally connected to the prefrontal and hippocampal seeds, respectively, regardless of the nature of the memory task. The medial prefrontal cortex (mPFC) receives direct projections from ventral CA1 and subiculum subfields of the hippocampus, and these synapses are plastic. During mnemonic tasks, theta rhythms may facilitate hippocampal inputs to the mPFC and may also integrate functionally relevant gamma-mediated cell assemblies in the mPFC (Colgin, 2011). During slow-wave sleep, temporal coordination of hippocampal sharp wave ripples—very high frequency oscillations—and medial prefrontal cortex spindles in the alpha-frequency range may be an important component of the process by which memories become hippocampus-independent (Colgin, 2011; Gordon, 2011).

2.5 Nested oscillations

2.5.1 Basics

Nested oscillations reflect a systematic relationship between the amplitude or power of one or more higher-frequency oscillation bands and the phase of a lower-frequency oscillation (Schroeder and Lakatos, 2009). As illustrated in Fig. 2.4, amplitude in the high gamma band (*dashed white box,*

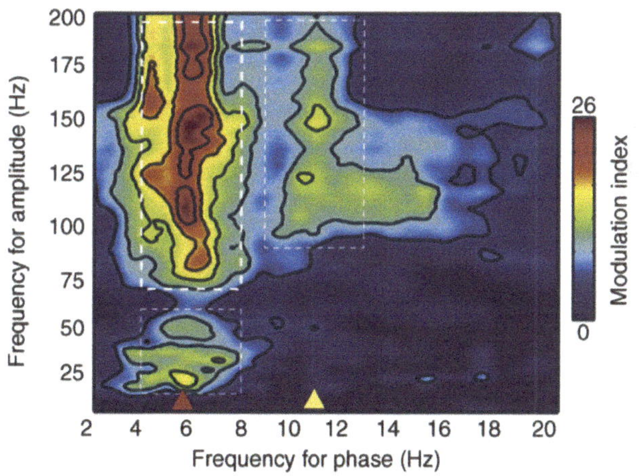

FIG. 2.4 Cross-frequency phase-amplitude coupling plot showing that high-gamma-band amplitude *(thick-dashed white box)* is grouped according to theta-band oscillatory phase *(red arrow)*. More subtle phase-amplitude coupling effects *(thin-dashed white boxes)* are also apparent between theta phase and gamma amplitude, and between – 10 Hz (alpha and mu) band phase *(yellow arrow)* and high gamma amplitude. *From Canolty, R., Edwards, E., Dalai, S., Soltani, M., Nagarajan, S., Kirseh, H., et al., 2006. High gamma power is phase-locked to theta oscillations in human neocortex. Science 313, 1626–1628. With permission from AAAS.*

upper left) is grouped by, or "nested within," the phase of a theta-band oscillation (*red arrowhead*, bottom). Coupling between theta and gamma (*dashed white box*, lower left) and between a ~ 10 Hz oscillation (*yellow arrowhead*, bottom) and high gamma (*thin-dashed white box*, middle) is also apparent (Canolty et al., 2006). Cross-frequency coupling has elements of hierarchical organization; that is, as gamma amplitude is coupled with theta phase, theta amplitude is coupled to delta phase. Coupling extends throughout the > 1 Hz frequency range typically studied and extends down into the "slow" (0.1–1 Hz) and "infra-slow" (0.01–0.1 Hz) range (Monto et al., 2008).

An important aspect that applies to all studies using neural oscillations is that filtering of even pure noise reveals pseudooscillations, albeit that they are no more than contents of the noise and not the results of an oscillatory process. This is even more important for cross-frequency coupling (CFC) where oscillations with different frequencies are compared. To validate such a procedure, clear peaks in the broadband spectrum is a prerequisite; especially the frequency component for defining the instantaneous phase should include one of the peaks (Aru et al., 2015; Hyafil, 2015; Young and Eggermont, 2009). CFC may also be affected by oscillation harmonics, so that one may test a fundamental component say alpha and one of its harmonics in the beta or gamma range (Hyafil, 2017). Consequently, wrong conclusions may result from decomposing the MEG/EEG into separate frequency bands before testing for a harmonic link between these bands, regardless of whether the distinct bands actually depend on separate generative mechanisms, as would be the case for genuine CFC (Hyafil, 2017).

2.5.2 Nested oscillations in perception and cognition

One of the most prominent roles assigned to neural oscillations is to mediate selective neural communication between areas, with in-phase regions communicating more efficiently than out-of-phase ones (Fries, 2009). Gamma rhythms have predominantly been implicated in this so-called "communication through coherence" mechanism, especially for bottom-up processes (Fries et al., 2007). However, fast oscillations only synchronize at a local scale. By contrast, slower rhythms (< 10 Hz) may synchronize between distant areas (Buzsáki and Draguhn 2004; Von Stein and Sarnthein, 2000). Coupling phase of fast oscillations to slower oscillations may be a way to selectively enhance communication between distant areas via long-distance synchronization between local gamma rhythms. In this way, coherence between distant slower oscillations associated with local slow/fast oscillations in posterior parietal cortex could bring synchrony to distant fast oscillations (Hyafil et al., 2015).

Nested oscillation could also reflect and underlying physiological mechanism. Jensen and Mazaheri (2010) proposed that information transfer between cortical regions is gated by inhibiting task-irrelevant regions, thus routing information to task-relevant regions. The functional inhibition maybe reflected in oscillatory activity in the alpha band (8–13 Hz), which provides pulsed inhibition and thereby reducing the processing capabilities of a given area (Chapter 3). Active processing in the engaged areas is reflected by increased neuronal synchronization in the gamma band (30–100 Hz) accompanied by an alpha band power decrease (Osipova et al., 2008). The importance of the theta frequency range in memory processing is well established (Section 2.4.4), and phase alignment of oscillations is considered to be necessary for synaptic plasticity. Intrathalamic LFP and surface EEG revealed that phase alignment in the theta rhythm was greater during successful compared with unsuccessful encoding, and this alignment was correlated with the theta-gamma cross-frequency coupling (Sweeney-Reed et al., 2015). Prestimulus theta (4–8 Hz) power in the hippocampus and neocortex predicts whether a memory for a subsequent event will be formed. Sweeney-Reed et al. (2016) found that greater prestimulus theta power in the right dorsomedial thalamic nucleus (DMTN) was associated with successful memory encoding, predicting both behavioral outcome and post-stimulus correlates of successful memory formation. In particular, significant correlations were observed between right DMTN theta power and both frontal theta and right anterior thalamic nucleus (ATN) gamma (32–50 Hz) phase alignment and frontal-ATN theta-gamma cross-frequency coupling. These results provide direct electrophysiological evidence in humans of a role for the DMTN and the ATN in memory formation.

2.6 Summary

The gamma rhythm has been implied in many cognitive functions such as perceptual binding attention, reaction times and initiation of movements. The gamma rhythm results from quasi-rhythmic firing of inhibitory neurons. Two major types of gamma exist. One is broadband and high frequency and is found in visual, auditory, and somatosensory cortex where it exerts an excitatory drive. The other type is narrow band and of lower frequency, is only found in the visual system, and exerts an inhibitory influence. The beta rhythm is dominant in the motor system and occurs in the auditory and visual systems; it is suppressed by movement and sound but has an important role in prediction of sensory and motor planning. The theta rhythm underlies memory processes in hippocampus and cortex. It also is important in chunking of speech and segregating concurrent sounds. One of its dominant roles in processing within brain networks is that it synchronizes phase-locked alpha and beta and activity across distributed cortical areas.

References

Adjamian, P., Holliday, I.E., Barnes, G.R., Hillebrand, A., Hadjipapas, A., Singh, K.D., 2004. Induced visual illusions and gamma oscillations in human primary visual cortex. Eur. J. Neurosci. 20 (2), 587–592.

Alavash, M., Daube, C., Wöstmann, M., Brandmeyer, A., Obleser, J., 2017. Large-scale network dynamics of beta-band oscillations underlie auditory perceptual decision-making. Netw. Neurosci. 1 (2), 166–191. https://doi.org/10.1162/netn_a_00009.

Arnal, L.H., Doelling, K.B., Poeppel, D., 2015. Delta-beta coupled oscillations underlie temporal prediction accuracy. Cereb. Cortex 25, 3077–3085.

Aru, J., Aru, J., Priesemann, V., Wibral, M., Lana, L., Pipa, G., et al., 2015. Untangling cross-frequency coupling in neuroscience. Curr. Opin. Neurobiol. 31, 51–61.

Aumann, T., Prut, Y., 2015. Do sensorimotor b-oscillations maintain muscle synergy representations in primary motor cortex? Trends Neurosci. 38 (2), 77–85.

Bartolo, R., Merchant, H., 2015. β Oscillations are linked to the initiation of sensory-cued movement sequences and the internal guidance of regular tapping in the monkey. J. Neurosci. 35 (11), 4635–4640.

Bartolo, R., Prado, L., Merchant, H., 2014. Information processing in the primate basal ganglia during sensory-guided and internally driven rhythmic tapping. J. Neurosci. 34 (11), 3910–3923.

Bauer, M., Oostenveld, R., Peeters, M., Fries, P., 2006. Tactile spatial attention enhances gamma-band activity in somatosensory cortex and reduces low-frequency activity in parieto-occipital areas. J. Neurosci. 26 (2), 490–501.

Breska, A., Deouell, L.Y., 2016. When synchronizing to rhythms is not a good thing: modulations of preparatory and post-target neural activity when shifting attention away from on-beat times of a distracting rhythm. J. Neurosci. 36 (27), 7154–7166.

Brovelli, A., Ding, M., Ledberg, A., Chen, Y., Nakamura, R., Bressler, S.L., 2004. Beta oscillations in a large-scale sensorimotor cortical network: directional influences revealed by granger causality. Proc. Natl. Acad. Sci. U. S. A. 101 (26), 9849–9854.

Buzsáki, G., 2002. Theta oscillations in the hippocampus. Neuron 33, 325–340.

Buzsáki, G., Draguhn, A., 2004. Neuronal oscillations in cortical networks. Science 304, 1926–1929.

Canolty, R., Edwards, E., Dalal, S., Soltani, M., Nagarajan, S., Kirsch, H., et al., 2006. High gamma power is phase-locked to theta oscillations in human neocortex. Science 313, 1626–1628.

Castelo-Branco, M., Neuenschwander, S., Singer, W., 1998. Synchronization of visual responses between the cortex, lateral geniculate nucleus, and retina in the anesthetized cat. J. Neurosci. 18 (16), 6395–6410.

Clouter, A., Shapiro, K.L., Hanslmayr, S., 2017. Theta phase synchronization is the glue that binds human associative memory. Curr. Biol. 27, 3143–3148.

Colgin, L.L., 2011. Oscillations and hippocampal–prefrontal synchrony. Curr. Opin. Neurobiol. 21, 467–474.

Cousijn, H., Haegens, S., Wallis, G., Near, J., Stokes, M.G., Harrison, P.J., et al., 2014. Resting GABA and glutamate concentrations do not predict visual gamma frequency or amplitude. Proc. Natl. Acad. Sci. U. S. A. 111 (25), 9301–9306.

Ding, N., Simon, J.Z., 2014. Cortical entrainment to continuous speech: functional roles and interpretations. Front. Hum. Neurosci. 8, 311. https://doi.org/10.3389/fnhum.2014.00311.

Eckhorn, R., Bauer, R., Jordan, W., Brosch, M., Kruse, W., Munk, M., et al., 1988. Coherent oscillations: a mechanism of feature linking in the visual cortex? Biol. Cybern. 60, 121–130.

Edwards, E., Soltani, M., Deouell, L.Y., Berger, M.S., Knight, R.T., 2005. High gamma activity in response to deviant auditory stimuli recorded directly from human cortex. J. Neurophysiol. 94, 4269–4280.

Eggermont, J.J., 2015a. Auditory Temporal Processing and its Disorders. Oxford University Press, Oxford, UK.
Eggermont, J.J., 2015b. Animal models of auditory temporal processing. Int. J. Psychophysiol. 95, 201–215.
Engel, A.K., König, P., Kreiter, A.K., Singer, W., 1991a. Interhemispheric synchronization of oscillatory neuronal responses in cat visual cortex. Science 252, 1177–1179.
Engel, A.K., Kreiter, A.K., König, P., Singer, W., 1991b. Synchronization of oscillatory neuronal responses between striate and extrastriate visual cortical areas of the cat. Proc. Natl. Acad. Sci. U. S. A. 88 (14), 6048–6052.
Florin, E., Vuvan, D., Peretz, I., Baillet, S., 2017. Pre-target neural oscillations predict variability in the detection of small pitch changes. PLoS One 12 (5). https://doi.org/10.1371/journal.pone.0177836.
Freeman, W.J., 1978. Spatial properties of an EEG event in the olfactory bulb and cortex. Electroencephalogr. Clin. Neurophysiol. 44 (5), 586–605.
Fries, P., 2009. Neuronal gamma-band synchronization as a fundamental process in cortical computation. Annu. Rev. Neurosci. 32, 209–224.
Fries, P., Nikolic, D., Singer, W., 2007. The gamma cycle. Trends Neurosci. 30, 309–316.
Fujioka, T., Trainor, L.J., Large, E.W., Ross, B., 2012. Internalized timing of isochronous sounds is represented in neuromagnetic beta oscillations. J. Neurosci. 32 (5), 1791–1802.
Gordon, J.A., 2011. Oscillations and hippocampal–prefrontal synchrony. Curr. Opin. Neurobiol. 21, 486–491.
Gray, C.M., Konig, P., Engel, A.K., Singer, W., 1989. Oscillatory responses in cat visual cortex exhibit inter-columnar synchronization which reflects global stimulus properties. Nature 338, 334–337.
Griffiths, T.D., Kumar, S., Sedley, W., Nourski, K.V., Kawasaki, H., Oya, H., et al., 2010. Direct recordings of pitch responses from human auditory cortex. Curr. Biol. 20, 1128–1132.
Gross, J., Schnitzler, A., Timmermann, L., Ploner, M., 2007. Gamma oscillations in human primary somatosensory cortex reflect pain perception. PLoS Biol. 5. https://doi.org/10.1371/journal.pbio.0050133.
Hipp, J.F., Engel, A.K., Siegel, M., 2011. Oscillatory synchronization in large-scale cortical networks predicts perception. Neuron 69, 387–396.
Howard, M.F., Poeppel, D., 2010. Discrimination of speech stimuli based on neuronal response phase patterns depends on acoustics but not comprehension. J. Neurophysiol. 104, 2500–2511.
Howard, M.F., Poeppel, D., 2012. The neuromagnetic response to spoken sentences: co-modulation of theta band amplitude and phase. NeuroImage 60, 2118–2127.
Hyafil, A., 2015. Misidentifications of specific forms of cross-frequency coupling: three warnings. Front. Neurosci. 9, 370. https://doi.org/10.3389/fnins.2015.00370.
Hyafil, A., 2017. Disharmony in neural oscillations. J. Neurophysiol. 118, 1–3.
Hyafil, A., Giraud, A.-L., Fontolan, L., Gutkin, B., 2015. Neural cross-frequency coupling: connecting architectures, mechanisms, and functions. Trends Neurosci. 8 (11), 725–740.
Igarashi, K.M., 2015. Plasticity in oscillatory coupling between hippocampus and cortex. Curr. Opin. Neurobiol. 35, 163–168.
Jensen, O., Mazaheri, A., 2010. Shaping functional architecture by oscillatory alpha activity: gating by inhibition. Front. Hum. Neurosci. 4, 186. https://doi.org/10.3389/fnhum.2010.00186.
Jensen, O., Kaiser, J., Lachaux, J.-P., 2007. Human gamma-frequency oscillations associated with attention and memory. Trends Neurosci. 30, 317–324.
Jia, X., Smith, M.A., Kohn, A., 2011. Stimulus selectivity and spatial coherence of gamma components of the local field potential. J. Neurosci. 31 (25), 9390–9403.
Jia, X., Tanabe, S., Kohn, A., 2013. Gamma and the coordination of spiking activity in early visual cortex. Neuron 77, 762–774.

Kayser, C., Montemurro, M.A., Logothetis, N., Panzeri, S., 2009. Spike-phase coding boosts and stabilizes the information carried by spatial and temporal spike patterns. Neuron 61, 597–608.

Kerlin, J.R., Shahin, A.J., Miller, L.M., 2010. Attentional gain control of ongoing cortical speech representations in a "cocktail party". J. Neurosci. 30 (2), 620–628.

Koepsell, K., Wang, X., Vaingankar, V., Wei, Y., Wang, Q., Rathbun, D.L., et al., 2009. Retinal oscillations carry visual information to cortex. Front. Syst. Neurosci. 3, 4. https://doi.org/10.3389/neuro.06.004.2009.

Kuki, T., Fujihara, K., Miwa, H., Tamamaki, N., Yanagawa, Y., Mushiake, H., 2015. Contribution of parvalbumin and somatostatin-expressing GABAergic neurons to slow oscillations and the balance in beta-gamma oscillations across cortical layers. Front. Neural Circuits 9, 6. https://doi.org/10.3389/fncir.2015.00006.

Luo, H., Poeppel, D., 2007. Phase patterns of neuronal responses reliably discriminate speech in human auditory cortex. Neuron 54, 1001–1010.

Mazaheri, A., Picton, T.W., 2005. EEG spectral dynamics during discrimination of auditory and visual targets. Cogn. Brain Res. 24, 81–96.

Miller, K.J., Hermes, D., Honey, C.J., Hebb, A.O., Ramsey, N.F., Knight, R.T., et al., 2012. Human motor cortical activity is selectively phase-entrained on underlying rhythms. PLoS Comput. Biol. 8 (9), e1002655. https://doi.org/10.1371/journal.pcbi.1002655.

Monto, S., Palva, S., Voipio, J., Palva, J.M., 2008. Very slow EEG fluctuations predict the dynamics of stimulus detection and oscillation amplitudes in humans. J. Neurosci. 28 (33), 8268–8272.

Muthukumaraswamy, S.D., Edden, R.A.E., Jones, D.K., Swettenham, J.B., 2009. Resting GABA concentration predicts peak gamma frequency and fMRI amplitude in response to visual stimulation in humans. Proc. Natl. Acad. Sci. U. S. A. 106 (20), 8356–8361.

Nourski, K.V., Steinschneider, M., Rhone, A.E., Kovach, C.K., Kawasaki, H., Howard III, M.A., 2019. Differential responses to spectrally degraded speech within human auditory cortex: an intracranial electrophysiology study. Hear. Res. 371, 53–65.

Nyhus, E., Curran, T., 2010. Functional role of gamma and theta oscillations in episodic memory. Neurosci. Biobehav. Rev. 34, 1023–1035.

Osipova, D., Hermes, D., Jensen, O., 2008. Gamma power is phase-locked to posterior alpha activity. PLoS One 3, e3990. https://doi.org/10.1371/journal.pone.0003990.

Perrenoud, Q., Pennartz, C.M., Gentet, L.J., 2016. Membrane potential dynamics of spontaneous and visually evoked gamma activity in V1 of awake mice. PLoS Biol. 14, e1002383. https://doi.org/10.1371/journal.pbio.1002383.

Pfurtscheller, G., Lopes da Silva, F.H., 1999. Event-related EEG/MEG synchronization and desynchronization: basic principles. Clin. Neurophysiol. 110, 1842–1857.

Raghavachari, S., Kahana, M., Rizzuto, D., Caplan, J., Kirschen, M., Bourgeois, B., et al., 2001. Gating of human theta oscillations by a working memory task. J. Neurosci. 21 (9), 3175–3183.

Ray, S., Maunsell, J.H.R., 2011. Different origins of gamma rhythm and high-gamma activity in macaque visual cortex. PLoS Biol. 9, e1000610. https://doi.org/10.1371/journal.pbio.1000610.

Ray, S., Maunsell, J.R.H., 2015. Do gamma oscillations play a role in cerebral cortex? Trends Cogn. Sci. 19 (2), 78–85.

Ray, S., Niebur, E., Hsiao, S.S., Sinai, A., Crone, N.E., 2008. High-frequency gamma activity (80–150 Hz) is increased in human cortex during selective attention. Clin. Neurophysiol. 119, 116–133.

Robin, J., Hirshhorn, M., Rosenbaum, R.S., Winocur, G., Moscovitch, M., Grady, C.L., 2015. Functional connectivity of hippocampal and prefrontal networks during episodic and spatial memory based on real-world environments. Hippocampus 25, 81–93.

Rodriguez, E., George, N., Lachaux, J.P., Martinerie, J., Renault, B., Varela, F.J., 1999. Perception's shadow: long-distance synchronization of human brain activity. Nature 397, 430–433.

Ross, B., Barat, M., Fujioka, T., 2017. Sound-making actions lead to immediate plastic changes of neuromagnetic evoked responses and induced b-band oscillations during perception. J. Neurosci. 37 (24), 5948–5959.

Salkoff, D.B., Zagha, E., Yüzgeç, Ö., McCormick, D.A., 2015. Synaptic mechanisms of tight spike synchrony at gamma frequency in cerebral cortex. J. Neurosci. 35 (28), 10236–10251.

Scheeringa, R., Fries, P., Petersson, K.-M., Oostenveld, R., Grothe, I., Norris, D.G., et al., 2011. Neuronal dynamics underlying high- and low-frequency EEG oscillations contribute independently to the human BOLD signal. Neuron 69, 572–583.

Schroeder, C.E., Lakatos, P., 2009. Low-frequency neuronal oscillations as instruments of sensory selection. Trends Neurosci. 32, 9–18.

Sedley, W., Cunningham, M.O., 2013. Do cortical gamma oscillations promote or suppress perception? An under-asked question with an over-assumed answer. Front. Hum. Neurosci. 7, 595. https://doi.org/10.3389/fnhum.2013.00595.

Sedley, W., Teki, S., Kumar, S., Overath, T., Barnes, G.R., Griffiths, T.D., 2012. Gamma band pitch responses in human auditory cortex measured with magnetoencephalography. NeuroImage 59, 1904–1911.

Singer, W., 1993. Synchronization of cortical activity and its putative role in information processing and learning. Annu. Rev. Physiol. 55, 349–374.

Steriade, M., 2000. Corticothalamic resonance, states of vigilance and mentation. Neuroscience 101 (2), 243–276.

Sweeney-Reed, C.M., Zaehle, T., Voges, J., Schmitt, F.C., Buentjen, L., Kopitzki, K., et al., 2015. Thalamic theta phase alignment predicts human memory formation and anterior thalamic cross-frequency coupling. eLife 4, e07578. https://doi.org/10.7554/eLife.07578.

Sweeney-Reed, C.M., Zaehle, T., Voges, J., Schmitt, F.C., Buentjen, L., Kopitzki, K., 2016. Prestimulus thalamic theta power predicts human memory formation. NeuroImage 138, 100–108.

Tallon-Baudry, C., Bertrand, O., 1999. Oscillatory gamma activity in humans and its role in object representation. Trends Cogn. Sci. 3, 151–162.

Tallon-Baudry, C., Bertrand, O., Fischer, C., 2001. Oscillatory synchrony between human extrastriate areas during visual short-term memory maintenance. J. Neurosci. 21 (20), RC177.

Teng, X., Tian, X., Rowland, J., Poeppel, D., 2017. Concurrent temporal channels for auditory processing: oscillatory neural entrainment reveals segregation of function at different scales. PLoS Biol. 15 (11), e2000812. https://doi.org/10.1371/journal.pbio.2000812.

Teng, X., Tian, X., Doelling, K., Poeppel, D., 2018. Theta band oscillations reflect more than entrainment: behavioral and neural evidence demonstrates an active chunking process. Eur. J. Neurosci. 48, 2770–2782.

Teng, X., Cogan, G.B., David Poeppel, D., 2019. Speech fine structure contains critical temporal cues to support speech segmentation. NeuroImage 202, 116152.

Tiesinga, P., Sejnowski, T.J., 2009. Cortical enlightenment: are attentional gamma oscillations driven by ING or PING? Neuron 63, 727–732.

Tóth, B., Kocsis, Z., Urbán, G., Winkler, I., 2016. Theta oscillations accompanying concurrent auditory stream segregation. Int. J. Psychophysiol. 106, 141–151.

Turella, L., Tucciarelli, R., Oosterhof, N.N., Weisz, N., Rumiati, R., Lingnau, A., 2016. Beta band modulations underlie action representations for movement planning. NeuroImage 136, 197–207.

Vinck, M., Bos, J.J., Van Mourik-Donga, L.A., Oplaat, K.T., Klein, G.A., Jackson, J.C., et al., 2016. Cell-type and state-dependent synchronization among rodent somatosensory, visual, perirhinal cortex, and hippocampus CA1. Front. Syst. Neurosci. 9, 187. https://doi.org/10.3389/fnsys.2015.00187.

Von Stein, A., Sarnthein, J., 2000. Different frequencies for different scales of cortical integration: from local gamma to long range alpha/theta synchronization. Int. J. Psychophysiol. 38, 301–313.

Wang, W., Viswanathan, S., Lee, T., Grafton, S.T., 2016. Coupling between theta oscillations and cognitive control network during cross-modal visual and auditory attention: supramodal vs modality-specific mechanisms. PLoS One 11 (7). https://doi.org/10.1371/journal.pone.0158465.
Welle, C.G., Contreras, D., 2016. Sensory-driven and spontaneous gamma oscillations engage distinct cortical circuitry. J. Neurophysiol. 115, 1821–1835.
Wyss, C., Tse, D.H.Y., Kometer, M., Dammers, J., Achermann, R., Shah, N.J., et al., 2017. GABA metabolism and its role in gamma-band oscillatory activity during auditory processing: an MRS and EEG study. Hum. Brain Mapp. 38, 3975–3987.
Young, C.K., Eggermont, J.J., 2009. Coupling of mesoscopic brain oscillations: recent advances in analytical and theoretical perspectives. Prog. Neurobiol. 89, 61–78.

CHAPTER

3

The alpha and delta rhythms and their interaction with other brain rhythms

3.1 The alpha rhythm

The alpha rhythm has long been considered as reflecting an idling state of the brain mainly because its amplitude was larger with eyes closed. The eyes open condition causing suppressed alpha could accompany visual preparatory attention, and it became clear that this suppression effect on alpha power was largest when the attention was allocated to spatial locations. Combined electroencephalography (EEG) and functional magnetic resonance imaging (fMRI) has shown that alpha is dominant in sensorimotor and attention networks. Visual alpha may function in several roles, inhibition of irrelevant neural activity, chunking information into relevant temporal windows, top-down control to facilitate prediction, and increase communication between frontal and posterior areas. In the somatosensory system, alpha (there called the mu rhythm) modulates attention and increases for both imperceptible and imperceptible stimuli. In the auditory cortex, entrainment of EEG rhythms by specific frequency components in speech or sound in general leads to resonance phenomena that engage alpha (tau rhythm) activity and influence stimulus detection. Recently a fast-growing body of literature has uncovered correlations of alpha-oscillations with cognitive processes, which is incompatible with the idling-brain view. These studies show that the power and the phase of alpha-band oscillations determine excitability and modulate signal processing in the underlying neural population, and thereby shape perception and cognition (Sadaghiani and Kleinschmidt, 2016).

https://doi.org/10.1016/B978-0-12-819818-6.00005-4

3.1.1 Basics of the alpha rhythm

Because of their large amplitude relative to other frequency bands, alpha oscillations in the EEG are easily detected. Most studies use a predefined alpha frequency range for all subjects, typically from 8 to 13 Hz. In some studies the alpha frequency band is defined subject-by-subject and based on the individual's peak frequency that typically lies between 9.5 and 12.5 Hz (Klimesch, 1999). We will indicate the frequency band for individual studies. The alpha rhythm is most pronounced over visual cortex, especially during relaxed alert states but is likewise an important functional rhythm across most of cortex such as the somatosensory cortex, auditory cortex, and association cortices (Sadaghiani and Kleinschmidt, 2016).

Intracranial recordings in macaque primary visual (V1), primary somatosensory (S1), and primary auditory (A1) cortex showed that alpha band activity has a similar spectral and laminar profile in each of these sensory areas. Alpha generators were present in each of the cortical layers, with a strong source in superficial layers. The fact that alpha is generated and modulates cellular excitability locally within supragranular and infragranular layers indicates that alpha is likely involved in feedforward and feedback processes from these layers, respectively (Haegens et al., 2015).

3.1.2 Alpha rhythms and spindles

Phenomenologically the alpha rhythm is most directly related to sleep spindles (Lopes da Silva, 1991). However, the conditions for the occurrence of the alpha rhythm are quite different from those necessary for sleep spindles, although the frequency range overlaps. Spindles appear during unconsciousness and are associated with blockade of synaptic transmission through the thalamus (Steriade and Llinâs, 1988), whereas alpha waves are involved in cognitive tasks. Manshanden et al. (2002) found that the intracortical sources of magnetencephalography (MEG) spindles, in contrast to those of the alpha and mu rhythms, were distributed over the centroparietal region.

Still, ketamine-induced spindles in the cat A1 have stimulus-dependent properties that resemble those of the alpha rhythm, for example, they are suppressed by sound. Britvina and Eggermont (2008a) recorded local field potentials (LFPs) and multiunit activity from cat A1 under ketamine anesthesia during successive "silence-stimulus-silence" conditions—each lasting 15 min—to investigate the effect of sound on cortical spindle oscillations (~ 8–10 Hz). Multifrequency sound suppressed the spindle waves, as shown by the decrease of spectral power within the spindle frequency range during stimulation when compared with the previous and following silent periods (Fig. 3.1). We also showed that the percentage suppression

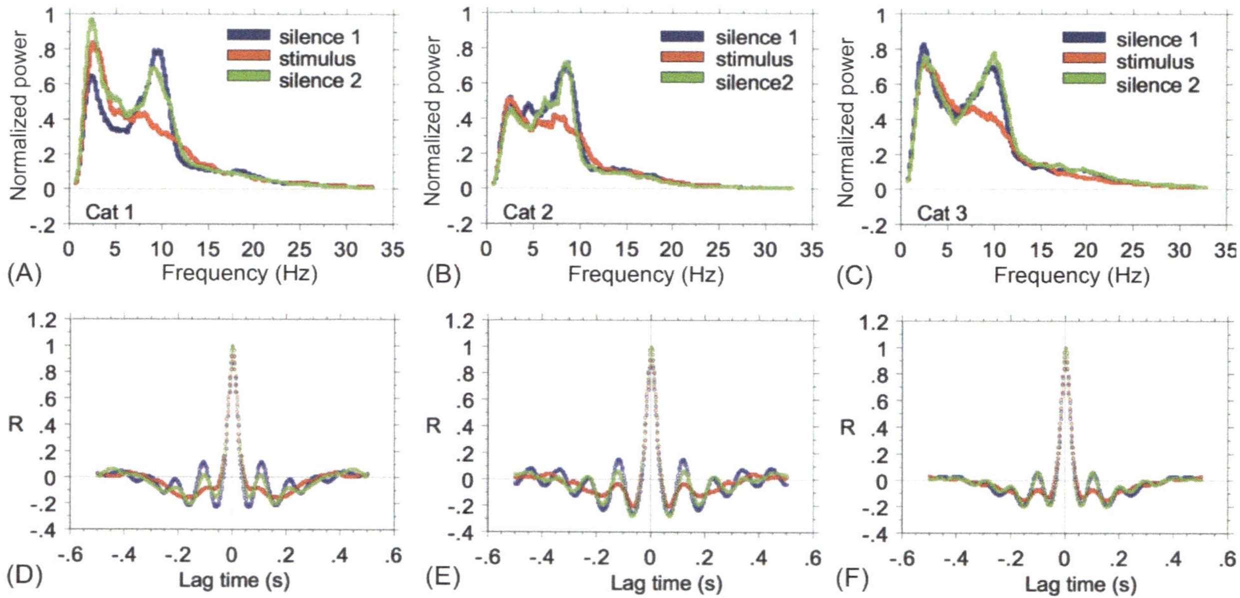

FIG. 3.1 Suppression of spindle wave power by multifrequency sound. The average local field potential (LFP) power spectra are shown for cat 1 (A), cat 2 (B), and cat 3 (C). The power during the first period of silence, the subsequent period, of multifrequency stimulation, and the second period of silence are represented by *blue*, *red*, and *green* lines, respectively. (D), (E), and (F) show the corresponding autocorrelation functions. *With permission from Britvina, T., Eggermont. J.J., 2008. Multi-frequency stimulation disrupts spindling activity in anesthetized animals. Neuroscience 151, 888–900. Elsevier.*

was independent of the power of the spindle waves during silence, and that the suppression of spindle power occurred within about one spindle period after stimulus onset. The suppression of spindle power during stimulation had no effect on the spatial correlation pattern of spindle waves. Firing rates increased under sound stimulation and spectrotemporal receptive fields could reliably be obtained (Britvina and Eggermont, 2008b) and were more pronounced in the spindle epochs than in the interspindle periods.

These findings question the validity of the "blockade of synaptic transmission" in the thalamus during spindling proposed by Steriade and Llinâs (1988), unless the sound-evoked activity reaches the auditory cortex through the midbrain reticular formation (MRF) and interlaminar thalamus—bypassing the auditory thalamus. Supporting this thalamic bypass are the fast maturing P_1 and P_2 components of the cortical auditory-evoked potential generated by the MRF (Ponton et al., 2000). However, the latencies thereof are much longer than the spike latencies in A1, indicating that they arrive through the standard afferent pathway.

3.1.3 Alpha modulates through inhibition

Alpha-band oscillations have inhibitory and gating roles that are closely linked to two-tone suppression and selection of attention (Klimesh, 2012). According to Jensen et al. (2012, 2014), posterior alpha oscillations (8–13Hz) provide an inhibitory mechanism for prioritizing and ordering unattended visual input according to "relevance." Interhemispheric coupling of auditory cortices is preferentially synchronized in the alpha frequency band (~ 7–12Hz) in response to auditory stimulation (Ghuman et al. 2013).

Sadaghiani et al. (2012) simultaneously recorded resting-state fMRI and EEG in 26 participants. They found that global synchrony fluctuations in the upper alpha band (individual peak alpha frequency + 2Hz) were positively correlated with activity in the dorsal attention network (DAN), cinguloopercular network (CON), and frontoparietal network (FPN). No other EEG frequency band showed equivalent results. This selective association supports an intrinsic relation between large-scale alpha phase synchrony and cognitive functions associated with the FPN (Chapter 4). Sadaghiani et al. (2012) suggested "a hierarchical relationship involving a role of alpha oscillations in tonic alertness, of their long-range phase locking in phasic alertness, and of their focal desynchronization in selective attention." Sadaghiani and Kleinschmidt (2016) noted that the neurophysiological importance of alpha-oscillations stems from their modulation of neural excitability. When the power of alpha-oscillations increases in a given brain region, overall neuronal activity in this brain region decreases (Haegens et al., 2011). This inhibitory effect is cyclic with its maximum

TABLE 3.1 Summary of the relation between cognitive control networks, alpha-band parameters, and cognitive functions.

	CON	DAN	FPN
Alpha-band parameter	Global alpha-power increase	Focal alpha-power decrease	Long-range alpha-phase locking
Effective mechanism	Noise clearance and information updating	(Dis-)engagement of regions	Distant communication
Cognitive function	Tonic alertness	Selective attention	Phasic adaptive-control

CON, cinguloopercular network; *DAN*, dorsal attention network; *FPN*, fronto parietal network.
Based on Sadaghiani, S., Kleinschmidt, A., 2016. Brain networks and a oscillations: structural and functional foundations of cognitive control. Trends Cogn. Sci. 20, 805–817.

occurring at the peak of the alpha-oscillation cycle. In other words, neural activity as measured by spike-firing rates (Haegens et al., 2011) and high-frequency population activity (gamma band > 30 Hz) in the intracranial electrocorticogram (ECoG) or MEG (Osipova et al., 2008) and fMRI (Scheeringa et al., 2011) is greater during the trough of the alpha cycle. The impact on gamma-oscillations where the power of gamma is coupled to the phase of alpha-oscillations is referred to as cross-frequency phase-amplitude coupling (Jensen et al., 2014; Chapter 2). Conversely, reduced alpha power leads to a release from this inhibition.

Based on the role of alpha-oscillations in cognition as described earlier, and the selective relation of these oscillations with the three core cognitive control networks, Sadaghiani and Kleinschmidt (2013, 2016) suggested a model of cognitive control mechanisms (Table 3.1). This model links the changes in alpha and putative mechanisms with cognitive functions. CON-activity time series are correlated during speech recognition across its sub-networks: dorsal cingulate cortex, frontal operculum, and right anterior insula/frontal operculum. These regions are connected by callosal fibers, as well as the frontal aslant tract that connects medial and inferior frontal regions (Eckert et al., 2016). Areas located on the boundary between-network communities, particularly the DAN, facilitate the integration or segregation of diverse cognitive systems (Gu et al., 2015).

Using EEG in five human subjects and one macaque monkey, De Pesters et al. (2016) investigated the temporal and spatial relationships between electrode locations in power modulations in the alpha band and in the broadband gamma range (70–170 Hz) during auditory and motor tasks. Their results confirm that broadband gamma power accurately tracks task-related behavior and that alpha power decreases in task-related areas. They also demonstrated that alpha power suppression lags gamma activity in auditory areas during the auditory task but precedes it in motor

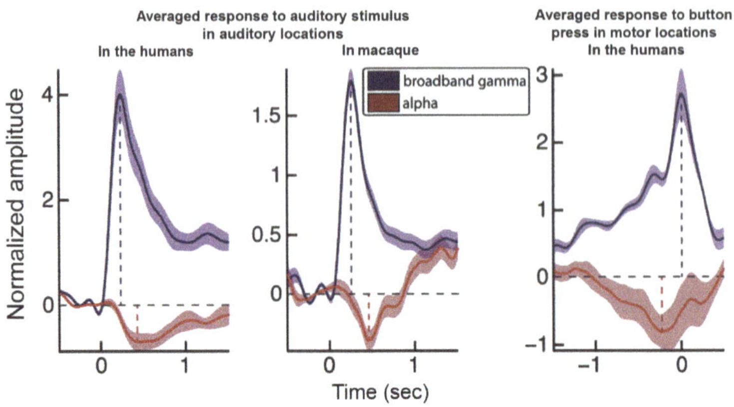

FIG. 3.2 Alpha power suppression lags broadband gamma power in auditory areas during the auditory task but precedes it in the motor areas during the motor task. The time courses depict the averaged responses in auditory locations during the auditory task in humans *(left panel)*, in macaque *(middle panel)*, and in motor locations during the motor task in humans *(right panel)* for broadband gamma *(blue)* and alpha *(red)* bands. Semitransparent shading represents the standard error. The *vertical dashed lines* indicate the timing of the positive peaks of the broadband gamma band *(blue)* and the negative peaks of alpha band *(red)* for each task. *With permission from De Pesters, A., Coon, W.G., Brunner, P., Gunduz, A., Ritaccio, A.L., Brunet, N.M., et al., 2016. Alpha power indexes task-related networks on large and small scales: A multimodal ECoG study in humans and a non-human primate. NeuroImage 134, 122–131. Elsevier.*

areas during the motor task (Fig. 3.2). De Pesters et al. (2016) showed that these differential modulations of alpha power could be observed not only across widely distributed systems (e.g., motor vs auditory system) but also within the auditory system. Specifically, alpha power was changed in 91% the locations within the auditory system that most robustly responded to particular sound stimuli.

Summarizing and extending the modulating role of alpha oscillations, Becker et al. (2018) noted that numerous EEG-fMRI studies have reported less metabolic demand during alpha activity (Becker et al., 2011; de Munck et al., 2007; Goldman et al., 2002; Laufs et al., 2003; Moosmann et al., 2003). This is seemingly compatible with the classical view of alpha activity being a passive, idling brain state. However, other viewpoints, specifically "gating-by-inhibition" (Jensen and Mazaheri, 2010), propose that alpha oscillations actively exert pulsed inhibition and that alpha desynchronization releases from this inhibition, thereby facilitating neural processing. Along the same lines, Sadaghiani and Kleinschmidt (2016) have suggested the metaphor of a "windshield wiper" proposing not only a suppressive but also updating mechanism that emphasizes current over accumulated earlier information. Becker et al. (2018) postulated that: "alpha oscillations, believed to exert active inhibitory gating, down-modulate the temporal width of long-range dependence in slower ongoing brain activity."

3.1.4 Alpha and speech coding

Monosyllabic pairs of words in which the first word ended with an/s/ and the second began with an/s/ (e.g., gas source) were chosen and paired together. In an EEG study where listeners had to report hearing one word or two words, Shahin and Pitt (2012) found that bursts of frontocentral alpha activity (9–14 Hz), following the onset of the physical /s/ and end of phrase, indexed speech segmentation. Furthermore, left-lateralized beta activity (14–18 Hz) following the end of phrase distinguished word (e.g., gas source) from nonword (e.g., nas sorf) segmentation.

As we have seen, enhanced alpha activity reflects inhibition of task-irrelevant neural populations. Strauß et al. (2015) used EEG to examine the influence of alpha phase on a lexical-decision task performed for stimuli (words and pseudoword counterparts) embedded in noise. Neural phase angles, relative to stimulus onset, were compared for correct versus incorrect lexical decisions using differences in mean phase angles and phase concentrations between correct and incorrect trials. Neural phase angles in the alpha frequency range (8–12 Hz) over right anterior sensors were approximately antiphase in a 100-ms prestimulus time window, and thus successfully distinguished between correct and incorrect lexical decisions. Moreover, alpha-band oscillations were again approximately antiphase across participants for correct versus incorrect trials during a later peristimulus time window (~ 500 ms poststimulus onset) at left-central electrodes. Lexical decision accuracy was not predicted by either event-related potentials or oscillatory power measures.

In MEG recordings from the left auditory cortex, Müller et al. (2015) showed that alpha suppression found when participants prepare to listen disappears when they expected a self-spoken sound. This suggests an inhibitory adjustment of auditory cortical activity already before sound onset. They also demonstrated that the medial prefrontal cortex, a region known for self-referential processes, mediates these condition-specific alpha power modulations (Fig. 3.3).

Billig et al. (2019) recorded ECoGs from patients to measure oscillations in the alpha range (7–14 Hz) in auditory brain regions at different levels of sensory processing. They found that these oscillations dominated in one particular nonprimary field, anterolateral Heschl's gyrus, and were suppressed when subjects listened to sentences. The suppression of alpha decreased with distance from anterolateral Heschl's gyrus throughout superior temporal cortex. Thus alpha oscillations have a differential manifestation and stimulus-sensitivity in primary and nonprimary auditory cortex.

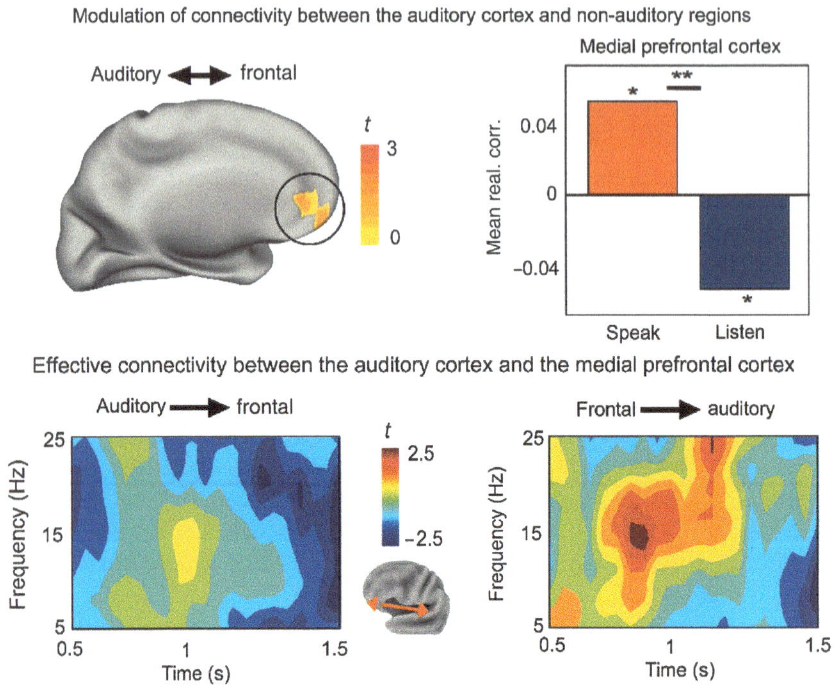

FIG. 3.3 Connectivity between medial prefrontal and auditory cortex. The *upper panel* shows the spatial dimension of the significant cluster derived from power-power correlations with the left auditory cortex. Alpha power in the auditory cortex was significantly stronger correlated with low-frequency power in the medial prefrontal cortex *(depicted by black circle)* when participants expected to listen to their self-spoken versus externally played-back voice (cluster $P<.05$, 1–1.5 s, 6–14 Hz, medial prefrontal cortex). The effect was due to a significant increase of power correlations compared with baseline when participants were expecting their self-produced voice ($P<.05$) and a significant decrease of power correlations when they expected their externally played-back voice ($P<.05$). The lower panel depicts effective connectivity between the left auditory cortex and the medial prefrontal cortex, quantified by Partial Directed Coherence. Information flow from the auditory cortex to the medial prefrontal cortex did not differ between conditions *(left)*. In contrast, information flow from the medial prefrontal cortex to the left auditory cortex was significantly stronger for the self-spoken versus externally played-back speech sounds *(right)*, with the effect being strongest for the time interval slightly preceding the main auditory alpha power modulations (0.7–1 s). *From Muller, N., Leske, S., Hartmann, T., Szebényi, S., Weisz, N., 2015. Listen to yourself: The medial prefrontal cortex modulates auditory alpha power during speech preparation. Cereb. Cortex 25, 4029–4037. By permission of Oxford University Press.*

3.2 Delta waves

Delta and theta waves (Chapter 2) are generated in superficial cortical layers and are synchronous across broad cortical areas. Considerable evidence now suggests that delta oscillations, besides occurring in slow-wave sleep (Amzica and Steriade, 1988) are also involved in the cognitive domain. Delta has been shown to aid decision-making in the somatosensory system by coherence between parietal and frontal brain areas (Nácher et al., 2013). In speech detection, delta waves that generally build up during perception appear biased toward correct responses.

3.2.1 Delta waves in decision-making

Stefanics et al. (2010) showed that performance changes, related to different levels of expectancy, originate in dynamic modulation of delta oscillation phase. They used detection tasks with rhythmic auditory targets, and found significant entrainment of the delta rhythm to the onset of the target tones with increasing phase synchronization at higher levels of predictability. Reaction times correlated with the phase of the delta band oscillation at target onset. The fastest reactions occurred during the delta phase that most commonly coincided with the target event in the high expectancy conditions. Stefanics et al. (2010) suggested that: "low-frequency oscillations play a functional role in human anticipatory mechanisms, presumably by modulating synchronized rhythmic fluctuations in the excitability of large neuronal populations and by facilitating efficient task-related neuronal communication among brain areas responsible for sensory processing and response execution."

Arnal and Giraud (2012) reviewed the neurophysiological bases of sensory predictions of "what" (predictive coding) and "when" (predictive timing), with an emphasis on low-level oscillatory mechanisms. They defined predictive timing as the process by which uncertainty about "when" events are likely to occur is minimized to facilitate their processing and detection (e.g., for periodic stimuli; Andreou et al., 2011). At the neurophysiological level, anticipating sensory events reset the phase of slow, delta-theta (2–8 Hz) activity before the stimulus occurs. The predictive alignment of delta-theta oscillations in an ideal excitability phase speeds up stimulus detection (Lakatos et al., 2008). When an expected event falls outside the expected temporal window, it receives reduced processing (Lakatos et al., 2008). If, conversely, an unexpected event (oddball) falls within the expected window (e.g., in mismatch negativity, MMN paradigms), then the "predicting what" scheme applies, whereby the response reflects the mismatch between expectation and input. Arnal and Giraud (2012) noted that the magnitude of neural responses depends on whether temporal predictions are tested in the presence or absence of explicit attention to the nature or location of those events (Sadaghiani et al., 2010). Attention interacts with predictive timing and enhances (rather than dampens) neural responses to stimuli that fall into its focus (Schroeder and Lakatos, 2009; Lakatos et al., 2008). However, when stimuli are implicitly expected based on their temporal regularity, early sensory responses are reduced (Costa-Faidella et al., 2011). In addition, entrained oscillations may become predictive by creating periodic temporal windows that higher-order regions can rely on to read out the encoded information.

3.2.2 Delta waves and speech coding

Short-duration speech cues with a high modulation frequency, typically in the ~ 30–50Hz gamma range, are associated with an important part of the speech fine structure, and correlate with attributes at the phonemic scale, such as formant transitions (e.g., /ba/ vs /da/), and the coding of voicing (e.g., /ba/ vs /pa/). Almost an order of magnitude slower, the acoustic envelope of naturalistic speech closely correlates with syllabic rate and has a modulation spectrum that peaks in the theta range between 4 and 7Hz. The intonation contour of an utterance occurs at yet a lower modulation rate, roughly 1–2Hz, that is, in the delta range. Although the temporal modulations on these three scales are not periodic, they are sufficiently rhythmic to elicit robust EEG/MEG regularities in the time domain, even in single utterances (Giraud and Poeppel, 2012).

Using MEG recordings, Doelling et al. (2014) found evidence that sharp events in the stimulus cause cortical rhythms to realign and parse the stimulus into syllable-sized chunks for further decoding. The frequency range of the neural effects that Doelling et al. (2014) observed was between 2 and 4Hz. They noted that this range matches the average syllabic rate and modal frequency of the temporal envelope of the stimuli (~ 3Hz). By removing temporal fluctuations in the stimulus that occur at the syllabic rate, envelope-tracking MEG activity, and intelligibility of the stimulus was reduced.

3.2.3 Delta coupling with other frequency bands

Arnal et al. (2015) hypothesized that coupled oscillations in the delta and beta bands would be instrumental in predicting the occurrence of auditory targets. While undergoing MEG recording, participants listened to brief rhythmic tone sequences (stimulus onset asynchrony of 800, 1000, or 1200ms) and detected target delays. Prior to target occurrence, Arnal et al. (2015) found that coupled delta (1–3Hz) and beta (18–22Hz) oscillations temporally aligned with upcoming targets and biased decisions toward correct responses, suggesting that delta-beta coupled oscillations underlie prediction accuracy (Fig. 3.4). Following target occurrence, participants updated their decisions using the magnitude of the alpha-band (10–14Hz) response as internal target timing (Mai et al., 2016).

Do neural oscillations alter local and global connectivity during audio-visual integration processing? Using EEG and an audio-visual speech-matching task, Ohki et al. (2016) measured amplitude-phase coupling and phase coherence between selected brain regions. They found

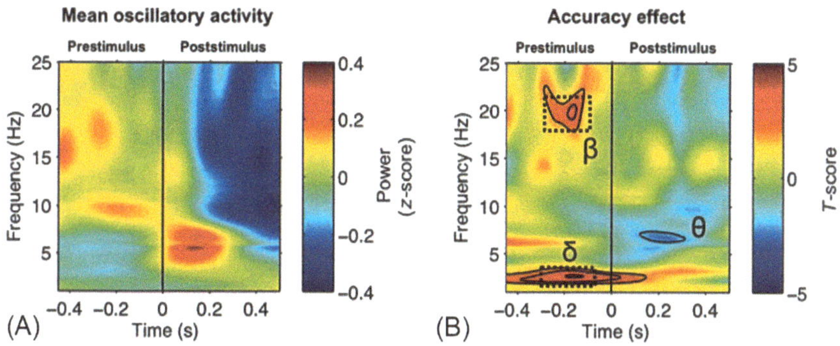

FIG. 3.4 Time-frequency analysis of prestimulus to poststimulus MEG activity and accuracy effects. (A) Mean oscillatory activity of responses to target tones in auditory sensors (selected individually, see Materials and Methods). Time is expressed relative to target onset *(black vertical line)*. (B) Accuracy effect (correct minus incorrect stimulus) on oscillatory activity. Effect is expressed in *t*-scores; contours indicate significance thresholds of $P<.05$, .01, and .001 for those clusters that survived the correction far multiple comparisons. *Black-dashed boxes* highlight prestimulus (− 300 to − 100 ms) accuracy effects in delta (1–3 Hz) and beta (18–22 Hz) power, corresponding to a stronger increase in these two frequency bands for targets that are subsequently correctly perceived. *From Arnal, L.H., Doelling, K.B., Poeppel, D., 2015. Delta–beta coupled oscillations underlie temporal prediction accuracy. Cereb. Cortex 25, 3077–3055. By permission of Oxford University Press.*

that, in the temporal pole, beta band amplitude was coupled with the phase of the task-related delta band activity. This supports the idea that beta oscillations in the temporal pole are related to processing combined audio-visual information. Furthermore, they demonstrated that the temporal pole and other multisensory areas (e.g., posterior parietal cortex and postcentral areas) are coordinated through delta-phase coherence. Thus both modulation of connectivity within the local network, and between the local and global network, play a role in audio-visual speech integration.

The aforementioned studies show that auditory cortex functions within a large network of other temporal and frontoparietal regions to provide the neural representations of speech. Using source-localized MEG data, Keitel et al. (2017) found that delta band entrainment interacted with the oscillatory activity in three distinct networks that control the parsing of continuous, natural speech by delta entrainment (Fig. 3.5). "In the left hemisphere, orbitofrontal areas influence the alignment of rhythmic auditory cortical activity to speech, possibly by providing predictive information. These predictions are also conveyed to the motor system, as shown by a significant cross-correlation between frontal beta power and central alpha. The motor system in turn interacts with delta-entrainment, possibly by exploiting the frontal predictions and utilizing its ability to process rhythms in a temporally precise manner. In the right hemisphere, parietal

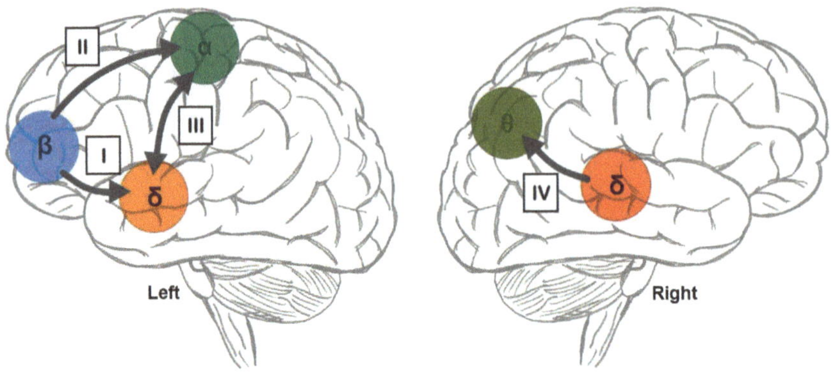

FIG. 3.5 Summary and time course of network-level effects of delta-entrainment in auditory areas. *Left hemisphere*: (I) Prefrontal predictions facilitate delta-speech tracking in aSTG. (II) Frontal predictions are conveyed to motor areas. (III) Motor areas interact with delta-entrainment in aSTG. *Right hemisphere*: (IV) Delta-speech tracking in pSTG engages parietal areas. *aSTG*, anterior superior temporal gyrus; *pSTG*, posterior superior temporal gyrus. *From Keitel, A., Ince, R.A.A., Gross, J., Kayser, C., 2017. Auditory cortical delta-entrainment interacts with oscillatory power in multiple fronto-parietal networks. NeuroImage 147, 32–42. Open Access.*

theta processes are particularly engaged in moments when auditory delta entrainment is weak." These results place the entrainment of rhythmic auditory cortical activity in the context of fronto-parietal speech relevant networks. They suggest functional and possibly mechanistic relations between auditory cortical entrainment, semantic, and lexical processes at higher stages of the auditory pathways and perception (Keitel et al., 2017).

3.3 Brain rhythms and event-related potentials

A prescient idea for auditory event-related potential (ERP) generation (Sayers and Beagley, 1974), more recently revived for visual ERPs (Jansen and Brandt, 1991; Makeig et al., 2001), suggests that ERPs could be based on the phase-synchronization of the EEG-rhythm generators. The idea of Sayers and Beagley (1974) was that stimulation with clicks or tone pips could synchronize the more or less random rhythmic activity of groups of neurons and make it visible at the scalp.

But this idea appears controversial. In contrast to the idea that averaged sensory ERPs are generated by partial phase resetting of ongoing EEG oscillations, Shah et al. (2004) proposed that ERPs predominantly result from stimulus-evoked neural responses, leaving open the possibility that phase resetting contributes to this process. Mäkinen et al. (2005) also approached this issue by examining single-trial auditory MEG data in humans. They demonstrated that phase coherence over trials can result from both a phase-coherent state of ongoing oscillations and from the presence of a phase-coherent event-related response which is additive to ongoing oscillations. Based on amplitude variance to establish the relationship between ongoing oscillations and event-related responses, Mäkinen et al. (2005) found that auditory stimuli do not give rise to phase reorganization of ongoing activity. They concluded that auditory ERPs are unique descriptors of neural information processing in humans, generated by processes separate from and additive to ongoing brain oscillations.

However, these two distinct mechanisms could look very similar in the case of rhythmic input. Doelling et al. (2019) ran numerical simulations of evoked and oscillatory computational models, showing that in the evoked case, phase lag is heavily rate-dependent, while the oscillatory model displays marked phase concentration across stimulation rates. They then compared these predictions with MEG data recorded while participants listened to music of varying note rates. Doelling et al. (2019) could show that the phase concentration of the experimental data is more in line with the oscillatory model than with the evoked model. This finding supports an auditory cortical signal that (1) contains components of both bottom-up evoked responses and internal oscillatory synchronization whose strengths are weighted by their appropriateness for particular stimulus types and (2) cannot be explained by evoked responses alone. Taken together, this argues for a model of auditory entrainment, whereby the auditory cortex actively synchronizes a low-frequency oscillator with the rhythms present in sound from 1 to 8 Hz.

3.4 Summary

Traditionally, alpha and delta waves have been associated with idling and sleep, respectively. More recently, it has been found that both rhythms play an important role in cognition. Alpha modulates by inhibiting activity of task irrelevant neural populations, and when desynchronized, that is, of lower amplitude, increases selective attention among others. Prestimulus alpha power suppression lags gamma activity in auditory tasks but precedes motor task gamma. Delta waves play a role in anticipation and predictive coding, as well as modulate reaction times. Delta entrained by speech interacts in distant brain networks, suggesting a functional relationship between this entrainment and semantic as well as lexical processes. Are evoked potentials and stimulus-induced brain rhythms—delta, theta, and alpha—independent or does stimulation reset the brain rhythms (by entrainment) to form evoked potentials? There is evidence for both depending on the specific stimulation, that is, transient or repetitive.

References

Amzica, F., Steriade, M., 1988. Electrophysiological correlates of sleep delta waves. Electroencephalogr. Clin. Neurophysiol. 107, 69–83.

Andreou, L.V., Kashino, M., Chait, M., 2011. The role of temporal regularity in auditory segregation. Hear. Res. 280, 228–235.

Arnal, L.H., Giraud, A.-L., 2012. Cortical oscillations and sensory predictions. Trends Cogn. Sci. 16 (7), 390–398.

Arnal, L.H., Doelling, K.B., Poeppel, D., 2015. Delta–beta coupled oscillations underlie temporal prediction accuracy. Cereb. Cortex 25, 3077–3085.

Becker, R., Reinacher, M., Freyer, F., Villringer, A., Ritter, P., 2011. How ongoing neuronal oscillations account for evoked fMRI variability. J. Neurosci. 31 (10), 11016–11027.

Becker, R., Van de Ville, D., Kleinschmidt, A., 2018. Alpha oscillations reduce temporal long-range dependence in spontaneous human brain activity. J. Neurosci. 38 (3), 755–764.

Billig, A.J., Herrmann, B., Rhone, A.E., Gander, P.E., Nourski, K.V., Snoad, B.F., et al., 2019. A sound-sensitive source of alpha oscillations in human non-primary auditory cortex. J. Neurosci. 39 (44), 8679–8689.

Britvina, T., Eggermont, J.J., 2008a. Multi-frequency stimulation disrupts spindling activity in anesthetized animals. Neuroscience 151, 888–900.

Britvina, T., Eggermont, J.J., 2008b. Spectro-temporal receptive fields during spindling and non-spindling epochs in cat primary auditory cortex. Neuroscience 154, 1576–1588.

Costa-Faidella, J., Baldeweg, T., Grimm, S., Escera, C., 2011. Interactions between "what" and 'when' in the auditory system: temporal predictability enhances repetition suppression. J. Neurosci. 31 (50), 18590–18597.

de Munck, J.C., Gonçalves, S.I., Huijboom, L., Kuijer, J.P., Pouwels, P.J., Heethaar, R.M., et al., 2007. The hemodynamic response of the alpha rhythm: an EEG/fMRI study. NeuroImage 35, 1142–1151.

De Pesters, A., Coon, W.G., Brunner, P., Gunduz, A., Ritaccio, A.L., Brunet, N.M., et al., 2016. Alpha power indexes task-related networks on large and small scales: a multimodal ECoG study in humans and a non-human primate. NeuroImage 134, 122–131.

Doelling, K.B., Arnal, L.H., Ghitza, O., Poeppel, D., 2014. Acoustic landmarks drive delta–theta oscillations to enable speech comprehension by facilitating perceptual parsing. NeuroImage 85, 761–768.

Doelling, K.B., Assaneo, M.F., Bevilacqua, D., Pesaran, B., Poeppel, D., 2019. An oscillator model better predicts cortical entrainment to music. Proc. Natl. Acad. Sci. U. S. A. 116 (20), 10113–10121.

Eckert, M.A., Teubner-Rhodes, S., Vaden, K.I., 2016. Is listening in noise worth it? The neurobiology of speech recognition in challenging listening conditions. Ear Hear. 37, 101S–110S.

Ghuman, A.S., van den Honert, R.N., Martin, A., 2013. Interregional neural synchrony has similar dynamics during spontaneous and stimulus-driven states. Sci. Rep. 3, 1481. https://doi.org/10.1038/srep01481.

Giraud, A.L., Poeppel, D., 2012. Cortical oscillations and speech processing: emerging computational principles and operations. Nat. Neurosci. 15, 511–517.

Goldman, R.I., Stern, J.M., Engel Jr., J., Cohen, M.S., 2002. Simultaneous EEG and fMRI of the alpha rhythm. Neuroreport 13, 2487–2492.

Gu, S., Pasqualetti, F., Cieslak, M., Telesford, Q.K., Yu, A.B., Kahn, A.E., et al., 2015. Controllability of structural brain networks. Nat. Commun. 6, 8414. https://doi.org/10.1038/ncomms9414.

Haegens, S., Händel, B.F., Jensen, O., 2011. Top-down controlled alpha band activity in somatosensory areas determines behavioral performance in a discrimination task. J. Neurosci. 31 (14), 5197–5204.

Haegens, S., Barczak, A., Musacchia, G., Lipton, M.L., Mehta, A.D., Lakatos, P., et al., 2015. Laminar profile and physiology of the a rhythm in primary visual, auditory, and somatosensory regions of neocortex. J. Neurosci. 35 (42), 14341–14352.

Jansen, B.H., Brandt, M.E., 1991. The effect of the phase of prestimulus alpha activity on the averaged visual evoked response. Electroencephalogr. Clin. Neurophysiol. 80, 241–250.

Jensen, O., Mazaheri, A., 2010. Shaping functional architecture by oscillatory alpha activity: gating by inhibition. Front. Hum. Neurosci. 4, 186. https://doi.org/10.3389/fnhum.2010.00186.

Jensen, O., Bonnefond, M., VanRullen, R., 2012. An oscillatory mechanism for prioritizing salient unattended stimuli. Trends Cogn. Sci. 16 (4), 200–2006.

Jensen, O., Gips, B., Bergmann, T.O., Bonnefond, M., 2014. Temporal coding organized by coupled alpha and gamma oscillations prioritize visual processing. Trends Neurosci. 37 (7), 357–369.

Keitel, A., Ince, R.A.A., Gross, J., Kayser, C., 2017. Auditory cortical delta-entrainment interacts with oscillatory power in multiple fronto-parietal networks. NeuroImage 147, 32–42.

Klimesch, W., 1999. EEG alpha and theta oscillations reflect cognitive and memory performance: a review and analysis. Brain Res. Rev. 29, 169–195.

Klimesh, W., 2012. Alpha-band oscillations, attention, and controlled access to stored information. Trends Cogn. Sci. 16 (12), 606–617.

Lakatos, P., Karmos, G., Mehta, A.D., Ulbert, I., Schroeder, C.E., 2008. Entrainment of neuronal oscillations as a mechanism of attentional selection. Science 320, 110–113.

Laufs, H., Kleinschmidt, A., Beyerle, A., Eger, E., Salek-Haddadi, A., Preibisch, C., et al., 2003. EEG-correlated fMRI of human alpha activity. NeuroImage 19, 1463–1476.

Lopes da Silva, F., 1991. Neural mechanisms underlying brain waves: from neural membranes to networks. Electroencephalogr. Clin. Neurophysiol. 79, 81–93.

Mai, G., Minett, J.W., Wang, W.S., 2016. Delta, theta, beta, and gamma brain oscillations index levels of auditory sentence processing. NeuroImage 133, 516–528.

Makeig, S., Westerfield, M., Jung, T.P., Enghoff, S., Townsend, J., Courchesne, E., et al., 2001. Dynamic brain sources of visual evoked responses. Science 295, 690–694.

Mäkinen, V., Tiitinen, H., May, P., 2005. Auditory event-related responses are generated independently of ongoing brain activity. NeuroImage 24, 961–968.

Manshanden, I., De Munck, J.C., Simon, N.R., Lopes da Silva, F.H., 2002. Source localization of MEG sleep spindles and the relation to sources of alpha band rhythms. Clin. Neurophysiol. 113 (12), 1937–1947.

Moosmann, M., Ritter, P., Krastel, I., Brink, A., Thees, S., Blankenburg, F., et al., 2003. Correlates of alpha rhythm in functional magnetic resonance imaging and near infrared spectroscopy. NeuroImage 20, 145–158.

Muller, N., Leske, S., Hartmann, T., Szebenyi, S., Weisz, N., 2015. Listen to yourself: the medial prefrontal cortex modulates auditory alpha power during speech preparation. Cereb. Cortex 25, 4029–4037.

Nácher, V., Ledberg, A., Deco, G., Romo, R., 2013. Coherent delta-band oscillations between cortical areas correlate with decision making. Proc. Natl. Acad. Sci. U. S. A. 110 (37), 15085–15090.

Ohki, T., Gunji, A., Takei, Y., Takahashi, H., Kaneko, Y., Kita, Y., et al., 2016. Neural oscillations in the temporal pole for a temporally congruent audio-visual speech detection task. Sci. Rep. 6, 37973. https://doi.org/10.1038/srep37973.

Osipova, D., Hermes, D., Jensen, O., 2008. Gamma power is phase-locked to posterior alpha activity. PLoS One 3. https://doi.org/10.1371/journal.pone.0003990.

Ponton, C.W., Eggermont, J.J., Kwong, B., Don, M., 2000. Maturation of human central auditory system activity: evidence from multi-channel evoked potentials. Clin. Neurophysiol. 111, 220–236.

Sadaghiani, S., Kleinschmidt, A., 2013. Functional interactions between intrinsic brain activity and behavior. NeuroImage 80, 379–386.
Sadaghiani, S., Kleinschmidt, A., 2016. Brain networks and a oscillations: structural and functional foundations of cognitive control. Trends Cogn. Sci. 20, 805–817.
Sadaghiani, S., Hesselmann, G., Friston, K.J., Kleinschmidt, A., 2010. The relation of ongoing brain activity, evoked neural responses and cognition. Front. Syst. Neurosci. 4, 20. https://doi.org/10.3389/fnsys.2010.00020.
Sadaghiani, S., Scheeringa, R., Lehongre, K., Morillon, B., Giraud, A.-L., D'Esposito, M., et al., 2012. Alpha-band phase synchrony is related to activity in the fronto-parietal adaptive control network. J. Neurosci. 32 (41), 14305–14310.
Sayers, B.M., Beagley, H.A., 1974. Objective evaluation of auditory evoked EEG responses. Nature 251, 608–609.
Scheeringa, R., Mazaheri, A., Bojak, I., Norris, D.G., Kleinschmidt, A., 2011. Modulation of visually evoked cortical fMRI responses by phase of ongoing occipital alpha oscillations. J. Neurosci. 31 (10), 3813–3820.
Schroeder, C.E., Lakatos, P., 2009. Low-frequency neuronal oscillations as instruments of sensory selection. Trends Neurosci. 32, 9–18.
Shah, A.S., Bressler, S.L., Knuth, K.H., Ding, M., Mehta, A.D., Ulbert, I., et al., 2004. Neural dynamics and the fundamental mechanisms of event-related brain potentials. Cereb. Cortex 14, 476–483.
Shahin, A.J., Pitt, M.A., 2012. Alpha activity marking word boundaries mediates speech segmentation. Eur. J. Neurosci. 36 (12), 3740–3748.
Stefanics, G., Hangya, B., Hernádi, I., Winkler, I., Lakatos, P., Ulbert, I., 2010. Phase entrainment of human delta oscillations can mediate the effects of expectation on reaction speed. J. Neurosci. 30 (41), 13578–13585.
Steriade, M., Llinâs, R.R., 1988. The functional states of the thalamus and the associated neuronal interplay. Physiol. Rev. 68 (3), 649–742.
Strauß, A., Henry, M.J., Scharinger, M., Obleser, J., 2015. Alpha phase determines successful lexical decision in noise. J. Neurosci. 35 (7), 3256–3262.

CHAPTER

4

Brain rhythms underlying perception and cognition

Conscious perception may require synchronous oscillatory activity in the gamma-band range of widely distributed cortical regions (Hipp et al., 2011). This implies that neural assemblies participating in a specific task-relevant network or processing features related to the same perception communicate through neural synchronization (Engel et al., 2001; Fries, 2009).

4.1 Rhythms in perception and cognition

4.1.1 Auditory perception

4.1.1.1 Binaural mechanisms

While the role of synchronized oscillatory activity in the gamma-band frequency range for conscious perception is well established in the visual domain (Engel and Singer, 2001; Chapter 2), there is less evidence concerning neurophysiological mechanisms in the various electroencephalogram (EEG)-frequency bands in conscious auditory perception (Steinmann et al., 2014). Recall that auditory processing is performed in a temporofrontal network stretching across the left and right hemispheres (Chapter 1). Apart from regional specialization, the integrity of interhemispheric interaction plays an important role in syllable perception (Pollmann et al., 2002) and speech comprehension (Friederici et al., 2007; Steinmann and Mulert, 2012). The anatomical link for this interaction is provided by the corpus callosum (CC); through the posterior third—isthmus and splenium—run neural pathways connecting right and left auditory cortices. Dichotic listening tasks—the simultaneous stimulation of the right and left ear by different sounds—show more frequent conscious perception of the speech syllable presented to the right ear. This is attributed to the stronger contralateral pathways running from the right ear to the speech-dominant left hemisphere. In contrast, the left ear input initially accesses the right

https://doi.org/10.1016/B978-0-12-819818-6.00012-1

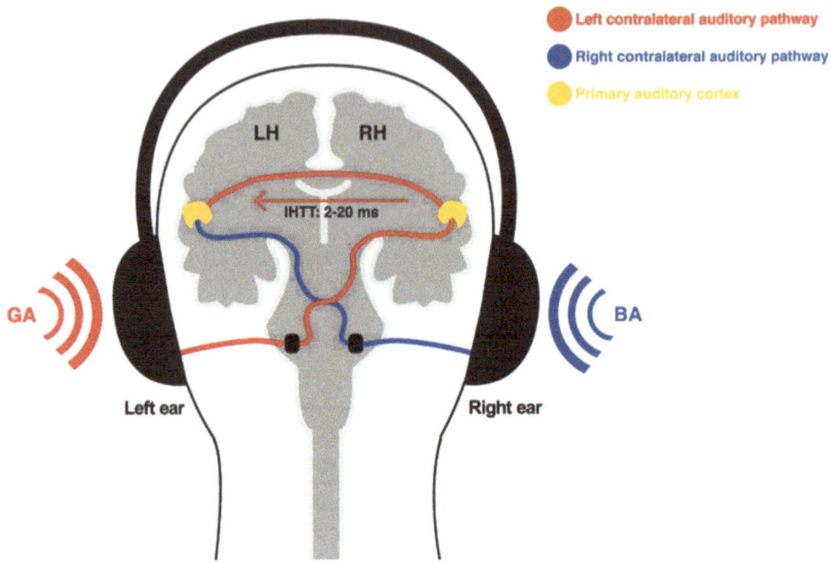

FIG. 4.1 Schematic illustration of the dichotic listening procedure in which two different syllables are presented simultaneously: one to the right ear (BA) and one to the left ear (GA). Although both ears are connected anatomically through ipsilateral and contralateral pathways with the auditory cortices, ipsilateral pathways are inhibited by the stronger contralateral activity during dichotic listening. The *blue line* indicates the contralateral pathway connecting the right ear-stimulus directly with the speech-dominant left hemisphere. In contrast, the syllable arriving from the left ear to the right hemisphere has to cross the CC to be efficiently processed *(red line)*. *Yellow*: Right and left primary auditory cortices. *IHTT*, inter-hemispheric transfer time; *LH*, left hemisphere; *RH*, right hemisphere. *Reprinted with permission from Steinmann, S., Leicht, G., Ertl, M., Andreou, C., Polomac, N., Westerhausen, R., et al., 2014. Conscious auditory perception related to long-range synchrony of gamma oscillations. NeuroImage 100, 435–443. Copyright 2014. Elsevier.*

hemisphere and needs additional transfer through interhemispheric pathways before it is processed in the left hemisphere (Fig. 4.1).

Using lagged-phase synchronization (LPS) analysis and source estimation, Steinmann et al. (2014) examined the functional connectivity between right and left primary (A1) and secondary (A2) auditory cortical areas in delta, theta, alpha, beta, and gamma bands during right/left ear reports. Interhemispheric LPS between right and left A1 and A2 was specifically increased in the gamma-band range, when participants consciously perceived the syllable presented to the left ear. Steinmann et al. (2014) suggested that enhanced synchronous gamma oscillations improve interhemispheric transfer and lead to a better perception of speech syllables presented to the left ear. In contrast, they (Steinmann et al., 2018) later found that during lower levels of gamma synchrony the perception of the dominant right ear is improved.

4.1.1.2 *Role of prestimulus brain rhythms*

Ongoing fluctuations in brain activity influence trial-by-trial perception of identical visual (Arieli et al., 1996) and tactile (Boly et al., 2007) stimuli where higher ongoing activity in the dorsal attention network (DAN) would facilitate perceptual performance, and higher activity in the default mode network (DMN) would deteriorate perceptual performance. Sadaghiani et al. (2009) tested whether these findings in the visual and somatosensory systems generalize to the auditory modality, and consequently to a nonspatial but memory-dependent paradigm. In a sparse event-related functional magnetic resonance imaging (fMRI) design, detection of auditory near-threshold stimuli was tested in relation to prestimulus baseline activity in early auditory cortex and in several distributed networks that were defined on the basis of resting state functional connectivity. The stimulus was a 500-ms noise burst with its frequency content modulated at 2 Hz—ranging from white noise to a narrower band of 0–5 kHz and back to white noise. Correct detections were associated with higher prestimulus activity in related early auditory cortex and in the anterior insula, anterior cingulate cortex (ACC), and thalamus. In contrast to findings in spatial modalities such as vision and somatosensory (Burton and Sinclair, 2000; Corbetta and Shulman, 2002; Robin, 2018) in the auditory modality higher prestimulus activity in the DAN of frontal and parietal cortex was biased toward misses, whereas higher activity in the DMN that includes posterior cingulate cortex (PCC) and precuneus was biased toward hits. This contradicts the simple dichotomic view on the function of the DAN and DMN for visual and somatosensory perception. Thus how ongoing activity fluctuations impact on perception depends on the relevant specific sensory (i.e., nonspatial) and cognitive (i.e., mnemonic) context (Sadaghiani et al., 2009).

4.1.2 Continuous versus discrete perception and attention

Perception is commonly regarded as a continuous process (VanRullen and Koch, 2003). However, in the visual system (Busch et al., 2009) and in the somatosensory system (Baumgarten et al., 2015), there is evidence for discrete visual sampling, which may be caused by modulatory cortical rhythms. Many behavioral measures of visual perception fluctuate continually in a rhythmic manner, reflecting the influence of endogenous brain oscillations, particularly theta (~ 4–7 Hz) and alpha (~ 8–12 Hz) rhythms (Benedetto et al., 2016; Fiebelkorn et al., 2013; Landau and Fries, 2012). These low-frequency rhythms provide a mechanism for parsing sensory information into "packets" and thus help to coordinate activity across ensembles in other (sub)cortical regions (Haegens et al., 2015). Cortical oscillators in nonauditory modalities exhibit frequency ranges that match their corresponding sensory channels. Somatosensory oscillators in the monkey have a peak frequency around 30 Hz (Ahissar and Vaadia, 1990), which is suitable for temporal information carried by rapidly adapting fibers (Johansson et al., 1982). Visual oscillations occur mostly in the alpha (~ 10 Hz) and gamma (40–100 Hz) ranges (Eckhorn et al., 1993; Gray and McCormick, 1996), matching the alpha and gamma ranges of retinal activation frequencies caused by the spontaneous miniature eye movements (Eizenman et al., 1985).

In contrast, these discrete perception cycles are not apparent in the auditory system. Zoefel and Heil (2013) examined in humans whether the detection of brief near-threshold sounds in quiet depends on the phase of EEG oscillations. When stimuli were presented at irregular intervals, they did not find a systematic relationship between detection probability and phase. When stimuli were presented at regular intervals (2 s), reaction times were significantly shorter and they observed phase entrainment of EEG oscillations corresponding to the frequency of stimulus presentation (0.5 Hz), revealing an adjustment of the system to the regular stimulation. Zoefel and VanRullen (2017) noted that a rhythmic component in sensory input is crucial for the auditory but not for the visual system. The alignment between oscillatory phase and rhythmic input (phase entrainment) is therefore an integral part of stimulus selection in the auditory system whereas the visual system merely adjusts its phase to upcoming events, without the need for any rhythmic component. The oscillatory entrainment to rhythmic stimulation seems to be a fundamental feature of the auditory system, probably evolved due the rhythmic nature of the auditory environment. Indeed, the tendency to synchronize with auditory rhythms is ubiquitous. When input is unpredictable, the visual system can maintain its oscillatory sampling, whereas the auditory system switches to a different, potentially internally oriented, "mode" of processing that might be characterized by alpha oscillations. However, see Ho et al. (2017) for a different perspective.

Related to discrete perception is rhythmic attention. From behavioral data, Fiebelkorn et al. (2013) determined how space- and object-based selection simultaneously evolve under conditions that promote sustained attention at a cued location, varying the cue-to-target interval from 300 to 1100 ms. They tracked visual-target detection at a cued location (i.e., space-based selection), at an uncued location that was part of the same object (i.e., object-based selection), and at an uncued location that was part of a different object (i.e., in the absence of space- and object-based selection). Fiebelkorn et al. (2013) demonstrated that even under static conditions, there is a moment-to-moment reweighting of attentional priorities based on object properties. This is reflected in rhythmic patterns of visual-target detection both within (at 8 Hz) and between (at 4 Hz) objects. Van Rullen et al. (2007) called this a blinking spotlight. A variant thereof was originally (Raymond et al., 1992) called the attentional blink, and recently firmly linked to brain rhythms (Shapiro et al., 2017): "the signature of the attentional blink—second target accuracy that increases with intertarget lag—is significantly larger for entraining streams that are in the alpha-beta frequency range, relative to streams that are slower (theta) or faster (gamma)." Although much evidence in humans points to a ~ 7 Hz attentional sampling rhythm, several authors have reported lower periodicities, ~ 3–4 Hz (e.g., Landau and Fries, 2012; Fiebelkorn et al., 2013). To reconcile these findings, it has been argued that in the latter studies the display and task instructions, either explicitly or implicitly, could have encouraged the possibility of dividing attention between two objects or locations. If attention does operate periodically, it could then have focused on each object or location during alternate cycles, with the result that the periodicity measured at any one location would actually be one-half of the ~ 7 Hz intrinsic rhythm of attention, as found by Fiebelkorn et al. (2013).

During anticipation of a visually cued auditory target and using MEG data, ElShafei et al. (2018) observed a decrease in alpha power ~ 9 Hz in the auditory cortices; and an increase ~ 13 Hz in the visual regions. Only this power increase in high alpha significantly correlated with behavior. Importantly, within the right auditory cortex, they showed a larger increase in high alpha power when attending an ipsilateral sound; and a stronger decrease in low alpha power when attending a contralateral sound. Thus facilitating and suppressing attentional mechanisms, with distinct timing in task-relevant and task-irrelevant brain areas, are differentially correlated to behavior and supported by distinct alpha subbands.

4.1.3 Brain rhythms and memory

4.1.3.1 The role of gamma

In a review paper, Miller et al. (2018) proposed that interactions between different rhythms in distinct cortical layers underlie maintenance and its volitional control (Fig. 4.2). Based on recordings from frontal cortex in awake monkeys, they concluded that: "synaptic weight changes between sparse bursts of spiking strengthen working memory maintenance. Volitional control acts via interplay between network oscillations in gamma (30–100 Hz) in superficial cortical layers (Layers 2 and 3) and alpha and beta (10–30 Hz) in deep cortical layers (layers 5 and 6). Deep-layer alpha and beta are associated with top-down information and inhibition, which regulates the flow of bottom-up sensory information associated with superficial layer gamma."

Working memory, which needs to temporary store short-term memory information, is characterized by persistent neural activity. This is reflected in increased power during memory retention and manipulation in the theta, beta, and gamma bands, and with increased alpha shielding

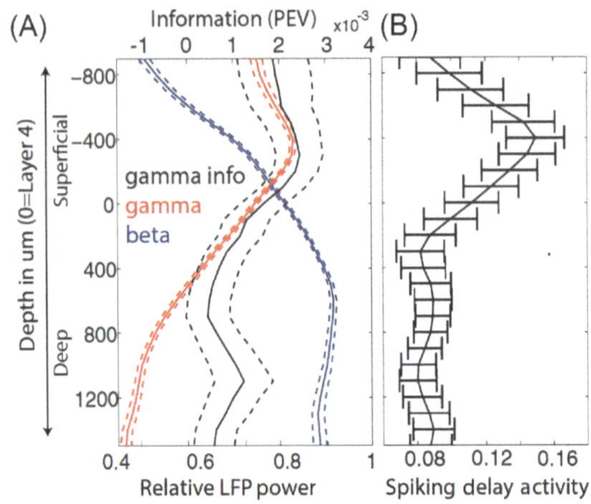

FIG. 4.2 Laminar organization of gamma/beta rhythms and spiking delay activity in awake monkey frontal cortex. (A) Gamma and alpha/beta power appear segregated into distinct layers. Gamma power peaks 400 mm above layer 4, whereas alpha/beta power peaks at 600 mm below layer 4. Gamma bursts in superficial, but not deep, layers carry significant information about the cued item during the working memory delay period (quantified by the percent explained variance, PEV statistic). Beta bursts do not carry significant information during the delay (not shown). *Dotted lines* are ± 1 SEM across sessions ($N=60$). (B) Spiking activity, quantified by multiunit change from baseline (a.u.) during the delay period, is strongest in superficial layers. The pattern of laminar pattern of delay activity correlates strongly with gamma and is strongly anticorrelated with alpha and beta. *Reprinted with permission from Miller, E.K., Lundqvist, M., Bastos, A.M., 2018. Working Memory 2.0. Neuron 100, 463–475. Copyright 2018. Elsevier.*

working memory from distractors. Gamma oscillations may be related to the comparison of stimulus-related information and memory contents, and subsequently with the utilization thereof (Herrmann et al., 2004). They proposed a model, termed the "match-and-utilization model," which assumed that the "early"-(evoked) gamma-band response reflects the matching of bottom-up signals with memory contents that occur rapidly (before 150 ms) after stimulus onset. Typically, the early gamma-band response should be localized and observed over the brain region computing the match. The model predicts that, whenever the matching process yields a positive result, the early gamma-band responses are enhanced. Bottom-up and top-down factors both influence synchronization of neural activity in the gamma band. Because readout processes such as action planning, behavioral control, or memory storage involve highly distributed brain regions, the "match-and-utilization model" predicts that the "late"-(induced) gamma-band response should be associated with large-scale coherence (Varela et al., 2001) reflecting the "broadcasting" of a synchronized neural message to other centers.

Zhang et al. (2018) measured memory recall in intracranial EEG recordings in patients with intractable epilepsy. Stimulus-specific activity was identified and then tracked during waking rest and sleep after encoding. Stimulus-specific gamma (30–90 Hz) activity during early (100–500 ms) and late (500–1200 ms) encoding was spontaneously reactivated during retrieval in waking state and sleep, independent of subsequent recalled forgotten items. Their findings showed that electrodes located in fusiform and inferior temporal lobe are more involved in stimulus-specific representations of early encoding activities while the medial temporal lobe seems to be more engaged during representations of late-encoding activity. Kucewicz et al. (2019) also used direct intracranial brain recordings from epilepsy patients performing free recall tasks. High-gamma frequency activity (65–115 Hz) showed consistent power responses during encoding of subsequently recalled and forgotten words on a subset of electrodes localized in 16 distinct cortical areas activated in the tasks. More of the high-gamma power during word encoding, and less power before and after the word presentation, was characteristic of successful recall and observed across multiple brain regions.

4.1.3.2 The role of theta

Do other bands besides alpha and gamma play a role in memory processes? Slow (~ 0.1 Hz) oscillatory UP-states (Chapter 1) may represent crucial time windows for memory reactivation and consolidation during sleep. Göldi et al. (2019) tested this assumption by utilizing closed-loop targeted memory reactivation: participants were reexposed to a prior learned foreign vocabulary during UP- and DOWN-states of slow EEG oscillations. Presenting memory cues during UP-states improved recall performance, whereas DOWN-state cueing did not result in a clear behavioral benefit. At the electrophysiological level, successful memory reactivation during UP-states was associated with a characteristic power increase in the theta and sleep-spindle band. No oscillatory changes were observable for DOWN-state cues. This effect was the strongest in the frontal cortex.

Wang et al. (2018) addressed whether momentary fluctuations in theta synchronization predict the likelihood of associative memory formation for multisensory events. They entrained the visual and auditory cortex at theta frequency (4 Hz) and in a synchronous or asynchronous manner by modulating the luminance and volume of movies and sounds at 4 Hz, with a phase offset at 0 or 180 degrees. EEG activity from human subjects was recorded while they memorized the association between a movie and a sound. They found that "associative memory performance was significantly enhanced in the 0° compared with the 180° condition. … Importantly, the strength of entrainment during encoding correlated with the efficacy of associative memory such that small phase differences between visual and auditory cortex predicted a high likelihood of correct retrieval in a later recall test. These findings suggest that theta oscillations serve a specific function in the episodic memory system: binding the contents of different modalities into coherent memory episodes."

4.2 Networks and cognition

4.2.1 Networks in attention and memory

Wianda and Ross (2019) recorded MEG in young adults, who performed a variant of a Sternberg working memory task. Time-frequency analysis revealed event-related decrease of alpha oscillation power during the encoding interval and an alpha power increase during memory retention. Alpha event-related decrease during encoding occurred at 10 Hz and alpha power increase during retention at 12 Hz, suggesting different alpha mechanisms. Analysis of alpha coherence and alpha-gamma cross-frequency synchrony showed that the connectivity between occipital and frontotemporal areas was correlated with alpha power increase during memory retention. Visualization of the resulting network in Fig. 4.3 showed coupling of anterior gamma amplitude (*arrow tail*) and frontotemporal alpha phase (*arrow head*). Likewise, temporal gamma amplitude was exclusively coupled to frontal alpha phase.

4.2.2 The frontoparietal network and task control

Neuronal activity in sensory and frontoparietal networks underlies the representation and attention control, respectively, of sensory information maintained in working memory. Within these regions, beta/gamma phase-synchronization supports the integration of sensory functions, while synchronization in theta/alpha bands supports the regulation of attention functions (Siebenhühner et al., 2016). They investigated whether such integration could be achieved by cross-frequency phase synchrony (CFS). Using simultaneous MEG and EEG, they found that CFS was enhanced between theta and alpha-gamma and between alpha and beta-gamma oscillations during visual working memory maintenance between the visual network, the frontoparietal network (FPN), and the DAN. CFS also connected the hubs of within-frequency-synchronized networks and its strength predicted individual visual working memory capacity. Siebenhühner et al. (2016) proposed that CFS integrates processing among synchronized neuronal networks from theta to gamma frequencies to link sensory and attention functions.

Graph theoretic analysis (Chapters 1 and 6) suggested that at least five distinct subnetworks are associated with cognitive control: the FPN, the cinguloopercular control network, the salience network, the ventral

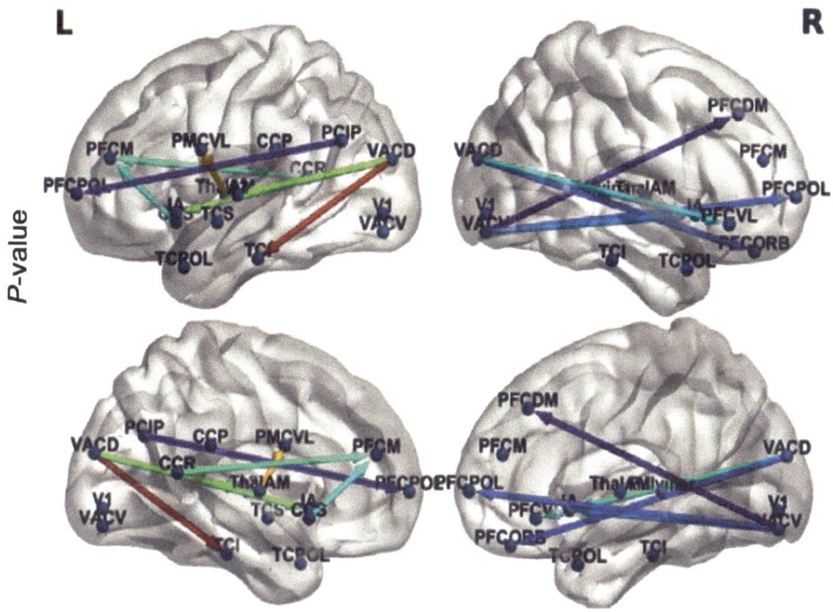

FIG. 4.3 Asymmetry of phase-amplitude coupling (PAC) as an indicator for directionality of communication within brain networks based of differences between local and distant asymmetry in their alpha-gamma PAC. The networks reveal predominant coupling between anterior gamma amplitude *(arrow tail)* and frontotemporal alpha phase *(arrow head)*. Temporal gamma amplitude was exclusively coupled to frontal alpha phase. *Blue* connections are strongly negative, and *red* connections are strongly positive. *Green, yellow,* and *turquoise* ones are not significantly directed. *A1*, primary auditory cortex; *A2*, secondary auditory cortex; *CCP*, posterior cingulate cortex; *CCR*, retrosplenial cingulate cortex; *CCS*, subgenual cingulate cortex; *M1*, primary motor cortex; *PCIP*, cortex of the intraparietal sulcus; *PFCCL*, centrolateral prefrontal cortex; *PFCDM*, dorsomedial prefrontal cortex; *PFCM*, medial prefrontal cortex; *PFCVL*, ventrolateral prefrontal cortex; *PMCM*, medial (supplementary) premotor cortex; *TCI*, inferior temporal cortex; *TCPOL*, polar temporal cortex; *TCS*, superior temporal cortex; *ThalAM*, thalamus; *V1*, primary visual cortex; *V2*, secondary visual cortex; *VACD*, anterior visual cortex dorsal; *VACV*, anterior visual cortex ventral. *From Wianda, E., Ross, B., 2019. The roles of alpha oscillation in working memory retention. Brain Behav. 9, e01263. https://doi.org/10.1002/brb3.1263. Open access.*

attention network, and the DAN. Cole et al. (2013) focused on the FPN for two reasons. First, the FPN includes the lateral prefrontal cortex (PFC; the focus of the guided activation theory) as a central component. Second, the flexible-hub theory predicts that most adaptive task-control flexible hubs likely exist in the FPN because the FPN is especially active (relative to other networks) during situations requiring highly adaptive task control. Using fMRI, Cole et al. (2013) performed experiments, which combined a set of rule components in many different ways, creating dozens of complex task sets that were each novel to participants. Cole et al. (2013) found that the FPN's brain-wide functional connectivity pattern shifted

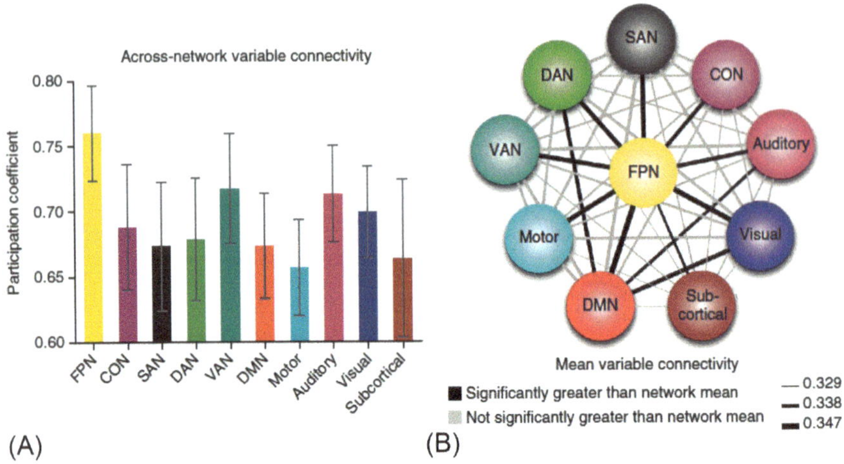

FIG. 4.4 Frontoparietal network's (FPN's) variable connectivity is truly global. (A) To rule out the possibility that FPN's high global variability coefficient (GVC) was driven by high variability of a small subset of connections, we estimated across-network variable connectivity (participation coefficient) for each network. Similar to GVC, FPN's participation coefficient of variable connectivity was highest ($P < .05$, FDR corrected) across the 10 networks. This suggests that FPN's variable connectivity is truly global. The error bars illustrate the intersubject SEM. (B) Pairwise mean variable connectivity between networks was examined. Variable connections are highlighted that were significantly ($P < .05$, FDR corrected) greater for one of the networks included in a given link than the GVC of the other network included in that link. For example, the FPN-DMN link was highlighted because FPN's mean variable connectivity with the DMN was significantly greater than the DMN's mean variable connectivity with the entire brain (i.e., its GVC). Note that variable connectivity from the Pearson correlation analysis was used for illustration given that this method provides connectivity estimates that are identical in both directions (i.e., to and from seed and target regions). The three lines in the legend are the minimum, median, and maximum variable connectivity strengths. The FPN's variable connectivity with each network was significantly greater than every network's GVC (i.e., the mean over each network's connections with all 264 regions), providing confirmation that this is truly a global effect. *CON*, cinguloopercular; *DAN*, dorsal attention; *DMN*, default-mode; *FPN*, frontoparietal; *SAN*, salience; *VAN*, ventral attention. *Reprinted with permission from Cole, M.W., Reynolds, J.R., Power, J.D., Repovs, G., Anticevic, A., Braver, T.S., 2013. Multi-task connectivity reveals flexible hubs for adaptive task control. Nat. Neurosci. 16(9), 1348–1355. Copyright 2013. Springer Nature.*

more than those of other networks across a variety of task states (Fig. 4.4). Note that both the auditory and visual networks are strongly and variably coupled to the DMN and the FPN. Furthermore, these patterns were consistent across practiced and novel tasks, suggesting that reuse of flexible hub connectivity patterns facilitates adaptive (novel) task performance. Together, these findings support a central role for frontoparietal flexible hubs in cognitive control and adaptive implementation of task demands.

Recording from patients with intracranial electrodes, Johnson et al. (2017) showed that working memory for everyday "what," "where," and "when" associations depends on multiplexed oscillatory systems. Signals of different frequencies simultaneously linked the PFC to parietooccipital and medial temporal regions, pointing to a complex web of bidirectional interactions. In patients performing a visuospatial working memory task, Johnson et al. (2017) found that task-induced oscillations in the same delta-theta (2–7 Hz) and alpha-beta (9–24 Hz) frequency ranges of PFC and parietal cortices carried information about the contents of working memory. Maintenance of memory was linked to directional connectivity from the parietal cortex to the PFC. However, presentation of the test prompt to cue identity, spatial, or temporal information changed delta-theta coordination from a unidirectional, parietal-led system to a bidirectional, FPN. Processing of spatiotemporal information was more bidirectional in the delta–theta range than was the processing of identity information, where alpha-beta connectivity did not exhibit sensitivity to the contents of working memory. These findings implicate a bidirectional delta-theta mechanism for FPN control over the contents of working memory (Johnson et al., 2017).

4.2.3 Fast and slow thinking

In a well-known book entitled *"Thinking, Fast and Slow,"* Kahneman (2013) elaborated on two brain systems that drive the way we think and make choices. "System One" is fast, intuitive, and emotional, whereas "System Two" is slower, more deliberate and more logical. Examining the underlying neural substrates, McClure et al. (2004, 2007) had identified several brain regions that comprise "System One" (Fig. 4.5A). These regions include the nucleus accumbens (NAc), the subgenual cingulate cortex (SGC), the medial orbitofrontal cortex (mOFC), the posterior cingulate cortex (PCC), and precuneus. All of these regions are within the limbic system, the paralimbic cortex, and the DMN, and have been directly implicated in reward processing. "System One" may also include a region in the dorsal ACC in the DAN. The ACC is frequently associated with the presence of response conflict. Finding ACC activity in this context may imply that choices involving an immediate reward option are associated with greater conflict. Such conflict is consistent with a two-system model of decision-making (McClure et al., 2007).

Regions that correlate significantly with "System Two" included both visual and motor areas, as well as regions commonly associated with higher-level cognitive functions. It is likely that areas of primary visual and premotor cortex [including premotor area (PMA) and supplementary motor area (SMA)] were activated because task performance required subjects to look at a visual display and press a button to indicate preference. The remaining areas, McClure et al. (2007) found to comprise "System Two" include a region in the PCC, bilateral areas in the posterior parietal cortex (PPC), bilateral areas in the anterior insula (Ant Ins), and several

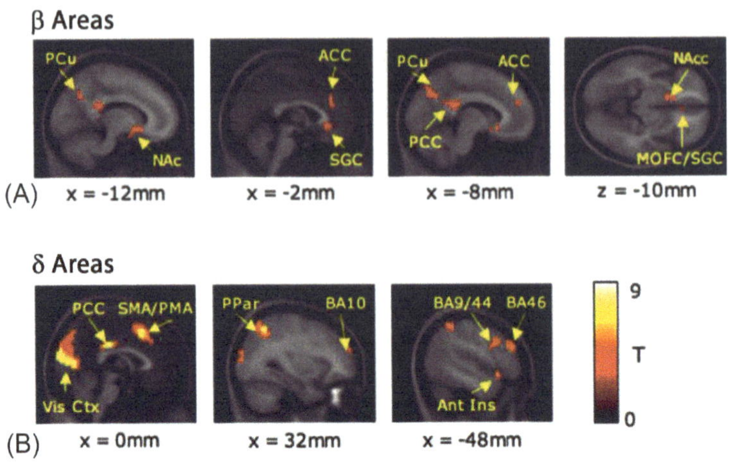

FIG. 4.5 System One (β) and System Two (δ) brain areas. (A) Brain areas that correlated with System One included a set of brain areas all closely linked with the mesolimbic dopamine system. These include the NAc, PCC, mOFC, and ACC ($P<.001$; minimum eight contiguous voxels). (B) System Two identified several brain areas identified with general cognitive processing, including the PPar bilaterally, the Ant Ins, and regions of the dorsolateral prefrontal cortex [Brodmann area 9 (BA9), BA44, BA46, and BA10: $P<.001$]. *ACC*, anterior cingulate cortex; *mOFC*, medial orbitofrontal cortex; *NAc*, nucleus accumbens; *PCC*, posterior cingulate cortex; *PCu*, precuneus; *PMA*, premotor area; *PPar*, posterior parietal cortex; *SMA*, supplementary motor area; *Vis Ctx*, visual cortex. *From McClure, S.M., Ericson, KM., Laibson, D.L, Loewenstein, G., Cohen, J.D., 2007. Time discounting for primary rewards. J. Neurosci. 27(21), 5796–5804. Copyright 2007, Society for Neuroscience.*

regions in the dorsolateral prefrontal cortex (DLPFC; Fig. 4.5B). Activity in DLPFC and PPC is observed commonly in tasks involving cognitive processes such as working memory, abstract problem-solving, and exertion of control in favor of long-term goals. The region in the PCC is located slightly rostral to the region that is comprised in "System One" but has no shared voxels (McClure et al., 2007).

Based on this and other studies (Kable and Glimcher, 2007; Peters and Buchel, 2011), Turner et al. (2019) recently expressed the accepted point of view that this "temporal discounting" depends on neural processes in the striatum and the ventromedial PFC related to the construction of subjective value. There is also agreement that the executive brain regions including the dorsolateral prefrontal cortex, dorsomedial frontal cortex, and posterior parietal cortex are associated with self-control.

4.3 Brain rhythms in speech and language

In previous chapters, we have seen that brain oscillations span the range from delta to high gamma. One could consider subthreshold oscillations—not accompanied by spikes—as putative resonance frequencies of the various structures. Since complex sounds such as speech contain specific envelope-frequency ranges related, for example, to phonemes, syllables, and prosody. These frequencies can entrain the subthreshold resonances and so represent these various frequency ranges in the speech envelope. Neural oscillations underlie speech processing in the segmentation and identification of linguistically meaningful units (Chapters 2 and 3). This happens on three timescales, such that the frequency of gamma-band oscillations synchronize with phonemes, theta-band oscillations synchronize with syllables, and delta-band oscillations synchronize with intonation-phrase boundaries (e.g., Giraud and Poeppel, 2012). The three bands form a hierarchical relationship, that is, low-frequency oscillations can top-down align higher-frequency oscillations, thereby amplifying bottom-up information extraction and establishing a coherent percept (Ahissar et al., 2001; Doelling et al., 2014; Luo and Poeppel, 2007). Temporal alignment might serve to implicitly align neural excitability with linguistically distinctive spectral information. In a response to Giraud and Poeppel (2012), albeit criticizing an overemphasis on speech envelope, Obleser et al. (2012) provide a strong argument about speech-entraining theta oscillations in auditory cortex: "we have ample evidence that slow brain oscillations become phase-locked to slow spectral regularities in an auditory signal, even in the absence of amplitude envelope fluctuations."

4.3.1 Speech

Most information conveyed by the temporal envelope of speech is confined to frequencies below 16 Hz, which roughly match strong evoked modulation rates of firing in cat A1 neurons (Eggermont, 1991) and local field potentials in human auditory cortex (Giraud et al., 2000; Chapter 7). To test the importance of cortical activity modulation rates for speech processing, Ahissar et al. (2001) manipulated the frequency of the temporal envelope of speech sentences and tested the effect on both speech comprehension and MEG-based cortical activity. Speech comprehension was degraded when sentences were presented in more compressed forms, that is, higher modulation rates. Ahissar et al. (2001) found that the average comprehension level, at each compression, correlated with the similarity between the frequencies of the temporal envelopes of the stimulus (< 20 Hz) and the subject's cortical resonances (stimulus-cortex frequency-matching) and the phase locking between the two temporal envelopes. Of these two correlates, phase locking was significantly more indicative for single-trial success. Their results suggest that the match between the speech rate and the a priori modulation capacities of the auditory cortex is a prerequisite for comprehension. This also suggests that the modulation capacity of cortical neurons determines the upper limit of speech modulation rate that can be followed by an individual, and that a comprehension of a sentence requires an online phase-locking to the temporal envelope of that sentence. This is likely underlying the age-related slowing of cortical oscillations and concomitant poor speech understanding (Eggermont, 2019).

Zion Golumbic et al. (2013) recorded the electrocorticogram (ECoG) from subdural grids implanted in six epilepsy patients. The patients viewed movie clips of two talkers reciting short narratives (9–12 s) presented simultaneously. In each trial, participants attended to one talker (visual and auditory), while ignoring the other one. Trials in which only one talker was presented served as a control. They found that brain activity dynamically tracks speech streams using both low-frequency (1–7 Hz) phase and high-frequency (75–150 Hz) amplitude fluctuations and that optimal encoding likely combines the two. In and near low-level auditory cortices, attention "modulates" the representation by enhancing cortical tracking of attended speech streams (Fig. 4.6), but ignored speech remains represented. In higher-order regions, the representation appears to become more "selective," in that there is no detectable tracking of ignored speech. This selectivity itself seems to sharpen as a sentence unfolds.

Kubanek et al. (2013) also recorded neural activity with high temporal resolution directly from the cortical surface in humans with intractable epilepsy who listened to a spoken story. They found that the high-gamma activity (75–115 Hz) in a single patient's auditory cortex robustly tracked the speech envelope. The effect could be observed during a single presentation of the spoken story to each subject. The effect is stronger in regions situated relatively early in the auditory pathway (belt areas) compared with other regions involved in speech processing, including the superior temporal gyrus (STG) and the posterior inferior frontal gyrus (Broca's region). Although the auditory belt regions also tracked melodic stimuli, STG and Brocas's region tracked the envelope of speech only, thus indicating that these regions encode speech envelope as a phonological—not purely acoustic—feature, and they do so only weakly. This is in line with previous suggestions that these regions specialize in more abstract, high-level (lexical/syntactic/semantic) analysis of speech.

Thus ongoing oscillations in auditory cortex modulate their phase to match the rhythm of temporally regular acoustic stimuli, increasing sensitivity to relevant environmental cues and improving detection accuracy (Peelle et al., 2013). What remains unclear is whether these phase-locked responses can be modulated by nonsensory information, which in the case of speech comprehension is by the linguistic content available in the speech signal. Using MEG, and processed stimuli with a noise-vocoding technique to vary intelligibility while keeping the temporal acoustic envelope consistent, Peelle et al. (2013) showed that the acoustic envelopes of sentences contain most power between 4 and 7 Hz and that it is in this

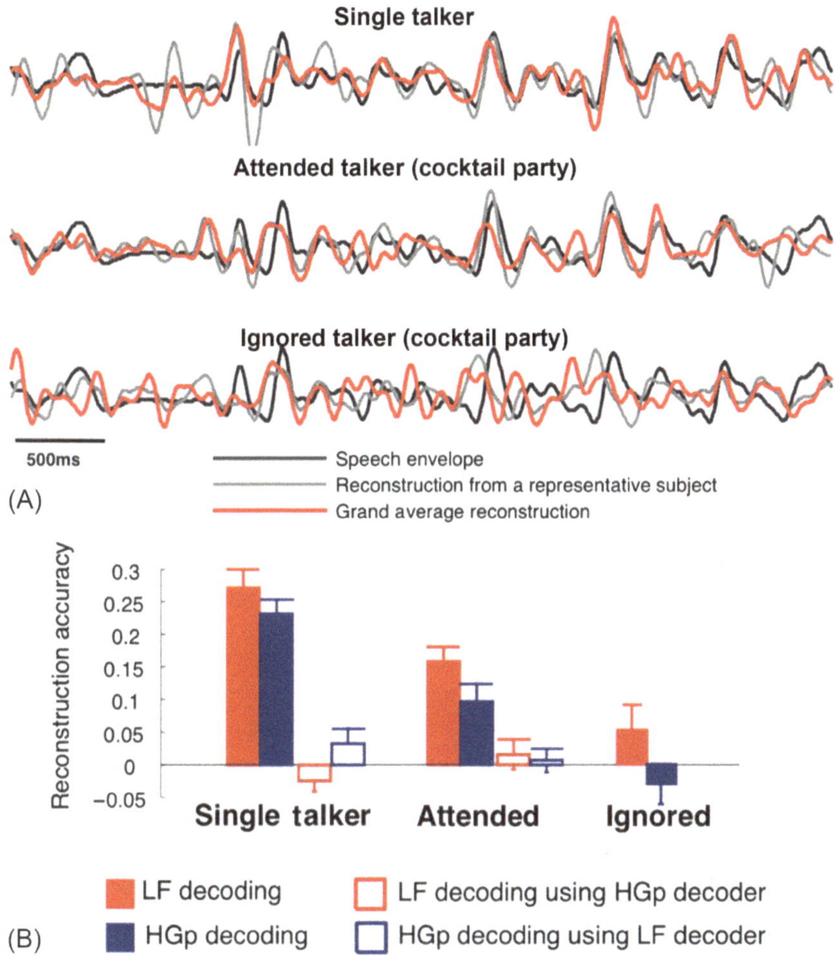

FIG. 4.6 Reconstruction of the speech envelope from the cortical activity. (A) A segment of the original speech envelope *(black)* compared with the reconstruction achieved using the low frequency (LF) signal from one participant *(gray)* and from all participants *(red)*. Reconstruction examples are shown for the Single Talker condition *(top)* as well as for the attended *(middle)* and ignored stimuli *(bottom)* in the Cocktail Party condition. (B) *Full bars*: Grand averaged of the reconstruction accuracy (i.e., the correlation-coefficient values between the actual and reconstructed time courses) across all participants using LF *(red)* or HGp *(blue)*. The Single Talker and the Attended Talker in the Cocktail Party condition could be reliably reconstructed using either measure, and in both cases, significantly better than the Ignored speaker. *Empty bars*: Envelope reconstruction accuracy obtained by applying each of the single-band decoders to data in the other band. As shown here, decoders constructed using either band performed poorly when applied to data in the other band. This implies that the two single-band decoders have systematically different features. *HGp*, high gamma range phase (75–150 Hz). *Reprinted with permission from Zion Golumbic, E.M., Ding, N., Bickel, S., Lakatos, P., Schevon, C.A., McKhann, G.M., et al., 2013. Mechanisms underlying selective neuronal tracking of attended speech at a "Cocktail Party". Neuron 77, 980–991. Copyright 2013. Elsevier.*

frequency band that phase locking between neural activity and envelopes is strongest. Bilateral oscillatory neural activity phase-locked to unintelligible speech, but this cortex-acoustic phase locking was enhanced when speech was intelligible. This enhanced phase locking was localized to left temporal cortex. Consequently, entrainment to connected speech does not only depend on acoustic characteristics but also is affected by listeners' ability to extract linguistic information. This suggests a biological framework for speech comprehension in which acoustic and linguistic cues reciprocally aid in stimulus prediction (Peelle et al., 2013).

4.3.2 Language

"When speech processing has succeeded, meaning must be decoded from the speech percept; language comprehension can proceed along two processing streams: The first stream is syntactic in nature, employing delta-band cycles to group words into syntactic phrases, alpha-band power modulations to store syntactic phrases in verbal working memory and theta-band power modulations to retrieve information from sentences' hierarchical working memory representations, as well as from long-term memory. The second processing stream is predictive in nature, employing beta-band oscillations to top-down predict the semantics of upcoming words that fit the cumulative meaning of the prior word sequence, and gamma-band oscillations to bottom-up assess the contextual semantic fit of incoming words" (Meyer, 2018).

Gamma-band (30–120 Hz) oscillations have been associated with lexical processing in speech (Tavabi et al., 2011). Word recognition in the context of acoustically degraded stimuli produces left-lateralized induced gamma-band activity of approximately 30–50 Hz with latency between 300 and 400 ms. This may reflect successful sound matching to auditory traces and lexical memory (Hannemann et al., 2007). Induced gamma-band modulation between 300 and 600 ms has also been reported during semantic violations in sentence comprehension (Hagoort et al., 2004), suggesting sentence-level semantic integration. Finally, Shahin et al. (2009) reported concomitant-induced gamma (30–60 Hz) enhancement and alpha (8–12 Hz) suppression in a lexical decision task.

4.4 Summary

Cognitive processes in general are considered similar for processing activity from the specific sensory modalities. However, several findings in the auditory modality contrast those in visual and somatosensory modalities. For example, discrete cycles of perception (based on theta and alpha rhythm chunking) demonstrated in the visual and somatosensory system are absent in the auditory, which uses oscillatory entrainment to rhythmic stimulation instead. Neural oscillations on three timescales—gamma-band oscillations synchronize with phonemes and underlie speech processing in the segmentation and identification of linguistically meaningful units. Theta-band oscillations synchronize with syllables and delta-band oscillations synchronize with intonation-phrase boundaries. A higher-order processing stream, using beta-band oscillations then predicts the semantics of upcoming words, and gamma-band oscillations to assess the contextual semantic fit of incoming words. High-gamma frequency activity is ubiquitous during encoding of recalled words. Theta oscillations serve a specific function in the episodic memory system: binding the contents of different modalities into coherent memory episodes. Maintenance of memory is linked to directional connectivity from the parietal cortex to the PFC, whereas bidirectional delta-theta mechanism allows FPN control over the contents of working memory. Thinking and making choices is done in two processing streams, one is fast, intuitive, and emotional, whereas the other is slower, more deliberate, and more logical. The fast one is implemented in the limbic system, the parahippocampus and the DMN, whereas the more deliberate one uses besides higher-order sensory systems, a region in the PCC, bilateral areas in the posterior parietal cortex, bilateral areas in the anterior insula, and several regions in the dorsolateral prefrontal cortex.

References

Ahissar, E., Vaadia, E., 1990. Oscillatory activity of single units in a somatosensory cortex of an awake monkey and their possible role in texture analysis. Proc. Natl. Acad. Sci. U. S. A. 87 (22), 8935–8939.

Ahissar, E., Nagarajan, S., Ahissar, M., Protopapas, A., Mahncke, H., Merzenich, M.M., 2001. Speech comprehension is correlated with temporal response patterns recorded from auditory cortex. Proc. Natl. Acad. Sci. U. S. A. 98, 13367–13372.

Arieli, A., Sterkin, A., Grinvald, A., Aertsen, A., 1996. Dynamics of ongoing activity: explanation of the large variability in evoked cortical responses. Science 273, 1868–1871.

Baumgarten, T.J., Schnitzler, A., Lange, J., 2015. Beta oscillations define discrete perceptual cycles in the somatosensory domain. Proc. Natl. Acad. Sci. U. S. A. 112 (39), 12187–12192.

Benedetto, A., Spinelli, D., Morrone, M.C., 2016. Rhythmic modulation of visual contrast discrimination triggered by action. Proc. Biol. Sci. 283, 20160692. https://doi.org/10.1098/rspb.2016.0692.

Boly, M., Balteau, E., Schnakers, C., Degueldre, C., Moonen, G., Luxen, A., et al., 2007. Baseline brain activity fluctuations predict somatosensory perception in humans. Proc. Natl. Acad. Sci. U. S. A. 104, 12187–12192.

Burton, H., Sinclair, R.J., 2000. Attending to and remembering tactile stimuli: a review of brain imaging data and single-neuron responses. J. Clin. Neurophysiol. 17, 575–591.

Busch, N.A., Dubois, J., VanRullen, R., 2009. The phase of ongoing EEG oscillations predicts visual perception. J. Neurosci. 29 (24), 7869–7876.

Cole, M.W., Reynolds, J.R., Power, J.D., Repovs, G., Anticevic, A., Braver, T.S., 2013. Multitask connectivity reveals flexible hubs for adaptive task control. Nat. Neurosci. 16 (9), 1348–1355.

Corbetta, M., Shulman, G.L., 2002. Control of goal-directed and stimulus-driven attention in the brain. Nat. Rev. Neurosci. 3, 201–215.

Doelling, K.B., Arnal, L.H., Ghitza, O., Poeppel, D., 2014. Acoustic landmarks drive delta–theta oscillations to enable speech comprehension by facilitating perceptual parsing. NeuroImage 85, 761–768.

Eckhorn, R., Frien, A., Bauer, R., Woelbern, T., Kehr, H., 1993. High frequency (60–90 Hz) oscillations in primary visual cortex of awake monkey. Neuroreport 4 (3), 243–246.

Eggermont, J.J., 1991. Rate and synchronization measures of periodicity coding in cat primary auditory cortex. Hear. Res. 56, 153–167.

Eggermont, J.J., 2019. The Auditory Brain and Age-Related Hearing Impairment. Academic Press, London, UK, ISBN: 978-0-12-815304-8.

Eizenman, M., Hallett, P.E., Frecker, R.C., 1985. Power spectra for ocular drift and tremor. Vis. Res. 25 (11), 1635–1640.

ElShafei, H.A., Bouet, R., Bertrand, O., Bidet-Caulet, A., 2018. Two sides of the same coin: distinct sub-bands in the a rhythm reflect facilitation and suppression mechanisms during auditory anticipatory attention. eNeuro 5 (4). https://doi.org/10.1523/ENEURO.0141-18.2018.

Engel, A.K., Singer, W., 2001. Temporal binding and the neural correlates of sensory awareness. Trends Cogn. Sci. 5 (1), 16–25.

Engel, A.K., Fries, P., Singer, W., 2001. Dynamic predictions: oscillations and synchrony in top-down processing. Nat. Rev. Neurosci. 2, 704–716.

Fiebelkorn, I.C., Saalmann, Y.B., Kastner, S., 2013. Rhythmic sampling within and between objects despite sustained attention at a cued location. Curr. Biol. 23, 2553–2558.

Friederici, A.D., von Cramon, Y.D., Kotz, S.A., 2007. Role of the corpus callosum in speech comprehension: interfacing syntax and prosody. Neuron 53, 135–145.

Fries, P., 2009. Neuronal gamma band synchronization as fundamental process of cortical computation. Annu. Rev. Neurosci. 32, 209–224.

Giraud, A.-L., Poeppel, D., 2012. Cortical oscillations and speech processing: emerging computational principles and operations. Nat. Neurosci. 15, 511–517.

Giraud, A.-L., Lorenzi, C., Ashburner, J., Wable, J., Johnsrude, I., Frackowiak, R., et al., 2000. Representation of the temporal envelope of sounds in the human brain. J. Neurophysiol. 84, 1588–1598.

Göldi, M., van Poppel, E.A.M., Rasch, B., Schreiner, T., 2019. Increased neuronal signatures of targeted memory reactivation during slow-wave up states. Sci. Rep. 9, 2715. https://doi.org/10.1038/s41598-019-39178-2.

Gray, C.M., McCormick, D.A., 1996. Chattering cells: superficial pyramidal neurons contributing to the generation of synchronous oscillations in the visual cortex. Science 274 (5284), 109–113.

Haegens, S., Barczak, A., Musacchia, G., Lipton, M.L., Mehta, A.D., Lakatos, P., et al., 2015. Laminar profile and physiology of the a rhythm in primary visual, auditory, and somatosensory regions of neocortex. J. Neurosci. 35 (42), 14341–14352.

Hagoort, P., Hald, L., Bastiaansen, M., Petersson, K.M., 2004. Integration of word meaning and world knowledge in language comprehension. Science 304, 438–441.

Hannemann, R., Obleser, J., Eulitz, C., 2007. Top-down knowledge supports the retrieval of lexical information from degraded speech. Brain Res. 1153, 134–143.

Herrmann, C.S., Munk, M.H.J., Engel, A.K., 2004. Cognitive functions of gamma-band activity: memory match and utilization. Trends Cogn. Sci. 8 (8), 347–355.

Hipp, J.F., Engel, A.K., Siegel, M., 2011. Oscillatory synchronization in large-scale cortical networks predicts perception. Neuron 69 (2), 387–396.

Ho, H.T., Leung, J., Burr, D.C., Alais, D., Morrone, M.C., 2017. Auditory sensitivity and decision criteria oscillate at different frequencies separately for the two ears. Curr. Biol. 27, 3643–3649.

Johansson, R.S., Landström, U., Lundström, R., 1982. Responses of mechanoreceptive afferent units in the glabrous skin of the human hand to sinusoidal skin displacements. Brain Res. 244 (1), 17–25.

Johnson, E.L., Dewar, C.D., Solbakk, A.K., Endestad, T., Meling, T.R., Knight, R.T., 2017. Bidirectional frontoparietal oscillatory systems support working memory. Curr. Biol. 27. 1829.e4–1835.e4.

Kable, J.W., Glimcher, P.W., 2007. The neural correlates of subjective value during intertemporal choice. Nat. Neurosci. 10, 1625–1633.

Kahneman, D., 2013. Thinking Fast and Slow. Farrar, Straus and Giroux, New York, NY, ISBN: 9780141033570.

Kubanek, J., Brunner, P., Gunduz, A., Poeppel, D., Schalk, G., 2013. The tracking of speech envelope in the human cortex. PLoS One 8. https://doi.org/10.1371/journal.pone.0053398.

Kucewicz, M.T., Saboo, K., Berry, B.M., Kremen, V., Miller, L.M., Khadjevand, F., et al., 2019. Human verbal memory encoding is hierarchically distributed in a continuous processing stream. eNeuro 6 (1). https://doi.org/10.1523/ENEURO.0214-18.2018.

Landau, A.N., Fries, P., 2012. Attention samples stimuli rhythmically. Curr. Biol. 22, 1000–1004.

Luo, H., Poeppel, D., 2007. Phase patterns of neuronal responses reliably discriminate speech in human auditory cortex. Neuron 54, 1001–1010.

McClure, S.M., Laibson, D.I., Loewenstein, G., Cohen, J.D., 2004. Separate neural systems value immediate and delayed monetary rewards. Science 306, 503–507.

McClure, S.M., Ericson, K.M., Laibson, D.I., Loewenstein, G., Cohen, J.D., 2007. Time discounting for primary rewards. J. Neurosci. 27 (21), 5796–5804.

Meyer, L., 2018. The neural oscillations of speech processing and language comprehension: state of the art and emerging mechanisms. Eur. J. Neurosci. 48 (7), 2609–2621.

Miller, E.K., Lundqvist, M., Bastos, A.M., 2018. Working memory 2.0. Neuron 100, 463–475.

Obleser, J., Herrmann, B., Henry, M.J., 2012. Neural oscillations in speech: don't be enslaved by the envelope. Front. Hum. Neurosci. 6, 250. https://doi.org/10.3389/fnhum.2012.00250.

Peelle, J.E., Gross, J., Davis, M.H., 2013. Phase-locked responses to speech in human auditory cortex are enhanced during comprehension. Cereb. Cortex 23 (6), 1378–1387.

Peters, J., Buchel, C., 2011. The neural mechanisms of intertemporal decision-making: understanding variability. Trends Cogn. Sci. 15, 227–239.

Pollmann, S., Maertens, M., von Cramon, D.Y., Lepsien, J., Hugdahl, K., 2002. Dichotic listening in patients with splenial and nonsplenial callosal lesions. Neuropsychology 16 (1), 56–64.

Raymond, J., Shapiro, K., Arnell, K., 1992. Temporary suppression of visual processing in an RSVP task: an attentional blink? J. Exp. Psychol.: Hum. Perc. Perform 18, 849–860.

Robin, J., 2018. Spatial scaffold effects in event memory and imagination. WIREs Cogn Sci. 9. https://doi.org/10.1002/wcs.1462.

Sadaghiani, S., Hesselmann, G., Kleinschmidt, A., 2009. Distributed and antagonistic contributions of ongoing activity fluctuations to auditory stimulus detection. J. Neurosci. 29 (42), 13410–13417.

Shahin, A.J., Picton, T.W., Miller, L.M., 2009. Brain oscillations during semantic evaluation of speech. Brain Cogn. 70, 259–266.

Shapiro, K.L., Hanslmayr, S., Enns, J.T., Lleras, A., 2017. Alpha, beta: the rhythm of the attentional blink. Psychon. Bull. Rev. https://doi.org/10.3758/s13423-017-1257-0.

Siebenhühner, F., Wang, S.H., Palva, J.M., Palva, S., 2016. Cross-frequency synchronization connects networks of fast and slow oscillations during visual working memory maintenance. eLife 5. https://doi.org/10.7554/eLife.13451.

Steinmann, S., Mulert, C., 2012. Functional relevance of interhemispheric fiber tracts in speech processing. J. Neurolinguistics 25 (1), 1–12.

Steinmann, S., Leicht, G., Ertl, M., Andreou, C., Polomac, N., Westerhausen, R., et al., 2014. Conscious auditory perception related to long-range synchrony of gamma oscillations. NeuroImage 100, 435–443.

Steinmann, S., Meier, J., Nolte, G., Engel, A.K., Leicht, G., Mulert, C., 2018. The callosal relay model of interhemispheric communication: new evidence from effective connectivity analysis. Brain Topogr. 31, 218–226.

Tavabi, K., Embick, D., Roberts, T.P.L., 2011. Spectral–temporal analysis of cortical oscillations during lexical processing. Neuroreport 22, 474–478.
Turner, B.M., Rodriguez, C.A., Liu, Q., Molloy, M.F., Hoogendijk, M., McClure, S.M., 2019. Cereb. Cortex 29 (2), 732–750.
Van Rullen, R., Koch, C., 2003. Is perception discrete or continuous? Trends Cogn. Sci. 7 (5), 207–213.
Van Rullen, R., Carlson, T., Cavanagh, P., 2007. The blinking spotlight of attention. Proc. Natl. Acad. Sci. U. S. A. 104, 19204–19209.
Varela, F., Lachaux, J.-P., Rodriguez, E., Martinerie, J., 2001. The brainweb: phase synchronization and large-scale integration. Nat. Rev. Neurosci. 2, 229–239.
Wang, D., Clouter, A., Chen, Q., Shapiro, K.L., Hanslmayr, S., 2018. Single-trial phase entrainment of theta oscillations in sensory regions predicts human associative memory performance. J. Neurosci. 38 (28), 6299–6309.
Wianda, E., Ross, B., 2019. The roles of alpha oscillation in working memory retention. Brain Behav. 9. https://doi.org/10.1002/brb3.1263.
Zhang, H., Fell, J., Axmacher, N., 2018. Electrophysiological mechanisms of human memory consolidation. Nat. Commun. 9, 4103. https://doi.org/10.1038/s41467-018-06553-y.
Zion Golumbic, E.M., Ding, N., Bickel, S., Lakatos, P., Schevon, C.A., McKhann, G.M., et al., 2013. Mechanisms underlying selective neuronal tracking of attended speech at a "cocktail party". Neuron 77, 980–991.
Zoefel, B., Heil, P., 2013. Detection of near-threshold sounds is independent of EEG phase in common frequency bands. Front. Psychol. 4, 262. https://doi.org/10.3389/fpsyg.2013.00262.
Zoefel, B., VanRullen, R., 2017. Oscillatory mechanisms of stimulus processing and selection in the visual and auditory systems: state-of-the-art, speculations and suggestions. Front. Neurosci. 11, 296. https://doi.org/10.3389/fnins.2017.00296.

CHAPTER

5

Neural spike-firing synchrony and population activity

As we have seen in the previous chapters, neural oscillations form rhythmic patterns in the central nervous system. In individual neurons, these patterns can appear either as oscillations in membrane potential or as rhythmic action potentials, which then produce oscillatory activation of postsynaptic neurons. At the level of neural populations, synchronized activity of large numbers of neurons can give rise to macroscopic oscillations, which can be observed in the electroencephalogram (EEG) or magnetoencephalogram (MEG). Oscillatory activity in groups of neurons generally arises from feedback connections between—or common-input activation of—the neurons that result in the synchronization of their firing patterns. The interaction between neurons can give rise to oscillations at a different frequency than the firing frequency of individual neurons.

5.1 What is spike-firing synchrony?

About 25 years ago (Eggermont, 1994), I wrote: "The cortex can be pictured as a statistical, weakly coupled, 'machine' (Braitenberg, 1978). Weak coupling implies that no single connection can dominate the firing rate of any neuron in the assembly. Braitenberg's assumption is supported by low numbers of direct connections between neurons, by high percentages of shared inputs, and by generally very small correlation strengths (Abeles, 1982). The main process in the cortex may therefore be its spontaneous activity, on which sensory input acts mainly as a modulator that causes a small perturbation of this spontaneous activity. ... As a weakly coupled 'machine' the cortex also shows a background synchronous activity resulting from its interconnectivity that has a broad temporal and spatial extension (e.g., Eggermont, 1992). ... Because specific thalamic inputs to the cortex diverge widely, the rather small effect of a stimulus on the effect of a single connection is greatly amplified by their synchronous

Brain Oscillations, Synchrony, and Plasticity
https://doi.org/10.1016/B978-0-12-819818-6.00004-2

action on a number of pyramidal cells. Under spontaneous, asynchronous conditions, ~ 300 synaptic inputs may be required to fire a pyramidal cell, whereas under stimulus, i.e., synchronous, conditions this number is at least a factor 10 smaller (Abeles, 1982). … Because of the multiplicative action of synchrony, small correlation strengths can still cause strong effects. The synchronizing action of neural connectivity is visible in the correlation of the firing times of the neurons involved, and this correlation can be assessed experimentally with the cross-coincidence histogram. Common inputs to neurons as well as synaptic interactions between them will give rise to spike-firing correlations (Aertsen et al., 1989)." Changes in the correlation structure of a neural circuit will be reflected in changes in its functional connectivity (Aertsen et al., 1989). The distinction between spike-firing synchrony and slower forms of correlated activity—for example, firing-rate covariance—can be related not only to the distinction between a temporal and a rate code (Eggermont, 1998) but also to the frequency of the modulating brain rhythm.

5.2 Multiple-spike train correlation in animal cortex

An elaborate early review (Eggermont, 1990) of spike-pair correlation in the central nervous system of mammals, subcortical and cortical, provides a comprehensive introduction to this topic. Here, I present some illustrative examples. Recent technical reviews are found in Eggermont (2010, 2019).

5.2.1 Spike-pair correlation

One way to study neuronal interaction is by simultaneously recording spike trains from two neurons and analyzing the temporal commonalities by cross-correlation of the spike firings (Moore et al., 1966). The peak of the cross-correlograms, also called cross-coincidence histograms, obtained from unit-pair firing in primary auditory cortex (A1) is typically found around zero-lag time (Fig. 5.1). This is usually taken as an indication of shared or common input—likely from the thalamus—by the two neurons involved. If the peak is displaced from the zero time point by one synaptic delay (~ 4 ms in A1) and is asymmetric, that is, shaped as the time-derivative of the excitatory evoked synaptic potential (EPSP), then the cross-correlogram can be interpreted as due to monosynaptic excitation. This is very rarely observed for extracellular recordings in A1, especially if the spike trains are obtained with separate microelectrodes. If the interaction is inhibitory, a delayed through is found in the cross-correlogram (Eggermont, 1990).

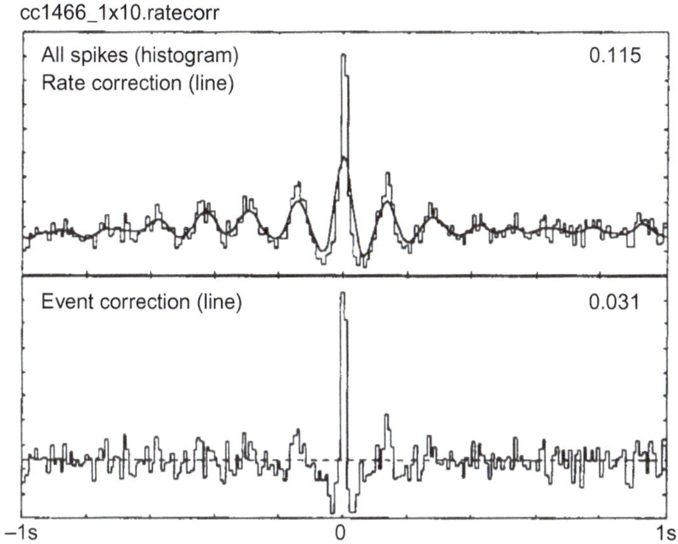

FIG. 5.1 Oscillatory cross-correlogram (10ms bin width) for which the secondary effects are largely due to rate covariation. The event-correlogram *(lower panel)* shows that only the first two side lobes of the oscillation pattern have some contribution from event correlation. *From Eggermont, J.J., Smith, G.M., 1995. Rate covariance dominates spontaneous cortical unit-pair correlograms. Neuroreport 6, 2125–2128. By Permission from Wolters Kluwer.*

The correlation between the firings of cortical neurons, either as a result of common input from the thalamus or resulting from the more numerous intracortical connections, is weak. In Abeles's (1982) recordings from cat A1, the average peak cross-correlation coefficient was only 0.06. In contrast the population activity in the resting-state cortex is widely synchronized and produces detectable signs of this synchrony in the EEG and MEG waves and reflected mostly as oscillations in the lower frequency bands (Chapters 3 and 4).

Interaction of local field potential (LFP)-spindles (here ~8Hz; Chapter 3) with spike firing can produce a periodic correlogram (Fig. 5.1, top panel). These secondary effects can be effectively removed by deconvolving the cross-correlogram with the auto-correlogram of the trigger unit (Eggermont and Smith, 1996). We found that secondary effects from burst firing, global synchrony related to LFPs and stimulus-related effects may largely obscure any underlying correlation produced by event correlation that would reflect direct connectivity (Eggermont and Smith, 1995, 1996). The advantage of using spontaneous or resting state condition is that stimulus effects on the cross-correlogram can be avoided. The effect of rate covariance on the correlogram can typically be predicted from the overall firing rates. Eggermont and Smith (1995, 1996) showed that cortical network activity, such as reflected in EEG rhythms, contributes considerably to the overall correlation and can mask the strength of direct anatomical connections, for example, from the thalamus, on the correlation of firing times (Fig. 5.1).

5.2.2 Epidural versus intracortical spike-based receptive fields

Epidural recordings in awake humans (Nir et al., 2007) have become insightful ways to localize rhythmic brain activity and relate it to cognitive functioning. So, it is important to establish what neural sources these recordings reflect. Bockhorst et al. (2018) used an array of 64 electrodes and combined large-coverage epicortical multiunit recordings in ferrets with intracortical recordings using laminar probes. Their data suggest that the epicortical electrodes selectively captured synchronized spiking of neurons in the underlying cortex. As a result of this synchronization, multiunit responses to sensory stimulation were more robust and less noisy compared with intracortical activity, and auditory and visual receptive field properties were well preserved in epicortical recordings (Fig. 5.2B). This is illustrated for auditory spectrotemporal receptive fields (STRFs) in Fig. 5.2. From simultaneous intracortical recordings in cat A1, Tomita and Eggermont (2005) also showed sharper STRFs for synchronized spiking activity, compared with single unit activity.

5.3 Spike-local field potential synchrony in animal cortex

LFPs are of growing importance in intracortical and epidural recordings. LFPs supplement action potential (spike) recordings and index activity relevant to EEG, MEG, and functional magnetic resonance imaging (fMRI) signals (Logothetis, 2002).

Eggermont et al. (2011) evaluated STRFs and frequency-tuning curves (FTCs) based on LFPs and spikes simultaneously recorded using multielectrode arrays in cat A1. They compared the characteristic frequency (CF) gradients for LFPs and spikes, and for the LFP-pair and spike-pair cross-correlation distances. The CF gradient for spike-based FTCs was about twice that for 2–40 Hz-filtered LFP-based FTCs, indicating greatly reduced frequency selectivity for LFP recordings. Visual comparisons of the data in Fig. 5.3 make this abundantly clear.

LFPs in visual cortex have been considered to reflect activity within very small domains of several hundred micrometers (Katzner et al., 2009). Kajikawa and Schroeder (2011) scrutinized this conclusion by comparing LFP, current source density (CSD), and multiunit activity signals in macaque auditory cortex. Estimated by FTC bandwidths, these signals' "listening areas" differ systematically between spikes and LFPs (as in Fig. 5.3)—with spike-based estimates providing the narrowest FTCs,

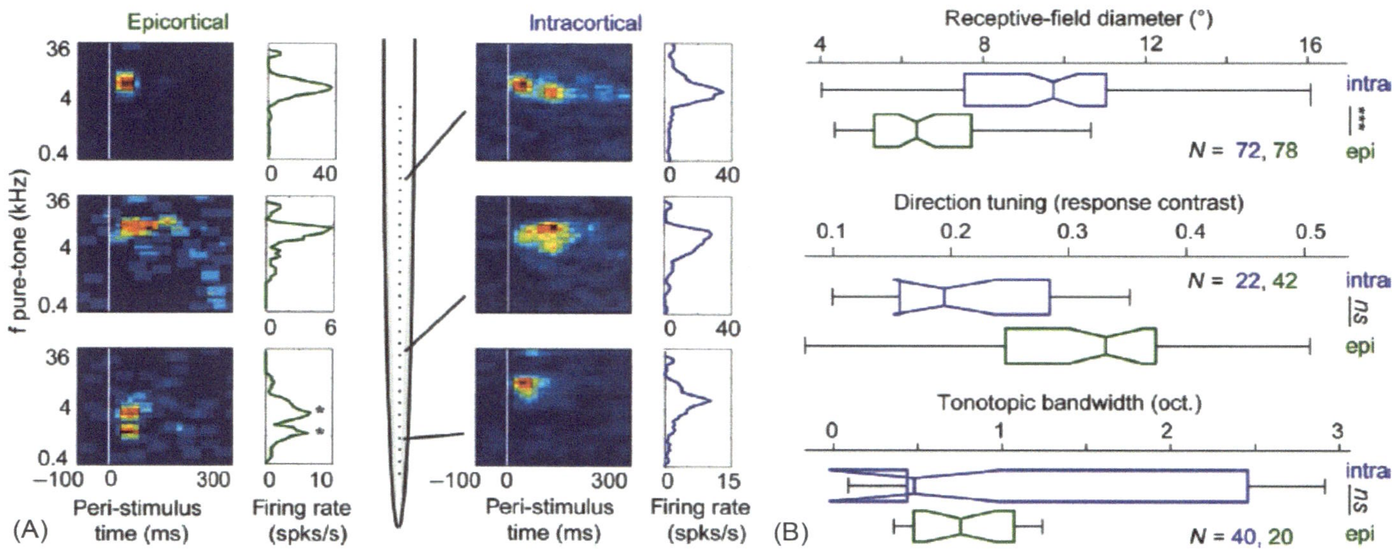

FIG. 5.2 Relation between intracortical and epicortical multiple-unit spiking: receptive fields and tuning to stimulus features. Comparisons include spatial (visual) and spectrotemporal (auditory) receptive fields and tuning to orientation of visual stimuli. (A) Example data from one experimental animal. Spectrotemporal receptive fields and frequency tuning-curves obtained from responses to auditory pure-tone stimulation. Asterisks in one plot of epicortical tuning indicate a double-peaked curve. (B) Population analysis of response properties for intracortical *(blue)* and epicortical *(green)* recordings. *N*, number of recordings. *(Top)* Visual receptive field diameter measured as full-width at half of the maximum (***$P < 10^{-5}$). *(Middle)* Tuning to movement direction of drifting Gabor patterns ($P = .06$). (Bottom) Bandwidth of auditory tuning curves measured as full-width at half of the maximum and expressed in octaves ($P > .05$). *Reprinted with permission from Bockhorst, T., Pieper, F., Engler, G., Stieglitz, T., Galindo-Leon, E., Engel, A.K., 2018. Synchrony surfacing: epicortical recording of correlated action potentials. Eur. J. Neurosci. 48(12), 3583–3596. Copyright 2018 Federation of European Neuroscience Societies.*

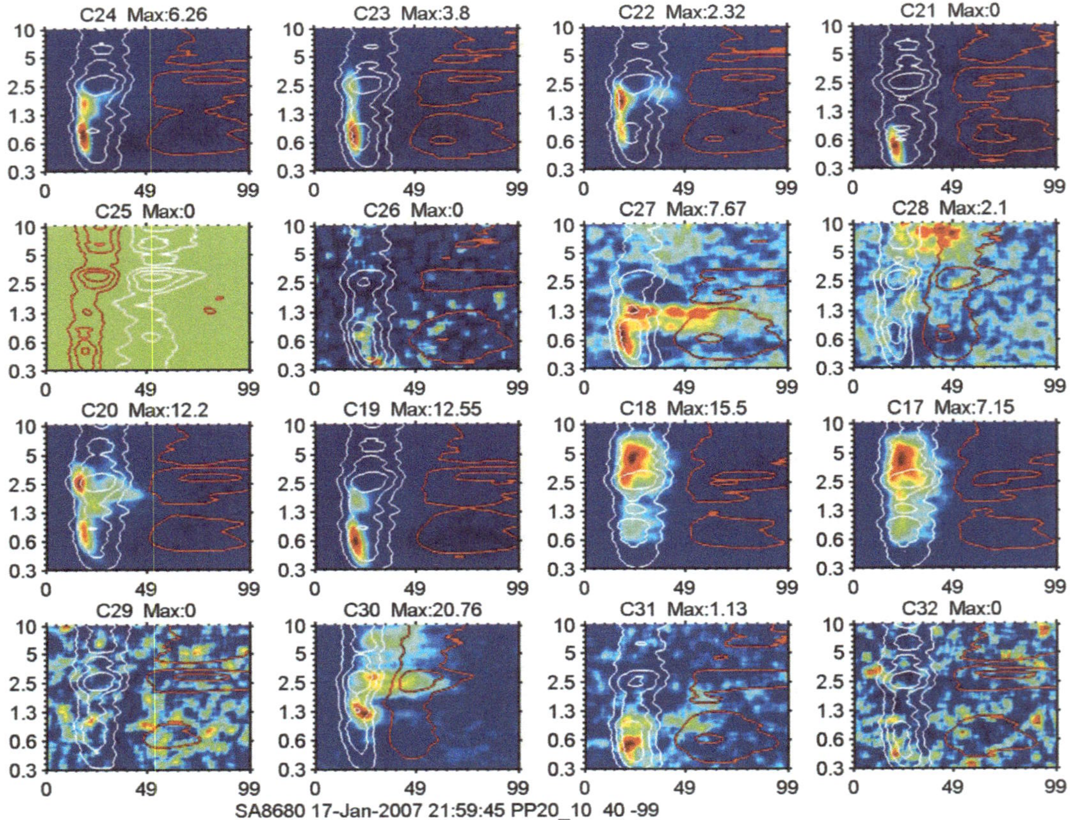

FIG. 5.3 Example of spike-based and local field potential (LFP)-based spectrotemporal receptive fields (STRFs) for one 16-electrode array recording. Each panel is scaled on its own extremes; red colors indicate high spike firing levels and blue colors low firing levels. Maximum values for the spike responses (number of spikes/bin/stimulus) are indicated above each panel. In some cases, this value is 0, whereas there is still some response visible. We use 0, when the value is < 0.01 spikes/bin/stimulus. Contour lines for LFP amplitude are white for negative values and *red* for positive values.

(Continued)

LFP-based FTCs the broadest and CSD-based FTCs in between. Albeit that observed LFPs receive local contributions, direct measurements indicate passive spread of LFPs to sites more than a centimeter from their origins. These findings were independent of the frequency content of the LFP. Kajikawa and Schroeder (2011) and Eggermont et al.'s (2011) results challenge the idea that LFP recordings typically integrate over extremely circumscribed local domains. Rather, LFPs appear as a mixture of genuine-local potentials with "volume conducted" potentials from distant sites (Herreras, 2016).

Snyder et al. (2018) measured LFPs with microelectrode arrays in two brain areas (extrastriate visual area V4 and dorsolateral prefrontal cortex) simultaneously with EEG at the nearby scalp in rhesus macaque monkeys. Although they found interarea coherence at both scales of measurement, they did not find that scalp-level coherence was reliably related to coherence between brain areas measured intracranially on a trial-to-trial basis. Thus Snyder et al. (2018) found no evidence that long-range coherence between scalp EEG electrodes relates to long-range coherence between LFPs.

5.4 Spike-LFP correlation in human cortex

To evaluate how individual neuron spikes couple to the LFP and blood oxygen level-dependent (BOLD) fMRI, Nir et al. (2007) simultaneous recorded single-unit activity and LFPs by using multiple electrodes in the human auditory cortex from three patients with pharmacologically intractable epilepsy. They found a wide range of correlations between individual neuron spiking and gamma-frequency LFPs (Fig. 5.4). However, this large variability could be predominantly explained by the degree of firing-rate correlations between neighboring neurons (cf. Riehle et al., 1997;

FIG. 5.3, CONT'D The frequency axis (log scale) runs from 300 Hz to 10 kHz and covers 5 octaves. The spatial orientation of the 16 electrodes in this figure is such that the top two rows are physically on the *left* of the *bottom two rows*. Within a row, the electrodes are separated by 250 mm and between rows by 500 mm. Although there is very little difference in the LFPs recorded on the various electrodes in the array, the spike STRFs are varying across the array. For example, the neighboring electrodes 19 and 20 show very different spike-STRFs and nearly identical LFP-STRFs. Here the LFP contours, especially the 25% of peak amplitude, cover about 5 octaves, whereas the neural activity covers 1–2 octaves. Spike firing occurs frequently at the short latency edge of the 50% contour *(e.g., top row)* but can also be within the 50% contour band *(third row, last two columns)*, or extend beyond the negative amplitude contours *(second row, last two columns)*. Channel 25 did not produce spikes but a clear LFP-based STRF starting with a positive phase and indicative of local hyperpolarization. *From Eggermont, J.J., Munguia, R., Pienkowski, M., Shaw, G., 2011. Comparison of LFP-based and spike-based spectro-temporal receptive fields and neural synchrony in cat primary auditory cortex. PLoS One 6(5): e20046. https://doi.org/10.1371/journal.pone.0020046. Open access.*

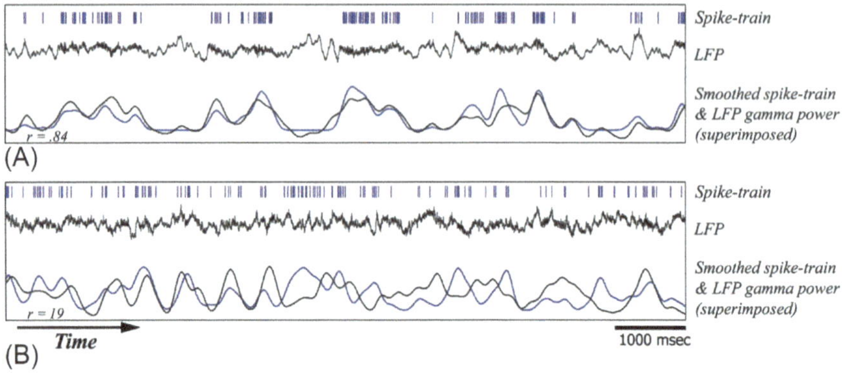

FIG. 5.4 Example of spike trains and LFPs. Data from one isolated neuron (mean firing rate = 14 Hz) and the simultaneously recorded LFP from the same microelectrode during segments of movie stimulation associated with strong or poor coupling of spikes to gamma LFP. (A) Strong coupling: Data acquired during a 10 s segment associated with strong spike-gamma coupling ($r = .84$). Rows *(top to bottom)* correspond to (l) spike train, (2) raw LFP filtered below 130 Hz, and (3) smoothed spike train and gamma-band power modulations superimposed; each time-course value represents the instantaneous firing rate *(blue)* or gamma power *(black)* at that time. (B) Poor coupling: Data from the same neuron and microelectrode acquired during a different 10 s segment associated with poor spike-gamma coupling ($r = .19$). Rows and colors are as shown in (A). Note that in this data segment, changes in firing rate are poorly correlated to changes in LFP gamma power *(bottom row). Reprinted with permission from Nir, Y., Fisch, L., Mukamel, R., Gelbard-Sagiv, H., Arieli, A, Fried, I., et al., 2007. Coupling between neuronal firing rate, gamma LFP, and BOLD fMRI is related to interneuronal correlations. Curr. Biol. 17, 1275–1285. Copyright 2007. Elsevier.*

Denker et al., 2011). Nir et al. (2007) found that gamma LFP was well coupled to BOLD responses measured across different individuals (Fig. 5.4). By contrast the coupling of single-unit activity to BOLD was highly variable and, again, tightly related to interneuronal firing-rate correlations. This suggested to Nir et al. (2007) that: "LFPs are coupled to the correlated firing rate in a local population but not necessarily to the firing rate of single units, which may often be uncorrelated with the averaged behavior of the local population."

Rey et al. (2014) recorded from the human medial temporal lobe (MTL) in five patients with pharmacologically intractable epilepsy, who were implanted with intracranial electrodes for clinical reasons. They analyzed the synchrony in simultaneously recorded single-cell spikes and LFPs during a visual recognition task. Patients were shown a sequence of briefly presented pictures and had to report whether they recognized the picture. Single-neuron responses to consciously perceived images were preceded by a global LFP deflection in the theta range (Fig. 5.5). In addition, there was a local and stimulus-specific increase in the single-trial gamma power. The timing of the neurons' firing was phase locked to these LFP responses. The unit responses upon picture recognition were correlated with an

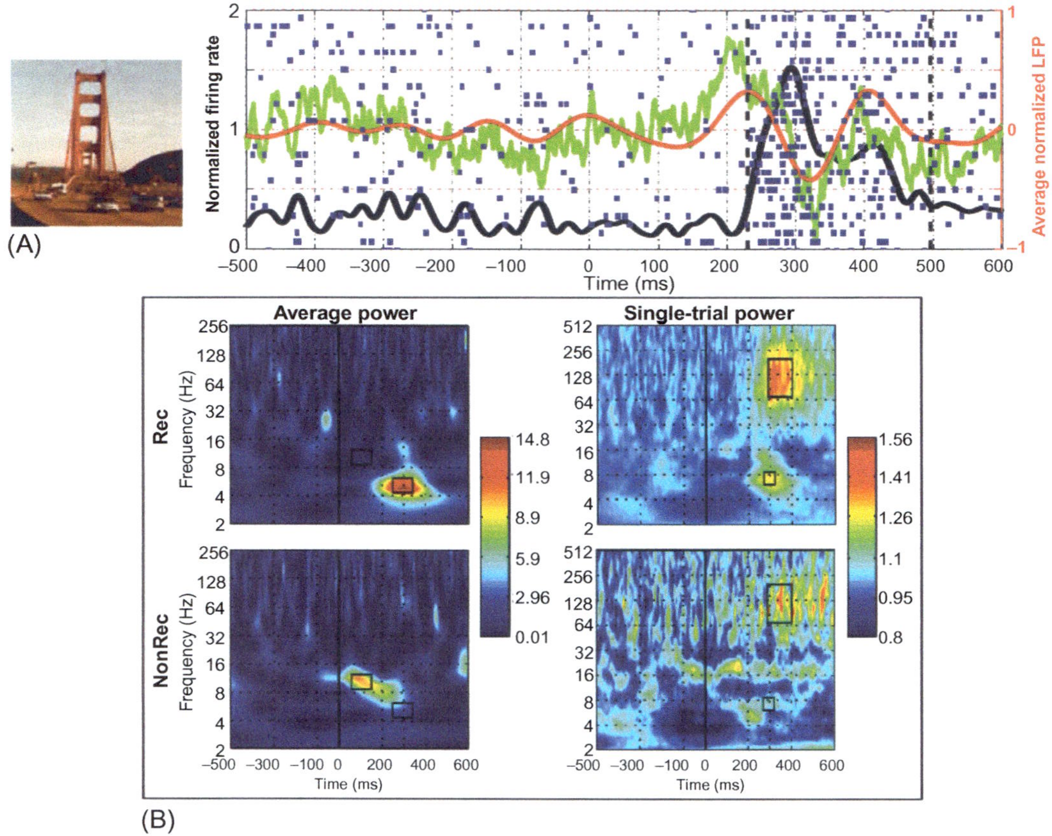

FIG. 5.5 LFP power in recognized and nonrecognized trials. (A) Example of a neuron in the left entorhinal cortex that responded to a picture of the Golden Gate Bridge. The raster plot, the instantaneous firing rate *(solid black line)*, onset and offset of the spiking response *(dashed vertical lines)*, raw average local field potential (LFP, *green line*), and average LFP in the theta band (4–8 Hz; *red line*) are shown. Only the 29 recognized trials were used for computing firing rate and average LFP. (B) *Left*: time-frequency plot of the average LFP (evoked) power for recognized (Rec) and nonrecognized (NonRec) trials.

(Continued)

increase in the evoked theta and the single-trial gamma power in the LFP signals (Fig. 5.5). Rey et al. (2014) proposed that: "whereas the gamma phase locking reflects the activation of local networks encoding particular recognized stimuli, the theta phase locking reflects a global activation that provides a temporal window for processing consciously perceived stimuli in the MTL."

5.5 Spikes, LFPs and EEG/MEG

Synchronized oscillations among large numbers of neurons are evident in electrocorticograms, LFP, EEG, and MEG recordings and are generally understood to depend on inhibition that produces alternating temporal windows of reduced and increased excitability in spiking neurons (Mathalon and Sohal, 2015). LFP oscillations in the cerebral cortex and thalamus are reflected as rhythmic waves in the EEG. The spontaneous EEG activity represents the continuously fluctuating resting state of the brain. Early on, it was already proposed that such EEG oscillations reflect correlated synaptic activity in cortical neurons (Klee et al., 1965; Creutzfeldt et al., 1966a,b).

Are these slow oscillations synchronous across the majority or the minority of brain regions, that is, are they a global or local phenomenon? To examine this, Nir et al. (2011) recorded scalp EEG, depth EEG, and unit firing in multiple brain regions of neurosurgical patients. A tight relationship between EEG slow waves in sleep and underlying unit activity was evident in all monitored brain structures. Although there was considerable variability among individual neurons, in each brain region, the activity of a substantial portion of neurons was phase locked with these slow waves. Moreover, the amplitude of EEG slow waves reflected the degree of modulation in neuronal firing. By examining simultaneous depth EEG and single-unit recordings in multiple regions of the human brain, Nir et al. (2011) showed that ON and OFF periods of spiking activity and corresponding EEG slow waves typically involve a limited number of brain regions. Similarly, sleep spindles were mostly local, whereas slow waves had a tendency to propagate from prefrontal cortex to the medial temporal

FIG. 5.5, CONT'D The regions of interest (ROIs) used for statistical comparisons are indicated by *black rectangles*. There is a significant increase in the theta band for Rec trials ($P=2\times10^{-3}$) and in the alpha band for NonRec trials ($P=.02$). *Right*: grand median time-frequency plot of the single-trial LFP power for Rec and NonRec trials. There is a significant increase in theta ($P<10^{-12}$) and gamma ($P<10^{-23}$) bands for Rec trials, but not for NonRec trials ($P>.05$). *Reprinted with permission from Rey, H.G., Fried, L, Quiroga, R.Q., 2014. Timing of single-neuron and local field potential responses in the human medial temporal lobe. Curr. Biol. 24, 299–304. Copyright 2014. Elsevier.*

lobe and hippocampus. Nir et al. (2011) also found that activity onsets in individual waves reflected the afferent synaptic drive to a given region. When focusing on the highest amplitude waves in each channel (top 20%), positive and negative peaks in depth EEG were associated with marked decreases and increases in unit discharges, respectively. In each subject and in each brain region, individual neurons whose activity was highly modulated by slow waves were identified. Such neurons were found not only in neocortex but also in limbic structures such as hippocampus and amygdala. Thus Nir et al. (2011) demonstrated that within specific brain structures, sleep slow waves in depth EEG reliably reflect synchronous transitions between ON and OFF periods among many neurons.

5.6 Summary

The distinction between spike-firing synchrony and slower forms of correlated activity, for example, firing-rate covariance can be related to the frequency of the modulating brain rhythm. The correlation between the firings of cortical neurons, either as a result of common input from the thalamus or resulting from the more numerous intracortical connections, is weak. In contrast the population activity in the resting-state cortex is widely synchronized and produces detectable signs of this synchrony in the EEG and MEG waves. Cortical network activity contributes considerably to the overall correlation and can mask the strength of direct anatomical connections. LFP oscillations in the cerebral cortex and thalamus are reflected as rhythmic waves in the EEG/MEG. LFPs appear as a mixture of genuine-local potentials with "volume conducted" potentials from distant sites. LFPs are coupled to the correlated firing rate in a local population but not necessarily to the firing rate of single units, which may often be uncorrelated with the averaged behavior of the local population.

References

Abeles, M., 1982. Local Cortical Circuits. An Electrophysiological Study. Springer Verlag, Berlin.

Aertsen, A.M.H.J., Gerstein, G.L., Habib, M.K., Palm, G., 1989. Dynamics of neuronal firing correlation: modulation of "effective connectivity". J. Neurophysiol. 61, 900–917.

Bockhorst, T., Pieper, F., Engler, G., Stieglitz, T., Galindo-Leon, E., Engel, A.K., 2018. Synchrony surfacing: epicortical recording of correlated action potentials Eur. J. Neurosci. 48 (12), 3583–3596.

Braitenberg, V., 1978. Cell assemblies in the cerebral cortex. In: Heim, R., Palm, G. (Eds.), Theoretical Approaches to Complex Systems. Springer-Verlag, Berlin, pp. 171–188.

Creutzfeldt, O.D., Watanabe, S., Lux, H.D., 1966a. Relations between EEG phenomena and potentials of single cortical cells. I. Evoked responses after thalamic and epicortical stimulation. Electroencephalogr. Clin. Neurophysiol. 20, 1–18.

Creutzfeldt, O.D., Watanabe, S., Lux, H.D., 1966b. Relations between EEG phenomena and potentials of single cortical cells. II. Spontaneous and convulsoid activity. Electroencephalogr. Clin. Neurophysiol. 20, 19–37.

Denker, D., Roux, S., Lindén, H., Diesmann, M., Riehle, A., Grün, S., 2011. The local field potential reflects surplus spike synchrony. Cereb. Cortex 21, 2681–2695.

Eggermont, J.J., 1990. The Correlative Brain. Theory and Experiment in Neural Interaction. Springer Verlag, Berlin, ISBN: 3-540-52326-X.

Eggermont, J.J., 1992. Neural interaction in cat primary auditory cortex. Dependence on recording depth, electrode separation and age. J. Neurophysiol. 68, 1216–1228.

Eggermont, J.J., 1994. Neural interaction in cat primary auditory cortex II. Effects of sound stimulation. J. Neurophysiol. 71, 246–270.

Eggermont, J.J., 1998. Is there a neural code? Neurosci. Biobehav. Rev. 22, 355–370.

Eggermont, J.J., 2010. Pair-correlation in the time and frequency domain. In: Grün, S., Rotter, S. (Eds.), Analysis of Parallel Spike Trains. Springer, New York, pp. 77–102.

Eggermont, J.J., 2019. Correlation analysis of parallel spike trains. In: Jaeger, D., Jung, R. (Eds.), Encyclopedia of Computational Neuroscience. Springer, https://doi.org/10.1007/978-1-4614-7320-6_390-1.

Eggermont, J.J., Smith, G.M., 1995. Rate covariance dominates spontaneous cortical unit-pair correlograms. Neuroreport 6, 2125–2128.

Eggermont, J.J., Smith, G.M., 1996. Neural connectivity only accounts for a small part of neural correlation in auditory cortex. Exp. Brain Res. 110, 379–392.

Eggermont, J.J., Munguia, R., Pienkowski, M., Shaw, G., 2011. Comparison of LFP-based and spike-based spectro-temporal receptive fields and neural synchrony in cat primary auditory cortex. PLoS One 6 (5). https://doi.org/10.1371/journal.pone.0020046.

Herreras, O., 2016. Local field potentials: myths and misunderstandings. Front. Neural Circuits 10, 101. https://doi.org/10.3389/fncir.2016.00101.

Kajikawa, Y., Schroeder, C.E., 2011. How local is the local field potential? Neuron 72, 847–858.

Katzner, S., Nauhaus, I., Benucci, A., Bonin, V., Ringach, D.L., Carandini, M., 2009. Local origin of field potentials in visual cortex. Neuron 61, 35–41.

Klee, M.R., Offenloch, K., Tigges, H., 1965. Cross-correlation of electroencephalographic potentials and slow membrane transients. Science 147, 519–521.

Logothetis, N.K., 2002. The neural basis of the blood-oxygen-level-dependent functional magnetic resonance imaging signal. Phil. Trans. R. Soc. Lond. B 357, 1003–1037.

Mathalon, D.H., Sohal, V.S., 2015. Neural oscillations and synchrony in brain dysfunction and neuropsychiatric disorders. It's about time. JAMA Psychiatry 72 (8), 840–844. https://doi.org/10.1001/jamapsychiatry.2015.0483.

Moore, G.P., Perkel, D.H., Segundo, J.P., 1966. Statistical analysis and functional interpretation of neuronal spike data. Annu. Rev. Physiol. 28, 493–522.

Nir, Y., Fisch, L., Mukamel, R., Gelbard-Sagiv, H., Arieli, A., Fried, I., et al., 2007. Coupling between neuronal firing rate, gamma LFP, and BOLD fMRI is related to interneuronal correlations. Curr. Biol. 17, 1275–1285.

Nir, Y., Staba, R.J., Andrillon, T., Vyazovskiy, V.V., Cirelli, C., Fried, I., et al., 2011. Regional slow waves and spindles in human sleep. Neuron 70, 153–169.

Rey, H.G., Fried, I., Quiroga, R.Q., 2014. Timing of single-neuron and local field potential responses in the human medial temporal lobe. Curr. Biol. 24, 299–304.

Riehle, A., Grün, S., Diesmann, M., Aertsen, A., 1997. Spike synchronization and rate modulation differentially involved in motor cortical function. Science 278, 1950–1953.

Snyder, A.C., Issar, D., Smith, M.A., 2018. What does scalp electroencephalogram coherence tell us about long-range cortical networks? Eur. J. Neurosci. 48, 2466–2481.

Tomita, M., Eggermont, J.J., 2005. Cross-correlation and joint spectro-temporal receptive field properties in auditory cortex. J. Neurophysiol. 93, 378–392.

CHAPTER

6

From oscillation synchrony to neural networks

The concept of the brain as consisting of networks goes back to Ramón y Cajal (from histological tracing using Golgi stains; Finger, 2000) and Hebb (1949; neural assembly concept) with theoretical analysis among others by von der Malsburg and Willshaw (1981). Recently, it has been brought back into the attention by functional magnetic resonance imaging (fMRI) studies by Fox et al. (2005), who coined large networks of brain areas based on correlations in ultraslow fluctuations in the signal, specifically the default mode network (DMN) and the dorsal attention network (DAN), which signals were in turn anticorrelated with each other. Brain rhythms from delta to gamma have been demonstrated to underlie the distinction between neurons in subnetworks synchronized for specific frequency bands. Since then, the idea of local- and long-distance synchronization of brain areas and their dyssynchrony in neurological and psychiatric disorders has taken off.

Networks based on oscillations in the electroencephalogram (EEG) and magnetoencephalogram (MEG) use frequencies that are much higher than the ultraslow (~ 0.1 Hz) fluctuations on which fMRI networks are based (Chapter 1). The various brain rhythms are forming a hierarchically structured conglomerate (Lakatos et al., 2005; Steriade, 2006). In this chapter, we assess whether networks based on the correlation structure of the various EEG/MEG rhythms can be considered as subnetworks of the fMRI-based ones.

6.1 Basics of EEG/MEG-based brain networks

6.1.1 Networks as graphs

The consideration of neural networks of the brain as computational tools likely started with von Neuman (1958) who envisioned the brain as "a simple, perhaps random wired network," and came to fruition in the graph-theoretical work of Watts and Strogatz (1998). Small-world

https://doi.org/10.1016/B978-0-12-819818-6.00007-8

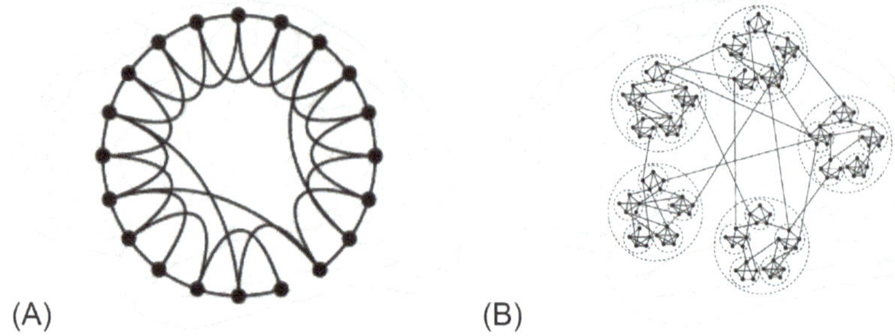

FIG. 6.1 Classical small-world network (A) versus hierarchical modular network (B). Classical small-world networks can be derived by partial random rewiring of regular networks, which results in high clustering and relatively short path lengths. While hierarchical modular networks may also possess these features, they can also be large-world networks with a finite topological dimension. This aspect makes them an intriguing model for brain networks. *From Hilgetag, C.C., Goulas, A., 2016. Is the brain really a small-world network? Brain Struct. Funct. 221, 2361–2366. Open access.*

networks (Watts, 1999), in which "most nodes are not neighbors of one another, but the neighbors of any given node are likely neighbors of each other and most nodes can be reached from every other node by a small number of hops or steps." Since then, the approach to brain networks using graph theoretical methods centers around the brain as a collection of small-world networks (Bullmore and Sporns, 2009; Hilgetag and Goulas, 2016). Such a hierarchical modular network comprises a grouping of small-world networks contrasting some concepts of the brain as one large small-world network (Fig. 6.1). Brain networks, ranging from simple neuron circuits to macroscopic networks of brain areas display the typical features of complex systems: high clustering, short path lengths (both typical of "small-world" networks; Reijneveld et al., 2007), presence of hubs, and the presence of modules (Chapter 1).

The basic ideas underlying the steps from oscillations, through synchrony, to brain networks are described in Chapter 1. We briefly recall these here. Network analysis requires that the original empirical data be represented in the form of a graph (Fig. 6.2). Graphs consist of nodes and edges; in the case of fMRI data, a node can be a voxel or a group of voxels corresponding to an anatomical brain region. In the case of EEG or MEG data, a node can be an electrode or a magnetic sensor but also an intracranial source estimated from the scalp recordings. Edges (links between nodes) can also be defined in various ways: as white-matter tracts (nerves), as correlations in the cortical thickness of structures between brain areas, or as a measure of synchronization between two fMRI blood oxygen level-dependent (BOLD) signals, two EEG signals or two MEG signals. The empirical data can be represented as an $N \times N$ connectivity matrix (cf. Fig. 1.2), with N being the number of nodes and with each

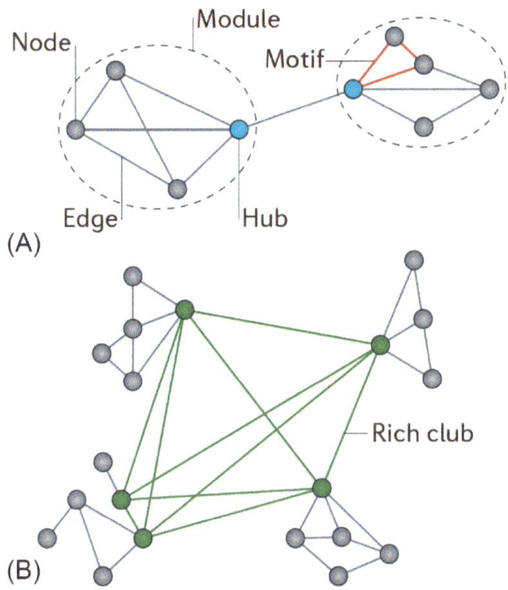

FIG. 6.2 Characteristic of graphs. (A) Modules consist of nodes, edges, and hubs and can be subdivided in motifs. (B) The connectivity structure of a set of hubs is called a rich club network. *Reprinted with permission from Stam, C.J., 2014. Modern network science of neurological disorders. Nat. Rev. Neurosci. 15 (10), 683–695. Copyright 2014. Springer Nature.*

matrix cell providing information about the presence and strength of any relation. Such a matrix can be analyzed as a weighted—by the correlation strength—graph, but more commonly a threshold is applied such that only the cells in the matrix with a suprathreshold value become links in the now unweighted graph. The fraction of edges (of the total number of possible edges, i.e., matrix cells) in the graph is called the density. Often, the values of the computed measures of a graph (such as clustering or path length) are compared with average values of random versions of the same graph, in which all of the node connections are shuffled. Graphs are undirected when links indicate symmetric relationships and directed when links correspond to unidirected relationships (Stam, 2014).

Complex networks can display structure at multiple scales. All nodes that are connected to a node by a direct link are called the neighbors of this node. The number of links connected to a node is called its degree (Stam, 2014). Connected nodes form a module when its nodes are more connected to each other than to nodes outside their module. Small subnetworks are referred to as motifs (Fig. 6.2A). Nodes that have high centrality (large number of edges with other modules) are called hubs. A set of highly interconnected hubs is referred to as a "rich club" (Fig. 6.2B). Rich club regions in the brain include the superior frontal gyrus, precuneus,

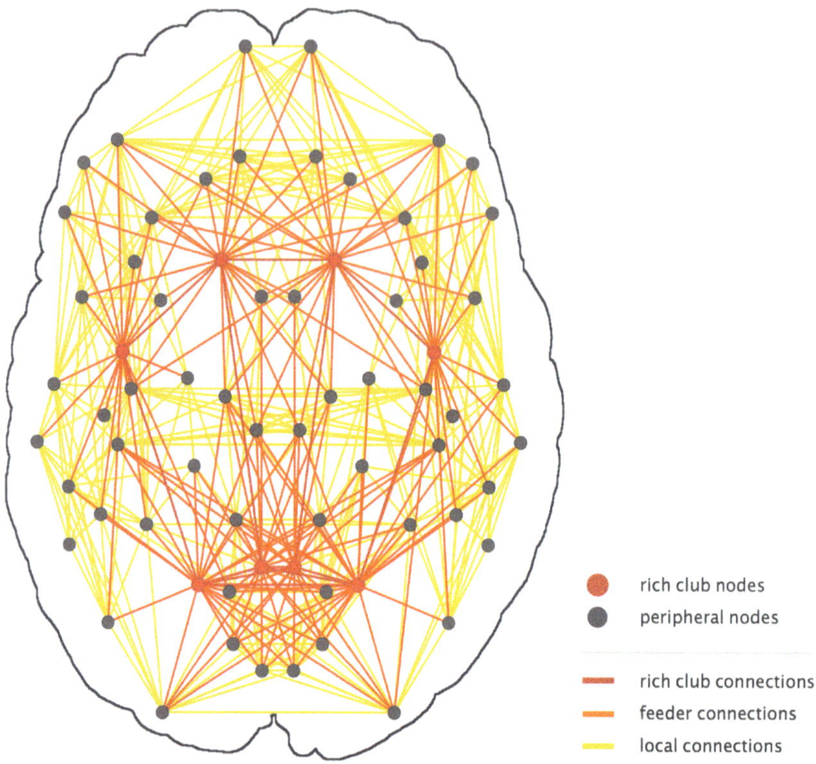

FIG. 6.3 Rich club. Schematic representation of a group-averaged reconstructed structural brain network. Nodes are categorized into rich club and nonrich club peripheral nodes and connections are color coded to indicate rich club, feeder, or local connections. *From Collin, G., Kahn, R.S., de Reus, M.A., Cahn, W., van den Heuvel, M.P., 2014. Impaired rich club connectivity in unaffected siblings of schizophrenia patients. Schizophr. Bull. 40 (2), 438–448. Open access.*

superior parietal gyrus, and insula, all bilateral (in total ~ 12% of the network nodes). Nodes of the network can then be classified into "rich club" nodes and "peripheral" (i.e., nonrich club) nodes (Collin et al., 2014). Rich club connections are those edges that link members of the rich club; feeder connections, are the edges that link rich club nodes to peripheral nodes; and local connections are those edges that interconnect peripheral nodes (Collin et al., 2014; Fig. 6.3).

In this view, the brain consists of scale-free networks whose degree (the number of connections to a node, e.g., voxel or source) distribution follows a power law, that is, linear on a double logarithmic scale (cf. Fig. 6.4C). This means that there are a large number of simple connected nodes, but fewer and fewer nodes with higher number of connections (hubs). Emergence of hubs is a consequence of a scale-free property of networks.

Hubs have a significant impact on the network topology. Hubs in the brain are metabolically high-cost, but allow efficient and fast connectivity. In brain disorders, hubs are often the main target resulting in decreased connectivity, that is, synchrony, between various (sub-)networks often including the DMN. In Alzheimer's disease the hubs have been identified as regions with high amyloid-β deposition, a hallmark of its progression (Chapter 12).

6.1.2 EEG/MEG-based and fMRI-based networks

6.1.2.1 Correlation between BOLD response and EEG rhythms

Broad-band EEG shows prominent infraslow fluctuations (0.01–0.1 Hz) during the execution of a somatosensory detection task, comparable with those in the BOLD response. Detection of sensory stimuli appears strongly correlated with the phase but not with the amplitude of these infraslow EEG fluctuations (Fig. 6.4; Monto et al., 2008; Lowet et al., 2016).

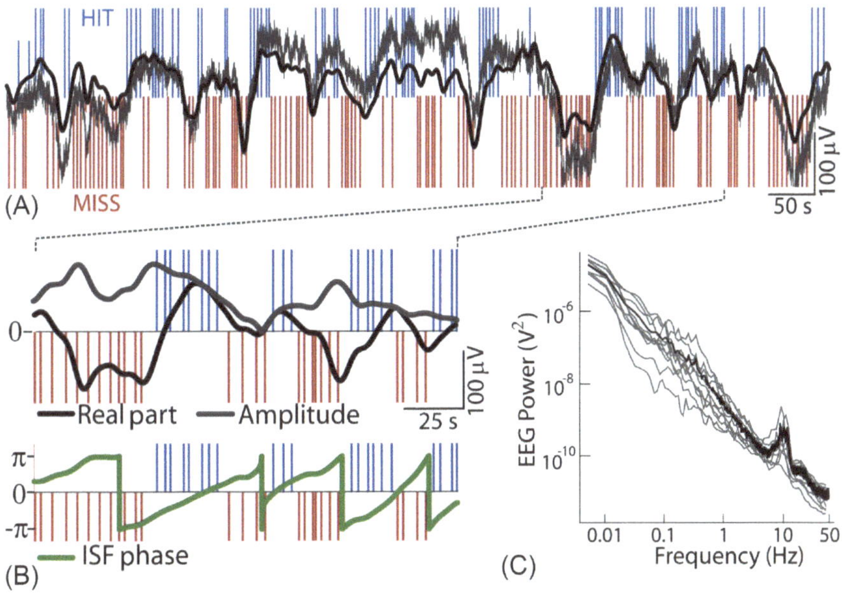

FIG. 6.4 Infraslow EEG is correlated with task performance. (A) Hits *(blue)* and misses *(red)* overlaid on full-band EEG data (0–200 Hz; *gray*) and band-pass-filtered infraslow fluctuations (ISF; 0.01–0.1 Hz; black) from Cz. (B) The ISF real part, amplitude *(top)*, and phase *(bottom)* are shown for part of the data in (A). (C) Individual power spectra *(gray)* and the mean power spectrum *(black)* from Cz in double-logarithmic coordinates. *From Monto, S., Palva, S., Voipio, J., Palva, J.M., 2008. Very slow EEG fluctuations predict the dynamics of stimulus detection and oscillation amplitudes in humans. J. Neurosci. 28, 8268–8272. Copyright 2008, Society for Neuroscience.*

The amplitudes in the 1–40 Hz bands were also strongly correlated with the phase of the infraslow fluctuations, suggesting that: "ongoing 0.01–0.1 Hz EEG fluctuations are prominent and functionally significant during execution of cognitive tasks."

Most of fMRI data identify static network configurations as they typically are used in averaging procedures to calculate their correlations and/or synchrony and make use of its high-spatial but low-temporal resolution. Especially the methods with high temporal resolution, MEG/EEG, have promoted a dynamic description of networks (Calhoun et al., 2014), where the connections evolve over time scales of tens to hundreds of milliseconds, and may reflect rapidly changing brain states that correlate with the time scales of cognitive processes (Section 6.3). Time averaging over these dynamic states then results in steady-state configurations that if obtained without sensory stimuli, with sensory stimuli, or cognitive tasks are called resting-state networks, evoked, or task-related networks, respectively.

For each of five patients undergoing surgical treatment for intractable epilepsy, He et al. (2008) computed a voxelwise BOLD-response correlation map of the sensorimotor network. Implanted electrocorticography (ECoG) electrodes then were localized according to the underlying BOLD response correlation map. Slow cortical activity (≤ 4 Hz) recorded by ECoG demonstrated a correlation structure similar to that of spontaneous BOLD response fluctuations across wakefulness, slow-wave sleep, and rapid-eye movement (REM) sleep. Gamma frequency power (≥ 50 Hz) also showed a similar correlation structure but only during wakefulness and REM sleep. These results connect the large-scale brain networks revealed by spontaneous BOLD signals and their underlying neurophysiology. Brookes et al. (2011) addressed the compatibility of network analysis based on functional connectivity (FC) in resting-state fMRI and MEG. They used envelope correlation and spectral coherence for source-space projected MEG signals. Brookes et al. (2011) showed good spatial agreement between FC measured independently using MEG and fMRI. FC between sensorimotor cortices was observed using both modalities, with the best spatial agreement when MEG data were filtered into the beta band (Fig. 6.5). Note that decreased MEG-beta power corresponds to increased BOLD signal. This negative correlation between beta activity and BOLD response resulting from decreased beta power during movement (Pfurtscheller, 1981; Chapter 2) is well known, in contrast to beta power and the BOLD response increases during movement preparation (Kilavik et al., 2013).

In a review comparing BOLD and electrophysiological connectivity between the DMN and DAN regions, Whitman et al. (2013) noted that alpha (~ 10 Hz) oscillations are phase-locked between the superior parietal lobe (in the DAN) and inferior occipital gyrus contralateral to attended locations.

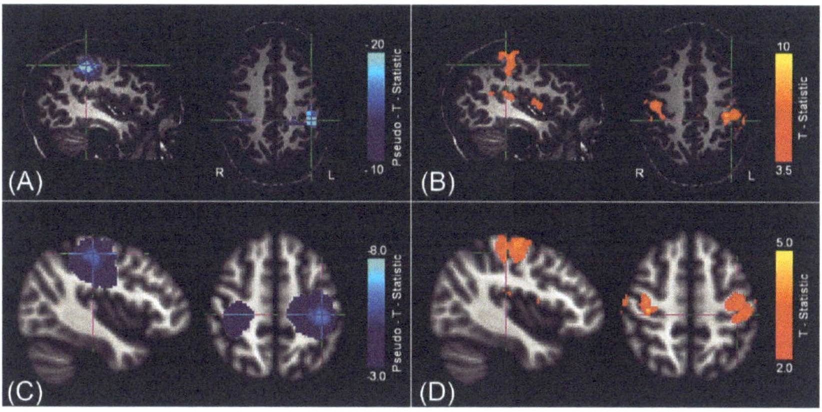

FIG. 6.5 Results of the MEG and fMRI localization experiments. (A) Beta-hand power decrease in bilateral motor cortex in response to the finger movement task (single subject); (B) corresponding increase in BOLD contrast in bilateral motor cortex (single subject). (C) Beta-band power decrease averaged across subjects and overlaid onto the MNI brain; (D) corresponding increase in BOLD also averaged across subjects. All images are shown according to the radiological convention. Cross hairs are placed at the peak beta-band power decrease. *From Brookes, M.J., Hale, J.R., Zumer, JM., Stevenson, C.M., Francis, S.T., Barnes, G.R., et al., 2011. Measuring functional connectivity using MEG: methodology and comparison with fc-MRI. NeuroImage 56, 1082–1104. Open access.*

Beta (~ 20 Hz) oscillations synchronize activity between the frontal eye fields, the intraparietal sulcus (both in DAN) and occipitotemporal cortices (Hipp et al., 2011; Fig. 2.2). Theta and gamma oscillations synchronize frontal regions such as the dorsolateral prefrontal cortex and superior frontal gyrus with the precentral gyrus and precuneus (all in DMN). As we have seen, slower oscillations in EEG/MEG/ECoG data occur at a rate comparable with the spontaneous resting-state oscillations measurable in the BOLD signal. Whitman et al. (2013) concluded that no single frequency band drives the BOLD response. Instead, several different frequencies of oscillation will likely work in concert to connect the regions belonging to a given fMRI network, particularly if that network involves many regions (cf. Magri et al., 2012; Wang et al., 2012).

6.1.2.2 Anticorrelated networks

Resting state networks (RSNs) exhibit a dichotomy between two anticorrelated systems (Fox et al., 2005). One system—extrinsic—engages with the environment and includes the visual, auditory, and sensorimotor (SMN) networks and the DAN, which controls spatial attention. The other system—intrinsic—includes the DMN and the frontoparietal network (FPN; cf. Fig. 4.4), which underlie episodic memory and executive control, respectively. Hacker et al. (2017) measured the spatial correspondence between ECoG band power and resting-state fMRI networks in awake human subjects. They showed that, while gamma power correspondence was common throughout the brain, theta (4–8 Hz) power correspondence was stronger in the DMN and FPN, whereas alpha (8–12 Hz) correspondence was stronger in the sensorimotor network (SMN) and DAN (Fig. 6.6A and B). A quantitative summary of these findings is presented in Fig. 6.6C, which shows the ECoG:fMRI correspondence spectrum (across carrier × modulation frequencies) that was modestly dependent on modulation frequency but strongly dependent on carrier frequency. Thus band-limited power spectral specificity appears in the carrier Í modulation display as broad horizontal bands with peaks within the high gamma (50–100 Hz) and theta (4–8 Hz) carrier frequencies. Fig. 6.6D and E illustrates complementary results, obtained with an electrode seed overlying the postcentral gyrus (PoCG), a locus within the SMN, which is a component of the extrinsic system. This locus showed strong ECoG:fMRI correspondence with alpha (8–12 Hz) and gamma but not with theta frequencies (Fig. 6.6F). Thus whereas both seeds exhibited similar resting state fMRI topographies and high gamma frequencies, intrinsic versus extrinsic spectral specificity was observed at theta and alpha frequencies, respectively.

6.2 Frequency specificity of network connectivity

From the Hacker et al. (2017) study, one may conclude that the human brain, at rest, exhibits frequency specific electrophysiology, respecting both the spectral structure of task responses and the hierarchical organization of resting-state networks.

6.2.1 Technical considerations

The use of brain rhythms as a basis for network analysis implicitly assumes that the rhythms are constituted by oscillations generated by a linear system and that the resulting correlations from which connectivity is deduced also reflect linearity. Frequently, Fourier analysis is used to decompose a certain frequency band, for example, alpha, into subbands then labeled alpha1, alpha2, and alpha3. The point here is that any broadband signal can be decomposed this way, but the physiological mechanisms underlying these subbands may be hard to define or are just not present. Especially for long-distance correlations, lower frequencies, for example, 6–8 Hz, may synchronize more easily, whereas higher frequencies, for example, 10–12 Hz, may not. In the alternative, peak correlations for higher frequency oscillations may be sharper and less noisy due to averaging a larger number of periods in the analysis window. Considering that brain rhythms are likely generated by relaxation oscillators (Buzsáki, 2006), which are definitely nonlinear, averaging to obtain cross-correlations or coherences is not warranted.

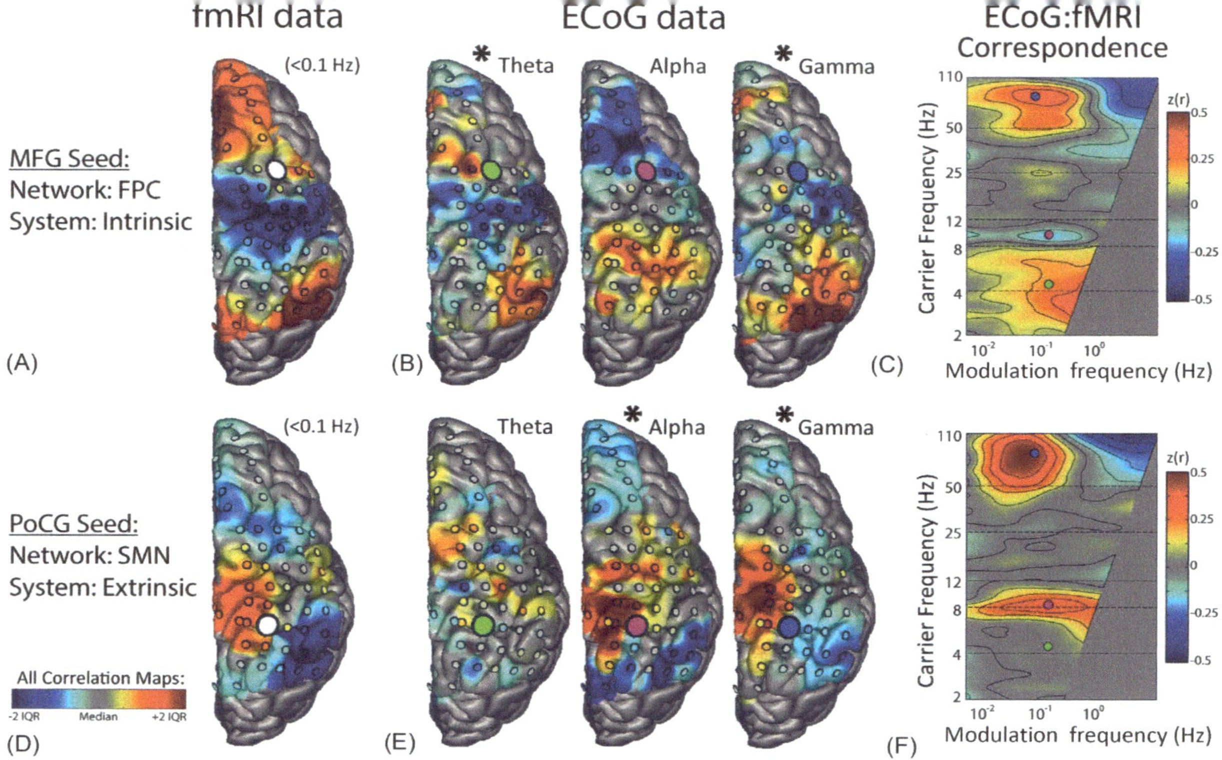

FIG. 6.6 Spatial correspondence of ECoG and fMRI correlation maps in a single subject. (A) Seed-based fMRI correlation map for a seed region *(white circle)* overlying the middle frontal gyrus (MFG), within the frontoparietal control system (FPC). The correlation map rendering is limited to regions with electrode coverage. (B) Seed-based ECoG band power correlation maps for the same seed location as in (A). (C) ECoG:fMRI correspondence assessed by spatial correlation of (A) and (B), parametric in ECoG band power carrier and modulation frequencies. Colored circles indicate frequencies for exemplars in (B). Note that ECoG correlation maps for theta and gamma frequencies *(asterisks)* for this seed region in (B) are similar to the fMRI correlation map in (A). *Bottom half* (D, E, F) illustrates results as per (A, B, C) for a seed region in the postcentral gyms (PoCG), within the sensorimotor network (SMN). Note that correlation maps for the SMN seed at alpha (rather than theta) and gamma frequencies (*asterisks* in E) are similar to the fMRI correlation map in (D). Thus the frequency-specific ECoG:fMRI similarity produces complimentary peaks of spatial correlation in the theta and alpha ranges in (C) and (F), respectively. *IQR*, inter quartile range. *Reprinted with permission from Hacker, CD., Snyder, A.Z., Pahwa, M., Corbetta, M., Leuthardt, E.C., 2017. Frequency-specific electrophysiologic correlates of resting state fMRI networks NeuroImage 149, 446–457. Copyright 2017. Elsevier.*

When assessing electrophysiological changes underlying RSNs, EEG/MEG facilitates a more direct inference on electrophysiological processes compared with fMRI, which measures metabolic consequences of this electrical activity and vasculature changes. The frequency bands of the EEG/MEG signal, by definition neural, allow one to study the hierarchy of functional connections across a range of spatial and temporal scales. These include the phase and envelope connectivity metrics mentioned above which may occur within different frequency ranges, and also cross frequency (Monto, 2012) and nonlinear coupling mechanisms. This brings me to single event analysis, indicated by a stimulus or task onset, so that a preevent window for a certain frequency band can be identified and its amplitude, and phase spectrum can be calculated. This is then compared with either a "hit or fail" behavioral response to a stimulus (e.g., threshold detection) or task. This method, which is still based on linear correlation, can overcome the averaging over brain states and used as predictive for the task (Chapter 3).

6.2.2 Direction of information flow

It is often overlooked that correlations and coherences, linear or not, identify only nondirected connections between brain areas, voxels, MEG/EEG traces, or sources. Directed techniques include Granger causality, information- or phase-transfer entropy, and dynamic causal modeling. The results of these procedures may be similar to structural connectivity resulting from diffusion tensor imaging tractography with seeded starting points. The undirected functional connectivity does not require connectivity through nerve tracts and often results from shared or common inputs (Damoiseaux and Greicius, 2009).

Using MEG, Hillebrand et al. (2016) inferred directionality from time series of intracortical sources of neuronal activation, using phase transfer entropy. Phase transfer entropy quantifies the transfer entropy between time-series of phase extracted from neuronal signals, for example, by filtering (Lobier et al., 2014). See also Gourévitch and Eggermont (2007) for application of transfer entropy to multielectrode single-unit recordings in animal auditory cortex. Hillebrand et al. (2016) observed well-organized posterior-to-anterior patterns of information flow in the higher-frequency bands (alpha1, alpha2, and beta band) dominated by regions in the visual cortex and posterior DMN. Opposite patterns of anterior-to-posterior flow were found in the theta band, involving mainly regions in the frontal lobe that were sending information to a more distributed network. Many strong information-sending regions in the theta band were also frequent receivers of information in the alpha2 band, and vice versa (Horschig et al., 2015). These results suggest that large-scale, higher-frequency band, resting-state patterns of information flow in the human brain from parietooccipital cortex to integrative frontal areas.

6.2.3 Single frequency-band analysis

Nir et al. (2008) simultaneously recorded bilateral single-unit, local field potentials (LFPs) and intracranial ECoG from auditory cortex in patients undergoing clinical monitoring. They found slow (< 0.1 Hz) spontaneous fluctuations of neuronal activity with significant interhemispheric correlations. These fluctuations were evident mainly in neuronal firing rates and in gamma (40–100 Hz) LFP-power modulations. Notably the interhemispheric correlations were enhanced during REM and stage 2 sleep. Multiple intracranial ECoG recordings revealed selectivity for functional networks in the spontaneous gamma LFP-power modulations. In particular, spatial selectivity was expressed in the depth data as higher correlation between auditory cortices than between auditory cortex and other cortical sites. Nir et al. (2008) pointed to slow spontaneous modulations in firing rate and gamma LFP as the likely correlates of spontaneous fMRI fluctuations in the human sensory cortex.

In participants performing a visual response task during high-density EEG recording, Kirschner et al. (2012) found that during mind wandering there was significantly more neural phase synchronization between brain regions associated with the DMN, whereas during periods when subjects were focused on performing the visual task there was significantly more neural phase synchrony within a task-positive brain network (TPN, cf. Fig. 1.1) that shared some of the same brain regions. These differences in network synchrony occurred in each of theta, alpha, and gamma frequency bands. Using MEG, Ghuman et al. (2013) found that, in response to somatosensory stimulation, interhemispheric coupling of somatosensory cortical regions (S1 and S2) is preferentially synchronized in the high-beta frequency band (~ 20–30 Hz). In contrast, interhemispheric coupling of auditory cortices (cf. Fig. 4.1) is preferentially synchronized in the alpha frequency band (~ 7–12 Hz) in response to auditory stimulation—500 ms duration click trains at 15–40 clicks/s, repeated every 2 s. These stimulus-driven synchronization frequencies were also selective to their interregional interactions during the resting-state. This similarity between stimulus-driven and spontaneous states suggests that frequency-specific oscillatory dynamics are intrinsic to the interactions between the nodes of these brain networks.

6.2.4 Nested oscillations

6.2.4.1 Artifacts?

Gerber et al. (2016) provided two empirical examples from intracranial recordings where cross-frequency-coupling (CFC) is driven by the shape of a periodic waveform rather than by a functional interaction between distinct sources. This arises when sharp waveforms (e.g., cortical-evoked potentials) occur throughout parts of the data, in particular whether they occur rhythmically. Gerber et al. (2016) concluded that such behavior of the data, which seems to be present in various cortical signals, might not be interpreted as reflecting functional modulation between distinct neural sources without additional evidence. In addition, even low amplitude periodic potentials that cannot be readily observed or controlled for, are sufficient for significant CFC to occur. However, Jensen et al. (2016) gave concrete counter examples illustrating cases when the identification of cross-frequency coupling can be trusted. See also Section 6.2.1.

6.2.4.2 *Variety of coupling*

Neuronal oscillations within and across brain regions have been proposed to play a crucial role in the communication of brain regions (Chapter 2). Communication between two regions may be established by phase synchronization of oscillations at lower frequencies (< 25 Hz), which serve as a temporal reference frame for information carried by high-frequency activity (> 40 Hz). This framework assumes (Bonnefond et al., 2017) that CFCs are essential for understanding how large-scale cognitive and perceptual networks operate. Palva and Palva (2018) noted that: "Several lines of research have shown that cognitive tasks engage neuronal oscillations simultaneously in multiple frequency bands that have distinct functional roles in cognitive processing. Gamma oscillations are associated with bottom-up processing, while slower oscillations in delta, theta, alpha and beta frequency bands may have roles in executive or top-down controlling functions."

In the macaque, Márton et al. (2019) examined bidirectional cross-frequency coupling across four sectors of the auditory hierarchy. They computed two measures of cross-frequency coupling, phase–amplitude coupling (PAC) and amplitude-amplitude coupling (AAC). Márton et al. (2019) found distinct patterns for bottom-up and top-down information processing among cross-frequency interactions. Both top-down and bottom-up interactions made prominent use of low frequencies: low-to-low-frequency (theta, alpha, and beta) and low-frequency-to-high-gamma couplings were predominant top–down, while low-frequency-to-low-gamma couplings were predominant bottom-up. These patterns were largely preserved across coupling types (PAC and AAC) and across stimulus types (natural and synthetic auditory stimuli), suggesting that they are a general feature of information processing in auditory cortex. Márton et al. (2019) suggest that: "the modulatory effect of low frequencies on gamma-rhythms in distant regions is important for bidirectional information transfer. The finding of low-frequency-to-low-gamma interactions in the bottom-up direction suggests that nonlinearities may also play a role in feedforward message passing." These findings suggest that cross-frequency interaction also play a prominent role in top-down auditory processing.

6.3 Dynamic networks

The current emphasis to dynamic coupling in networks (Calhoun et al., 2014; Hutchison et al., 2013)—functional networks that form and dissolve over short (down to subsecond) time frames—means that EEG/MEG, which exhibits millisecond temporal resolution, offers advantages over fMRI where the measured BOLD signal has a temporal resolution in the seconds range. These arguments, coupled with good spatial resolution of MEG—better than EEG but clearly less then fMRI—which can be 3mm or better in brain regions with a high signal-to-noise ratio, suggest that MEG should be a method of choice for investigation of the human connectome (O'Neill et al. (2015). However, a number of significant technical challenges exist, in particular the MEG inverse problem—inferring equivalent source distributions in the cortex of moment-to-moment change in neural current based only on the recorded extra cranial magnetic fields—is ill posed. This means that estimated time courses of brain current at spatially separate regions are not necessarily independent. As a result, estimated functional connectivity between regions can be artifactually inflated. This significant confound makes connectivity modeling using MEG nontrivial (Brookes et al., 2011). The same implies even stronger to EEG recordings, which represent in addition radial dipole activity and subcortical activity.

Current correlation-based approaches not only average over time but also over functional connectivity between regions, but these averages are less meaningful for regions that are part of multiple networks. There are two ways to proceed. One way is to allow a region's activity pattern to reflect one network's activity some of the time, and another network's activity at other times (Smith et al., 2012). So here spatial invariance is sacrificed and temporal stationarity is assumed. The other way is to assume spatial invariance and sacrifice temporal stationarity. Allen et al. (2014) used this to highlight particularly dynamic connections between regions in lateral parietal and cingulate cortex and assumed that these subnetworks are spatially fixed over time. They argued against a labeling scheme (as in Smith et al., 2012) where such regions are treated as separate and antagonistic entities. Thus, Allen et al. (2014) focused on the variability of temporal interactions between subnetworks, whereas Smith et al. (2012) focused on the spatial complexity and overlap between networks, and implicitly assumed that temporal relationships are consistent over time.

Calhoun et al. (2014) extracted this time varying connectivity in the chronnectome, a model of the brain in which nodal activity and connectivity patterns are changing in fundamental ways through time. In the context of this paper, "dynamics" is thus referring to intrinsic nonstationarities to describe a focus on identifying time-varying, but reoccurring, patterns of coupling among brain regions.

Kabbara et al. (2017) used EEG recorded during task-free paradigm to track the fast temporal dynamics of resting-state brain networks. They found a functional dynamic core network consisting of a set of key brain regions (mainly located in the cingulate and the medial frontal cortex) that ensure segregation and integration functions. Most of the identified hubs were found to belong to the DMN. These central regions were found to alternate dynamically and play the role of either provincial (local) or connector (global) hubs. Lin et al. (2017) also found that DMN topology changes over time with different patterns associated with different brain states induced by task performance. The nodal and global topological organization can be rebuilt by such different brain states. This indicates that the posttask, resting-state topology of the brain network is dynamically altered as a function of immediately prior cognitive experience.

Wens et al. (2019) investigated spontaneous synchronization patterns of dynamic FC with RSN nodes. These modes of dynamic coupling corresponded to recurrent and sporadic transient events of FC changes in varied and widespread coupling patterns. They found that the alpha (8–12 Hz) band disclosed longer-lived FC events with larger deviations, whereas the beta (12–21 Hz)-band modes involved spatially more extended coupling patterns. The DMN modes showed predominant cross-network integration that was more common in the alpha and the beta bands. The dynamic integration of the DMN with other RSNs involved mostly the sensorimotor network and visual network in the alpha band, most RSNs including the auditory network in the theta (8–12 Hz) and the high beta (21–30 Hz) bands, and all of them in the beta band. In contrast with static FC, dynamical competition among these cross DMN-RSN couplings was ubiquitous.

Recent application of the dynamic brain connectome method has become frequent. I selected two examples one related to the link between intelligence and the brain dynamics, and the other correlating recovery of traumatic brain injury to time spent in dynamic brain states. As is known (Stam, 2010), network topology appears predictive of cognitive functioning. Hilger et al. (2019) showed that higher intelligence is associated with higher temporal stability (lower temporal variability) of brain network modularity, in that subjects with higher intelligence scores engage in fewer periods of extremely high modularity. They identified the DAN as primary locus of the observed association between intelligence and temporal brain network stability. Van der Horn et al. (2019) investigated the dynamic functional connectome in relation to long-term recovery after mild-to-moderate traumatic brain injury (TBI). They collected longitudinal resting-state functional MRI data, 1 and 3 months after injury from patients with TBI (92% mild TBI) and healthy subjects. The dynamic neural blueprint of incomplete recovery was characterized by an increased global connectivity pattern. Patients with incomplete recovery at 3 months spent less time in a dynamic brain state that was characterized by the presence of two main functional subdomains, one for sensory and motor function, and one for higher cognitive processing.

6.4 Summary

We assessed, based on the correlation structure of the various EEG/MEG rhythms, whether these networks can be considered subnetworks of the fMRI-based ones. Correlations and coherences, linear or not, identify only nondirected connections between brain areas, voxels, MEG/EEG traces, or sources. Techniques such as Granger causality, information- or phase-transfer entropy, and dynamic causal modeling result in directed connectivity networks. These may be similar to structural connectivity resulting from diffusion tensor imaging tractography. The undirected functional connectivity does not require connectivity through nerve tracts and often results from shared or common inputs. Most of fMRI data identify static network configurations as they typically result from averaging procedures to calculate their correlations and/or synchrony and make use of its high-spatial but low-temporal resolution. MEG and EEG have high temporal resolution and allow a dynamic description of networks, where the connections evolve over time scales of tens to hundreds of milliseconds. Dynamic networks may reflect rapidly changing brain states that correlate with the time scales of cognitive processes. These networks have functional dynamic core consisting of a set of key brain regions (mainly located in the cingulate and the medial frontal cortex) that ensure segregation and integration functions. Their identified hubs were mostly found to belong to the DMN. We described two applications, one related to the link between intelligence and the brain dynamics, and the other correlating recovery of traumatic brain injury to time spent in dynamic brain states.

References

Allen, E.A., Damaraju, E., Plis, S.M., Erhardt, E.B., Eichele, T., Calhoun, V.D., 2014. Tracking whole-brain connectivity dynamics in the resting state. Cereb. Cortex 24, 663–676.

Bonnefond, M., Kastner, S., Jensen, O., 2017. Communication between brain areas based on nested oscillations. eNeuro 4 (2). https://doi.org/10.1523/ENEURO.0153-16.2017.

Brookes, M.J., Hale, J.R., Zumer, J.M., Stevenson, C.M., Francis, S.T., Barnes, G.R., et al., 2011. Measuring functional connectivity using MEG: methodology and comparison with fcMRI. NeuroImage 56, 1082–1104.

Bullmore, E., Sporns, O., 2009. Complex brain networks: graph theoretical analysis of structural and functional systems. Nat. Rev. Neurosci. 10, 186–198.

Buzsáki, G., 2006. Rhythms of the Brain. Oxford University Press, Oxford, UK.

Calhoun, V.D., Miller, R., Pearlson, G., Adali, T., 2014. The chronnectome: time-varying connectivity networks as the next frontier in fMRI data discovery. Neuron 84, 262–274.

Collin, G., Kahn, R.S., de Reus, M.A., Cahn, W., van den Heuvel, M.P., 2014. Impaired rich club connectivity in unaffected siblings of schizophrenia patients. Schizophr. Bull. 40 (2), 438–448.
Damoiseaux, J.S., Greicius, M.D., 2009. Greater than the sum of its parts: a review of studies combining structural connectivity and resting-state functional connectivity. Brain Struct. Funct. 213, 525–533.
Finger, S., 2000. Santiago Ramón y Cajal. From nerve nets to neuron doctrine. In: Minds Behind the Brain: A History of the Pioneers and their Discoveries. Oxford University Press, New York, ISBN: 0-19-508571-X, pp. 197–216.
Fox, M.D., Snyder, A.Z., Vincent, J.L., Corbetta, M., Van Essen, D.C., Raichle, M.E., 2005. The human brain is intrinsically organized into dynamic, anticorrelated functional networks. Proc. Natl. Acad. Sci. U. S. A. 102, 9673–9678.
Gerber, E.M., Sadeh, B., Ward, A., Knight, R.T., Deouell, L.Y., 2016. Non-sinusoidal activity can produce cross-frequency coupling in cortical signals in the absence of functional interaction between neural sources. PLoS One 11 (12). https://doi.org/10.1371/journal.pone.0167351.
Ghuman, A.S., van den Honert, R.N., Martin, A., 2013. Interregional neural synchrony has similar dynamics during spontaneous and stimulus-driven states. Sci. Rep. 3, 1481. https://doi.org/10.1038/srep01481.
Gourévitch, B., Eggermont, J.J., 2007. Evaluating information transfer between auditory cortical neurons. J. Neurophysiol. 97, 2533–2543.
Hacker, C.D., Snyder, A.Z., Pahwa, M., Corbetta, M., Leuthardt, E.C., 2017. Frequency-specific electrophysiologic correlates of resting state fMRI networks. NeuroImage 149, 446–457.
He, B.J., Snyder, A.Z., Zempel, J.M., Smyth, M.D., Raichle, M.E., 2008. Electrophysiological correlates of the brain's intrinsic large-scale functional architecture. Proc. Natl. Acad. Sci. U. S. A. 105, 16039–16044.
Hebb, D.O., 1949. The Organization of Behavior. McGill University Press, Montreal, Canada.
Hilger, K., Fukushima, M., Sporns, O., Fiebach, C.J., 2019. Temporal stability of functional brain modules associated with human intelligence. Hum. Brain Mapp. 41 (2), 362–372.
Hilgetag, C.C., Goulas, A., 2016. Is the brain really a small-world network? Brain Struct. Funct. 221, 2361–2366.
Hillebrand, A., Tewarie, P., van Dellen, E., Yu, M., Carbo, E.W.S., Douw, L., et al., 2016. Direction of information flow in large-scale resting-state networks is frequency-dependent. Proc. Natl. Acad. Sci. U. S. A. 113 (14), 3867–3872.
Hipp, J.F., Engel, A.K., Siegel, M., 2011. Oscillatory synchronization in large-scale cortical networks predicts perception. Neuron 69, 387–396.
Horschig, J.M., Smolders, R., Bonnefond, M., Schoffelen, J.-M., van den Munckhof, P., Schuurman, P.R., et al., 2015. Directed communication between nucleus accumbens and neocortex in humans is differentially supported by synchronization in the theta and alpha band. PLoS ONE 10 (9), e0138685. https://doi.org/10.1371/journal.pone.0138685.
Hutchison, R.M., Womelsdorf, T., Allen, E.A., Bandettini, P.A., Calhoun, V.D., Corbetta, M., et al., 2013. Dynamic functional connectivity: promise, issues, and interpretations. NeuroImage 80, 360–378.
Jensen, O., Spaak, E., Park, H., 2016. Discriminating valid from spurious indices of phase-amplitude coupling. eNeuro 3 (6). https://doi.org/10.1523/ENEURO.0334-16.2016.
Kabbara, A., El Falou, W., Khalil, M., Wendling, F., Hassan, M., 2017. The dynamic functional core network of the human brain at rest. Sci. Rep. 7, 2936. https://doi.org/10.1038/s41598-017-03420-6.
Kilavik, B.E., Zaepffel, M., Brovelli, A., MacKay, W.A., Riehle, A., 2013. The ups and downs of beta oscillations in sensorimotor cortex. Exp. Neurol. 245, 15–26. https://doi.org/10.1016/j.expneurol.2012.09.014.

Kirschner, A., Kam, J.W.Y., Handy, T.C., Ward, L.M., 2012. Differential synchronization in default and task-specific networks of the human brain. Front. Hum. Neurosci. 6, 139. https://doi.org/10.3389/fnhum.2012.00139.

Lakatos, P., Shah, A.S., Knuth, K.H., Ulbert, I., Karmos, G., Schroeder, C.E., 2005. An oscillatory hierarchy controlling neuronal excitability and stimulus processing in the auditory cortex. J. Neurophysiol. 94, 1904–1911.

Lin, P., Yang, Y., Gao, J., De Pisapia, N., Ge, S., Wang, X., et al., 2017. Dynamic default mode network across different brain states. Sci. Rep. 7, 46088. https://doi.org/10.1038/srep46088.

Lobier, M., Siebenhuhner, F., Palva, S., Palva, J.M., 2014. Phase transfer entropy: a novel phase-based measure for directed connectivity in networks coupled by oscillatory interactions. NeuroImage 85, 853–872.

Lowet, E., Roberts, M.J., Bonizzi, P., Karel, J., De Weerd, P., 2016. Quantifying neural oscillatory synchronization: a comparison between spectral coherence and phase-locking value approaches. PLoS One 11 (1). https://doi.org/10.1371/journal.pone.0146443.

Magri, C., Schridde, U., Murayama, Y., Panzeri, S., Logothetis, N.K., 2012. The amplitude and timing of the BOLD signal reflects the relationship between local field potential power at different frequencies. J. Neurosci. 32, 1395–1407.

Márton, C.D., Fukushima, M., Camalier, C.R., Schultz, S.R., Averbeck, B.B., 2019. Signature patterns for top-down and bottom-up information processing via cross-frequency coupling in macaque auditory cortex. eNeuro 6 (2). https://doi.org/10.1523/ENEURO.0467-18.2019.

Monto, S., 2012. Nested synchrony—a novel cross-scale interaction among neuronal oscillations. Front. Physiol. 3, 384. https://doi.org/10.3389/fphys. 2012.00384.

Monto, S., Palva, S., Voipio, J., Palva, J.M., 2008. Very slow EEG fluctuations predict the dynamics of stimulus detection and oscillation amplitudes in humans. J. Neurosci. 28, 8268–8272.

Nir, Y., Mukamel, R., Dinstein, I., Privman, E., Harel, M., Fisch, L., et al., 2008. Interhemispheric correlations of slow spontaneous neuronal fluctuations revealed in human sensory cortex. Nat. Neurosci. 11 (9), 1100–1108.

O'Neill, G.C., Barratt, E.L., Hunt, B.A.E., Tewarie, T.K., Brookes, M.J., 2015. Measuring electrophysiological connectivity by power envelope correlation: a technical review on MEG methods. Phys. Med. Biol. 60, R271–R295.

Palva, J.M., Palva, S., 2018. Functional integration across oscillation frequencies by cross-frequency phase synchronization. Eur. J. Neurosci. 48 (7), 2399–2406.

Pfurtscheller, G., 1981. Central beta rhythm during sensorimotor activities in man. Electroencephalogr. Clin. Neurophysiol. 51, 253–264. https://doi.org/10.1016/0013-4694(81) 90139-5.

Reijneveld, J.C., Ponten, S.C., Berendse, H.W., Stam, C.J., 2007. The application of graph theoretical analysis to complex networks in the brain. Clin. Neurophysiol. 118, 2317–2331.

Smith, S.M., Miller, K.L., Moeller, S., Xu, J., Auerbach, E.J., Woolrich, M.W., et al., 2012. Temporally-independent functional modes of spontaneous brain activity. Proc. Natl. Acad. Sci. U. S. A. 109 (8), 3131–3136.

Stam, C.J., 2010. Characterization of anatomical and functional connectivity in the brain: a complex networks perspective. Int. J. Psychophysiol. 77, 186–194.

Stam, C.J., 2014. Modern network science of neurological disorders. Nat. Rev. Neurosci. 15 (10), 683–695.

Steriade, M., 2006. Grouping of brain rhythms in corticothalamic systems. Neuroscience 137, 1087–1106.

Van der Horn, H.J., Vergara, V.M., Espinoza, F.A., Calhoun, V.A., Mayer, A.R., Van der Naalt, J., 2019. Functional outcome is tied to dynamic brain states after mild to moderate traumatic brain injury. Hum. Brain Mapp. 41 (3), 617–631.

von der Malsburg, C., Willshaw, D., 1981. Co-operaticity and brain organization. Trends Neurosci. 4, 80–83.

von Neuman, J., 1958. The Computer and the Brain. Yale University Press, New Haven, CT.

Wang, L., Saalmann, Y.B., Pinsk, M.A., Arcaro, M.J., Kastner, S., 2012. Electrophysiological low-frequency coherence and cross-frequency coupling contribute to BOLD connectivity. Neuron 76, 1010–1020.

Watts, D., 1999. Small Worlds. The Dynamics of Networks Between Order and Chaos. Princeton University Press, NJ.

Watts, D.J., Strogatz, S.H., 1998. Collective dynamics of 'small-world' networks. Nature 393 (6684), 440–442.

Wens, V., Bourguignon, M., Vander Ghinst, M., Mary, A., Marty, B., Coquelet, N., et al., 2019. Synchrony, metastability, dynamic integration, and competition in the spontaneous functional connectivity of the human brain. NeuroImage 199, 313–324.

Whitman, J.C., Ward, L.M., Woodward, T.S., 2013. Patterns of cortical oscillations organize neural activity into whole-brain functional networks evident in the fMRI BOLD signal. Front. Hum. Neurosci. 7, 80. https://doi.org/10.3389/fnhum.2013.00080.

CHAPTER

7

Rhythms and networks in the hearing and deaf brain

7.1 Introduction

In a review of auditory temporal processing (Eggermont, 2015), I wrote: "Sound reflects the dynamic spectrum of pressure fluctuations in the air. ... The temporal aspects of sound as analyzed and represented in the auditory system have been thoroughly studied. Temporal mechanisms underlie pitch perception, which allows the enjoyment of music, and are required to make a distinction between voiced and unvoiced consonants. Speech encoding is likely the most important function of the human auditory system. The rapid changes in the spectral makeup of speech with time, e.g., as in formant transitions, are characteristic and shaped for use in communication. Speech also contains periodic elements resulting from the glottal pulses that determine the pitch of the human voice. Onsets and gaps determine the phonemes in speech. At least *under optimal listening conditions,* frequency information in the formants appears far less important than the temporal content of speech for recognition of phonemes and words in simple sentences (Shannon et al., 1995)."

7.2 Brain rhythms in cortical auditory and speech areas

7.2.1 Animal research

The first studies that showed auditory gamma in animals (Franowicz and Barth, 1995; Barth and MacDonald, 1996) localized this to primary (A1) and secondary (A2) auditory cortex of anesthetized rats. In the primary and caudomedial auditory cortex of anesthetized macaque monkeys, Brosch et al. (2002) explored neuronal activity in the gamma range. Their stimuli-induced gamma oscillations that were not phase-locked to the onset of stimulation, in the majority of multiunit clusters and sites at which local

https://doi.org/10.1016/B978-0-12-819818-6.00011-X

field potentials (LFPs) were recorded. The amplitudes of stimulus-induced gamma oscillations in LFPs were, on average, almost twice the amplitude of spontaneously occurring gamma oscillations. Gamma activity at different sites within the primary and the caudomedial auditory field could be synchronized at near-zero phase. They concluded that stimulus-induced gamma oscillations originate from sources in the auditory cortex.

Using voltage-sensitive dye imaging in anesthetized guinea pigs, Farley and Noreña (2013) measured frequency tuning to sound across several core and belt auditory cortical areas, and then recorded spontaneous activity across this defined network. They found that phase coherence in spontaneous delta-theta band activity was highest between regions of core and belt areas that had similar tone-frequency tuning, even if they were distant. Further, core and belt regions with high phase coherence showed differences in phase. These phase shifts observed during spontaneous LFP activity paralleled latency differences for evoked activity. Farley and Noreña (2013) then compared ongoing spontaneous activity traces derived from auditory cortical and surrounding fields. They found that the pattern of activity in the auditory cortex was dramatically different from activity in an adjacent nonauditory responsive region. Thus the cortical sensory modality is an important factor for predicting the structure of slow-wave spontaneous activity.

In macaque A1, (O'Connell et al., 2015) observed that for delta and gamma stimulation rates, the amplitude spectrum of the entrained neuronal activity is "peaked" compared with the spontaneous spectrum. However, lower amplitudes around the peak in the auditory stimulus stream-related spectra did not results in significant net integrated amplitude change in the overall delta and gamma bands (Fig. 7.1). The authors suggest that the peaks could therefore represent a reorganization of oscillatory activity to match the stimulus repetition rates, a characteristic of oscillatory entrainment (Haegens and Zion Golumbic, 2018).

7.2.2 Human research

Early on and using magnetoencephalography (MEG), (Pantev et al., 1991) discovered a ~40-Hz transient oscillatory response, evoked in the human brain by the onset of auditory stimuli, consisting of four or more cycles locked in phase to stimulus onset in approximately the 20- to 130-ms poststimulus interval. The response was localized to a source in the supratemporal auditory cortex, some millimeters deeper and anterior to the source of the larger-amplitude slow-wave M100 component of the evoked magnetic field. They considered the oscillatory cortical activation elicited by auditory stimuli similar to the gamma band cortical oscillations elicited by olfactory and visual stimuli and may represent an essential component of auditory perceptual processing.

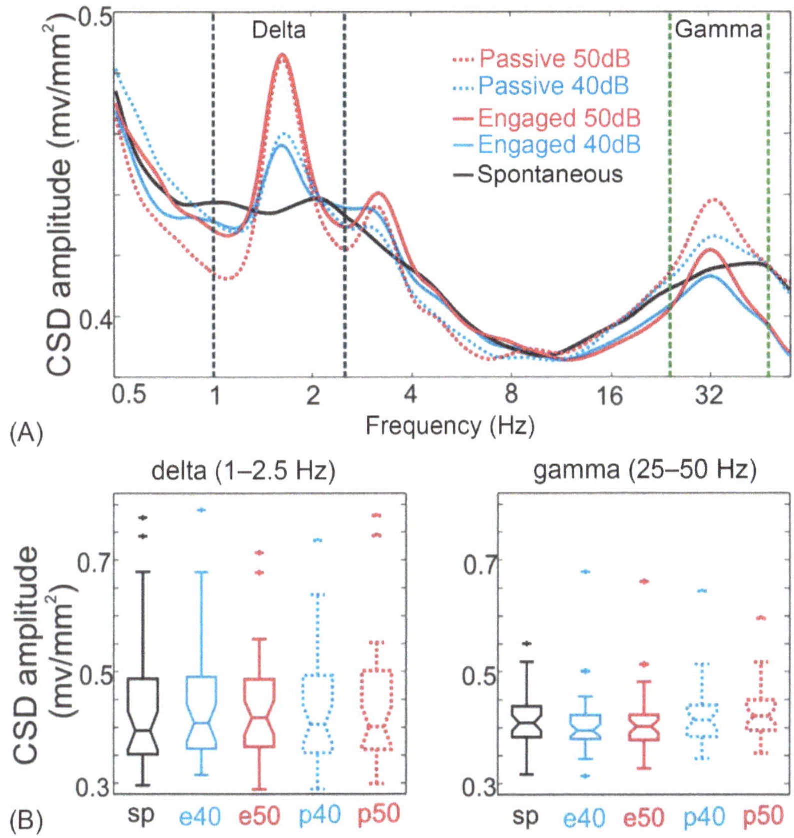

FIG. 7.1 Tie amplitude of spontaneous and auditory stimulus stream-related delta and gamma band neuronal activity. (A) The spectra averaged across all experiments ($n = 48$) show that the presentation of click-train streams results in discrete peaks that correspond to the repetition of clicks within click-trains (33 Hz) and the rate at which click-trains occurred in the stream (1.6 Hz). These peaks are larger in the case of louder streams. The *dotted lines* mark the delta (*black*, 1–2.5 Hz) and gamma (*green*, 25–50 Hz) frequency bands. (B) Boxplots show pooled amplitudes of neuronal activity averaged in the delta (*left*) and gamma (*right*) frequency bands. *From O'Connell, M.N., Barczak, A., Ross, D., McGinnis, T., Schroeder, C.E., Lakatos, P., 2015. Multi-scale entrainment of coupled neuronal oscillations in primary auditory cortex. Front. Hum. Neurosci. 9, 655. https://doi.org/10.3389/fnhum.2015.00655 Open access.*

Gourévitch et al. (2008) recorded depth-intracortical LFPs in humans in response to amplitude modulated (AM) noise to map activation in primary (A1), secondary (A2) and associative auditory areas and to study the propagation of the cortical activity between them. They showed that the following areas respond to AM in both hemispheres: A1, A2, and planum temporale (PT)), the insular gyrus, the Brodmann area (BA) 22 and the posterior part of T1 gyrus (T1Post) (Fig. 7.2). They observed a stronger response to AM in A2 and T1Post of the left hemisphere independent of the modulation

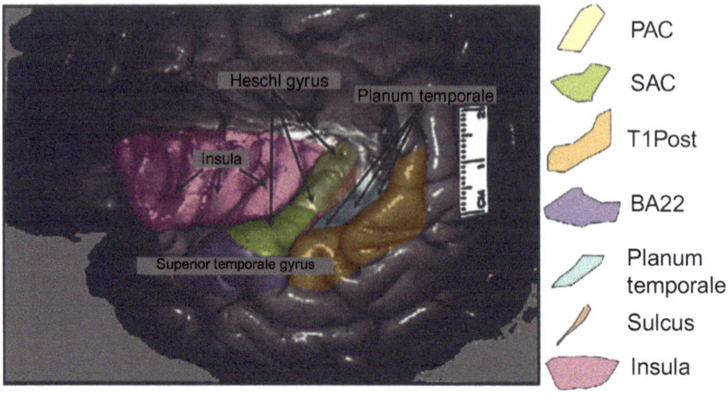

FIG. 7.2 Areas of the human cortex (left hemisphere here) showing an activity in response to amplitude modulated white noise. Part of the frontal lobe has been removed. *PAC*, primary auditory cortex; *SAC*, secondary auditory cortex. *Reprinted with permission from Gourévitch, B., Le Bouquin Jeannés, R., Faucon, G., Liégeois-Chauvel, C., 2008. Temporal envelope processing in the human auditory cortex: response and interconnections of auditory cortical areas. Hear. Res. 237, 1–18, Elsevier. Copyright 2008.*

frequency, and in the left BA22 for amplitude modulation frequencies of 8 and 16Hz, compared with those in the right. A sequential activation of A1, A2, and BA22 areas was clearly visible at all modulation frequencies, while other auditory areas were more involved in parallel processing upon a stream originating from A1, which thus acts as a distribution hub.

More recently and from electrocorticograms (ECoGs) in 10 human subjects with intractable epilepsy, (Potes et al., 2014) identified stimulus-related modulations in the alpha (8–12Hz) and high gamma (70–110Hz) bands at neuroanatomical locations implicated in auditory processing. Specifically, they found stimulus-related ECoG modulations in the alpha band in areas adjacent to tonotopically organized A1 (Chapter 1). They also identified stimulus-related ECoG modulations in the high-gamma band in areas close to A1 and in other perisylvian areas known to be involved in higher-order auditory processing and in superior premotor cortex. Across these areas, modulations in the high-gamma band preceded those in the alpha band by 280ms and activity in the high-gamma band predicted alpha activity, but not vice versa. Additionally, Granger causality analysis identified relationships of high-gamma activity between distinct locations in early auditory pathways within the superior temporal gyrus (STG) and posterior STG, between posterior STG (Wernicke's area) and inferior frontal cortex (Broca's area), and between STG and premotor cortex (Fig. 7.3). Evidence suggests that these relationships reflect direct corticocortical connections rather than common driving input from subcortical structures such as the medial geniculate body.

Fontolan et al. (2014) obtained human depth recordings from auditory cortical regions in both hemispheres, while subjects listen to sentences,

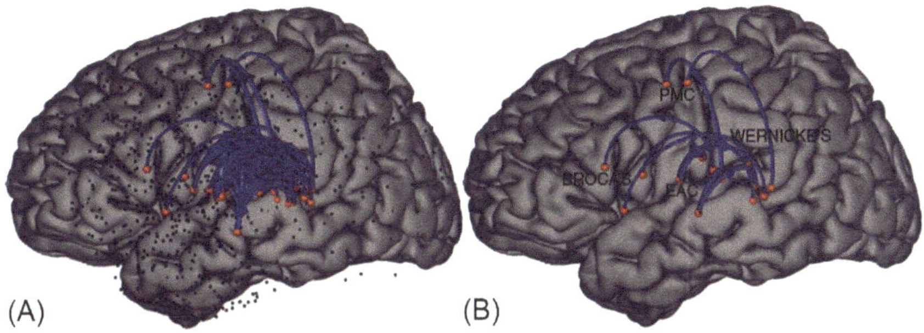

FIG. 7.3 Intersubject Granger causality analysis. Causal relationships of activity in the high gamma-hand recorded during listening to music and between different pairs of electrode locations. Across all possible 630,800 electrode combinations, only 68 connections were significant at $\alpha = 1.58e{-}8$ after Bonferroni correction (A). These connections were further reduced to the 10 most salient 10 connections (B). *Reprinted with permission from Potes, C., Brunner, P., Gunduz, A., Knight, R.T., Schalk, G., 2014. Spatial and temporal relationships of electrocorticographic alpha and gamma activity during auditory processing. NeuroImage 97, 188–195, Elsevier. Copyright (2014).*

and show that information travels in each direction using separate EEG-frequency channels. Bottom-up and top-down propagation dominates in gamma- and delta-beta (< 40 Hz) bands, respectively. "The predominance of low frequencies for top-down information transfer is confirmed by cross-regional frequency coupling, which indicates that the power of gamma-activity in A1 is modulated by the phase of delta–beta activity sampled from association auditory cortex. This cross-regional coupling effect is absent in the opposite direction."

7.3 Auditory memory and speech

7.3.1 Auditory memory

Memory comprises several subsystems. *Sensory memory* allows individuals to retain traces of sensory information after the original stimulus has terminated. It is strongly involved in the processing of sensory stimuli and orienting within the environment, and it is active in every moment of an organism's life. Sensory information is stored in sensory memory just long enough to be transferred to *short-term memory* and used in *working memory*. Working memory is a conscious cognitive system with limited capacity and responsible for the temporary holding, processing, and manipulation of information (Bonetti et al., 2018).

(Majerus et al., 2017) examined the role of the dorsal attention network (DAN) during working memory by manipulating encoding attentional load and memory control requirements during a short-term probe recognition task for sequences of auditory (digits, letters) or visual (lines, unfamiliar faces) stimuli. They found that "multivariate neural activity patterns in the DAN during working memory do not only represent top–down attentional control aspects but also attentional aspects involved in memory encoding. … These attentional functions are supported by the DAN in both auditory and visual modalities, but their implementation further requires modality-specific neural processes."

Using EEG, (Shahin et al., 2009) examined brain oscillations during a semantic speech-processing task that required both lexically mediated memory matching and selective attention. Participants listened to nouns spoken in male and female voices, and detected an animate target word (e.g., ant; $p=20\%$) in a train of inanimate standard words (e.g., chair; $p=80\%$) or vice versa. For a control task, subjects listened to the same words and detected a target male voice in standards of a female voice or vice versa. The standard trials of the semantic task showed enhanced upper beta (25–30 Hz) and gamma band (30–60 Hz) activity compared with the voice task. Upper beta and gamma enhancement were accompanied by a suppression of alpha (8–12 Hz) and lower to mid beta (13–20 Hz) activity, mainly localized to posterior electrodes. Enhancement of phase-locked theta activity peaking near 275 ms also occurred over the midline electrodes (e.g., Cz). (Shahin et al., 2009) concluded: "Theta, upper beta, and gamma band enhancement may reflect lexically mediated template matching in auditory memory, whereas the alpha and [lower] beta suppression likely indicate increased attention processes and memory demands."

Using high-density EEG in a musical delayed match-to-sample task, (Tseng et al., 2019) observed that low-frequency oscillations in the frontal, temporal, and parietal regions were increased during the processing of both target and distracting melodies. Cross-frequency coupling between low-frequency phases and high-frequency amplitudes was elevated in the frontal and parietal regions when the participants were distinguishing between the target and distractor, suggesting that the phase-amplitude coupling could be an indicator of neural mechanisms underlying memory retrieval. Finally, there was frontotemporal—anterior cingulate cortex and inferior-temporal gyrus, respectively—phase-locking in the theta band and frontoparietal phase-locking in the alpha band during the recognition of the two stimuli (Fig. 7.4). Furthermore the interregional phase-locking and phase-amplitude coupling among the frontal, temporal, and parietal regions occurred at the very beginning of musical memory retrieval, which might reflect the precise timing when cognitive resources were involved in the retrieval of targets and the rejection of similar distractors.

7.3.2 Oscillations and speech under difficult conditions

We have extensively dealt with this topic for speech perception under normal conditions in Chapter 4; here we illustrate some aspects of speech processing under difficult conditions. Such conditions are generally

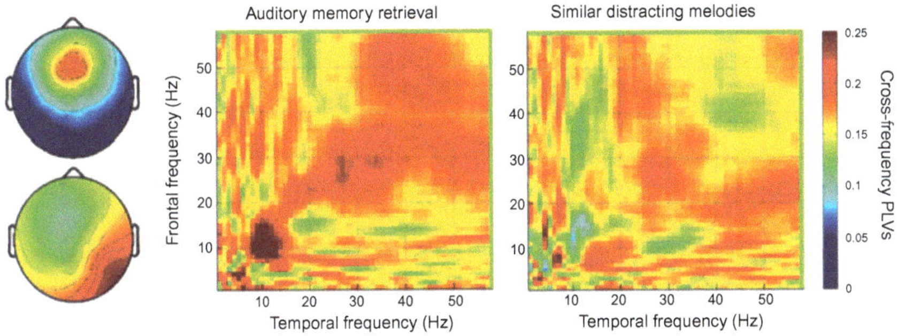

FIG. 7.4 An example of cross frequency phase lock value between 100 and 400ms after the stimulus onset between the frontal independent component close to ACC and right temporal independent component close to ITG of a single subject The phase synchronization between the frontal and temporal regions during memory retrieval and similar distracting conditions. *ACC*, Anterior cingulate cortex; *ITG*, inferior temporal gyrus. *From Tseng, Y.-L., Liu, H.-H., Liou, M., Tsai, A.C., Chien, V.S.C., Shyu, S.-T., et al., 2019. Lingering sound: event-related phase-amplitude coupling and phase-locking in fronto-temporo-parietal functional networks during memory retrieval of music melodies. Front. Hum. Neurosci. 13, 150. https://doi.org/10.3389/fnhum.2019.00150. Open access.*

experienced by people with hearing loss. Here, we look at how normal hearing people deal with this. Using MEG and a continuous listening task, (Vander Ghinst et al., 2016) evaluated the coupling between the listener's cortical activity and the temporal envelopes of different sounds in a multitalker auditory scene. They recorded from right-handed healthy adult humans who listened to five different recorded stories (attended speech streams), one without any multitalker background (No noise) and four mixed with a "cocktail party" multitalker background noise Coherence analysis revealed that the modulations of the attended speech stream in the no-noise condition, were coupled at ~ 0.5Hz to the activity of both superior temporal gyri, whereas the modulations at 4–8Hz were coupled to the activity of the right supratemporal auditory cortex. In cocktail party conditions, the coupling was at both frequencies stronger for the attended speech stream than for the unattended multitalker background. During the cocktail party conditions, the ~ 0.5Hz coupling became left-hemisphere dominant, compared with bilateral coupling without the multitalker background, whereas the 4–8Hz coupling remained right-hemisphere lateralized in both conditions. The results highlight the key role of listener's left superior temporal gyri in extracting the slow ~ 0.5Hz modulations, likely reflecting the prosody and intonation (Giraud and Poeppel, 2012) in the attended speech stream within a multitalker auditory scene.

(Pefkou et al., 2017) addressed the dynamics of auditory decoding by varying the difficulty of syllable tracking and speech decoding using comprehensible and incomprehensible time-compressed auditory sentences. They recorded EEGs in humans and found that neural activity in both theta and beta range was sensitive to syllabic rate. Phase patterns of slow neural activity consistently followed the syllabic rate (4–14 Hz), even when this rate went beyond the classical theta range (4–8 Hz). The power of theta activity increased linearly with syllabic rate but showed no sensitivity to comprehension. Conversely the power of beta (14–21 Hz) activity was insensitive to the syllabic rate, yet reflected comprehension on a single-trial basis. Thus speech comprehension is constrained by concurrent stimulus-driven theta and low-gamma activity, and by endogenous beta activity, but not primarily by the capacity of theta activity to track the syllabic rhythm (Pefkou et al., 2017).

7.4 The connectome in hearing loss

Note that most studies of resting-state EEG and/or magnetoencephalography (MEG) on hearing loss were obtained in tinnitus patients, who typically have hearing loss. These are discussed in Chapter 10. Here, I review the structural and functional magnetic resonance imaging (fMRI) data, starting with unilateral sensorineural hearing loss (UHL). Several recent papers that recorded structural and functional aspects of congenital and acquired sensorineural hearing loss will subsequently be discussed. Reference data for normal hearing are presented in Chapters 1 and 10. Connectome changes in age-related hearing loss were featured in my recent book (Eggermont, 2019).

7.4.1 Unilateral hearing loss

7.4.1.1 Sudden unilateral hearing loss

Sudden deafness is an unexplained, rapid loss of hearing either all at once or over a few days, and frequently affects only one ear. (Yang et al., 2014) investigated regional brain structural and functional changes in patients with sudden and persistent UHL in adulthood. Hearing ability was assessed by pure tone audiometry and showed pure tone averages ≥ 40 dB HL. Voxel-based morphometry (VBM) was performed to detect brain regions with changed gray matter volume in UHL. Compared with controls, UHL patients showed *decreased* gray matter volume in bilateral posterior cingulate gyrus (PCC) and precuneus, left superior/middle/inferior temporal gyrus, and right parahippocampal gyrus and lingual gyrus. Thus brain areas with atrophy not only comprised the auditory cortex but also other nonauditory regions. Patients also showed significantly *decreased* amplitude of blood oxygen level dependent (BOLD) response in bilateral precuneus, left inferior parietal lobule, and right inferior frontal gyrus and insula and *increased* amplitude of BOLD response in right inferior and middle temporal gyrus, suggesting functional deficits. (Xu et al., 2016) investigated two large samples of sudden-UHL patients and correlated the sudden-UHL characteristics with changes in the functional network properties. Pure tone audiometry was performed to assess hearing ability. All of the unilateral UHL patients had a pure-tone average (PTA) hearing threshold at the octave frequencies 0.5, 1, and 2 kHz > 30 dB in the hearing loss ear. Network changes were quantified with graph-theoretical means (Chapter 6). They found that the functional connectome of unilateral sudden-UHL patients is characterized by a shift toward small-world networks with a significant increase in the clustering coefficient, in the global efficiency and the local efficiency but a significant decrease in the characteristic path length. Sudden-UHL patients exhibited these changes in the limbic and paralimbic systems, and in the auditory network brain areas. These changes already occurred within the acute period at both the global and local level. Significantly, different areas were involved in left versus right sudden UHL.

7.4.1.2 Long-term unilateral hearing loss

Burton et al. (2012) studied BOLD responses in core, belt, and parabelt auditory cortex in adults with normal hearing and unilateral hearing loss (UHL). Regardless of ear of stimulation, normal hearing people showed larger contralateral responses in several auditory cortex fields. With right ear stimulation in UHL, ipsilateral responses were larger compared with normal hearing in core and belt auditory cortex (ACx). With left ear stimulation in UHL, only posterior ACx core showed larger ipsilateral responses, suggesting that most ACx in the left hemisphere had greater resilience against the reduced crossed inputs from a deafferented right ear. Parabelt regions located posterolateral to core and belt ACx showed reduced activation in UHL compared with normal hearing irrespective of ear of stimulation and lateralization of inputs. Thus the effect in UHL compared with normal hearing differed by ACx fields and ear of deafness. (Zhang et al., 2018) confirmed this ear specificity in long-term UHL patients. Summarizing, the functional changes in long-term (as well as in sudden) single-sided deafness are highly deafness-side dependent.

Pediatric UHL is associated with speech and language, and cognitive deficits. To identify functional network architecture differences in the brains of children (7–17 years of age) with UHL, (Jung et al., 2017) used resting-state fMRI. Compared with the normal hearing group, children with UHL showed *increased* connectivity patterns between networks such as the cinguloopercular network (CON) and visual processing centers. However, there were *decreased*, as well as aberrant connectivity patterns with the coactivation of the default mode network (DMN) and frontoparietal network, a relationship that usually is negatively correlated. These connectivity changes may therefore be adaptive and maladaptive.

(Zhang et al., 2015) analyzed resting-state fMRI data from normal hearing subjects and patients (age 27–60 years) with long-term postlingual UHL. The PTA (0.25–2 kHz) for the UHL patients was at least 70 dB in the deaf ear and less than 25 dB in the better hearing ear. More changes in the

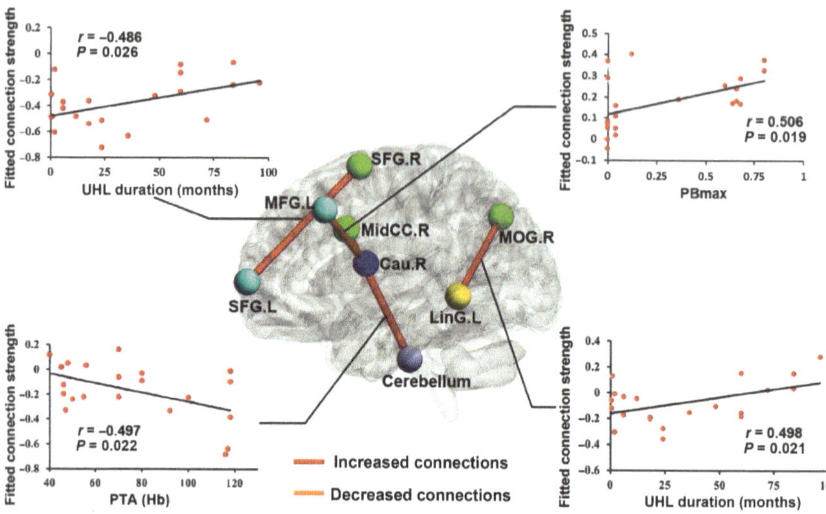

FIG. 7.5 Relationships between network measures and clinical performance in UHL. The UHL duration was significantly correlated with the connections strength of the left SFG and right SFG, and the connections strength of left lingual gyrus and right MOG. *Green nodes*, attention network; *cyan*, default mode network; *yellow*, visual network; *purple*, subcortical network; *gray*, cerebellum. There was a significant positive correlation between the connection strength of right middle cingulate gyrus and cerebellum with the PTA scores. There was also a significant positive correlation between the connection strength of left MFG and right caudate nucleus with the PBmax scores. *Cau*, caudate nucleus; *LinG*, lingual gyrus; *MidCC*, middle cingulate gyrus; *MOG*, middle occipital gyrus; *PBmax*, maximum phonetically balanced word recognition; *SFG*, superior frontal gyrus; *STG*, superior temporal gyrus. *From Zhang, Y., Mao, Z., Feng, S., Liu, X., Lan, L., Zhang, J., et al., 2018a. Altered functional networks in long-term unilateral hearing loss: a connectome analysis. Brain. Behav. 8, e00912. https://doi.org/10.1002/brb3.912. Open access.*

DMN were found in the left than right long-term UHL. There was no significant relationship between this asymmetry of gray matter volume and the duration of the disease, or time of onset. A different research group, (Zhang et al., 2018) acquired resting-state fMRI data from patients with long-term UHL caused by acoustic neuromas, and matched healthy controls. Both UHL patients and controls showed efficient small-world properties in their functional networks. Compared with normal controls, UHL patients showed *increased* nodal betweenness in many brain regions; most of these regions were components of the visual network, DMN and subcortical network. *Decreased* nodal betweenness in UHL was observed between the left middle frontal gyrus and the right caudate (Fig. 7.5). The changes in intranetwork connections in UHL were significantly correlated with disease duration, hearing level and word recognition. (Zhang et al., 2018) noted that *long-term* UHL patients and controls showed common small-world properties of the functional networks. This indicates that long-term

UHL patients still exhibit an optimized topological organization in brain networks despite the long-term deprivation of unilateral auditory input. Specifically, the results were consistent with the finding of a typical small-world network in *sudden*-UHL patients (Xu et al., 2016).

(Qiao et al., 2019) found significantly *decreased* resting-state fMRI activation in the auditory cortex of right UHL compared with normal hearing controls. The degree of involvement was correlated with residual hearing ability in right single-sided deafness. In addition, the frontoparietal cortex was less activated during the visual task in right UHL with increased functional connectivity between the frontoparietal cortex and the auditory cortex, suggesting frontoparietal resources may be recruited less by vision but more by hearing. These observations suggest that right UHL can lead to a downward cross-modal plasticity. The left UHL showed a similar alteration trend with right UHL, but without statistical significance.

7.4.2 Bilateral hearing loss

7.4.2.1 Sensorineural hearing loss

(Husain et al., 2014) demonstrated *decreased* connectivity between seed regions in the DAN with the insula and postcentral gyrus, and *increased* connectivity between seeds in the DMN and middle frontal gyrus in mild-to-moderate hearing-impaired participants (~ 40 dB loss for 4–8 kHz). Using fMRI and neuropsychology tests, (Xu et al., 2017) found that "Compared with normal hearing subjects, patients with long-term bilateral hearing loss presented weaker brain region activations in A1 and the language network; enhanced neural activities in the DMN [as in (Zhang et al., 2015) in UHL]. Poorer cognitive performance scores were always associated with higher hearing levels and weaker activated strengths in A1. However, the abnormal changes in the DMN did not directly affect the cognitive abilities of patients with bilateral hearing loss."

(Luan et al., 2020) obtained rs-fMRI data and neuropsychological test scores from patients (aged 31–69 years) with long-term bilateral sensorineural hearing loss (SNHL) and well-matched healthy controls (aged 32–71 years). They found widespread alterations of functional coupling with the subdivisions of the anterior cingulate cortex (ACC) in regions involved in cognitive, emotional, and multiple sensory processing, particularly within the CON, DMN, and auditory network in the SNHL patients. The functional coupling between the ACC and ACx during auditory-related demanding cognitive tasks was robust. Hence, the enhanced connectivity in the ACC may be a compensation of the integrative capacity of ACC for the loss of periphery hearing loss, and in turn this functional reorganization might affect the higher-order functions through the interactions between the ACC and ACx. (Luan et al., 2019) then investigated the role of the dorsolateral prefrontal cortex (DLPFC) in the cross-modal reorganization and cognitive participation of the ACx in long-term SNHL by combining functional and structural measurements. No differences in gray matter volume, resting-state neural activity, or diffusion characteristics in the DLPFC were found between the SNHL and control groups. The functional connectivity between the DLPFC and the ACx and visual areas, such as the cuneus, fusiform, lingual cortex, and calcarine sulcus was increased in patients with SNHL.

In an fMRI study in people with SNHL of which some had tinnitus, (Xu et al., 2019) found that the left anterior insula (AI) showed *decreased* connectivity with right precentral gyrus and superior frontal gyrus. The connections between the dorsal ACC and middle frontal gyrus, superior parietal gyrus, and supplementary motor area (SMA) were also *reduced*. Granger causality maps were generated for each subject. Compared with controls, the SNHL group showed *decreased* outflow from the left AI to the left cerebellum lobule VI, right middle frontal gyrus and right precentral gyrus. The inflow from right inferior temporal gyrus to the left AI was also *decreased*. Meanwhile, the outflow from the right AI to bilateral middle frontal gyrus was *reduced* in SNHL subjects. There was also *decreased* outflow from the dACC to the left precentral gyrus and right SMA, but *increased* inflow from right cerebellum lobule VIII. Furthermore, correlation analysis demonstrated that some of these atypical connectivity measures were correlated with performance of neuropsychological tests. These findings suggest that the inefficient modulation of the salience network might contribute to central effects of SNHL, as well as associated cognition and emotion deficits. This may also underlie the various comorbidities of tinnitus. (Xu et al., 2019) then investigated hearing loss-induced changes among neural networks within cerebellar subregions and the changes in cerebellar-cerebral connectivity patterns using resting-state fMRI. Voxelwise functional connectivity (FC) analysis demonstrated *decreased* couplings between the cerebellum and other cerebral areas, including the temporal pole (TP), insula, supramarginal gyrus, IFG, medial frontal gyrus, and thalamus, in long-term bilateral sensorineural hearing loss patients. Moreover, there was a negative correlation between anxiety and FC between the left cerebellar lobe VI and left insula. Hearing ability and anxiety scores were also correlated with FC between the left cerebellar lobe VI and left TP, as well as the right cerebellar lobule VI and left IFG. Thus sensorineural hearing loss disrupts cerebellar-cerebral circuits, some potentially linked to anxiety, and interregional cerebellar connectivity.

7.4.2.2 *Comparing prelingual and postlingual deafness*

There is a sensitive period (2–4 years after birth) allowing optimal central auditory development (Eggermont and Moore, 2012). Typically, hearing loss occurring before the critical period is called prelingual deafness. (Kim et al., 2014) evaluated morphological changes in A1 of prelingual deaf (mean age = 50.4 years), postlingual deaf (mean age = 50.9 years), and

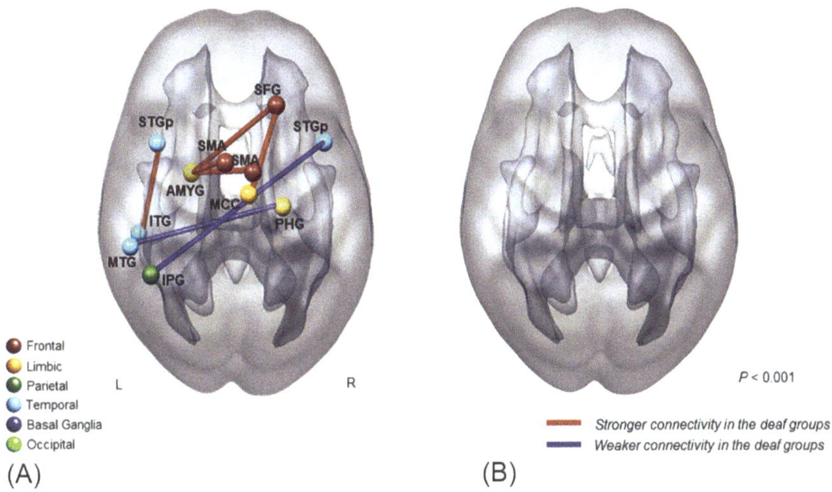

FIG. 7.6 Differences in the morphological covariance network between (A) pre-LD and hearing adults, and (B) post-LD and hearing adults. Using single linkage matrices and pseudo-matrices constructed by roaring the matrices of both groups, significantly different ($P<0.001$) connections were identified. *Red edges* indicate that the connection showed earlier merging during filtration (and thus closer coupling), and *blue edges* indicate that the connection showed later merging (and thus looser coupling), in the deaf groups than the hearing adults. While connectivity of the post-LD adults were the same as the hearing adults, the pre-LD group showed closer or looser coupling between certain nodes than the hearing adults, thus implying different cortical reorganization. *AMYG*, amygdala; *IPG*, inferior parietal gyrus; *ITG*, inferior temporal gyrus; *MCC*, median cingulate and paracingulate gyri; *MTG*, middle temporal gyrus; *PHG*, parahippocampal gyrus; *SFG*, superior frontal gyrus, dorsolateral; *SMA*, supplementary motor area; *STGp*, temporal pole of superior temporal gyrus. *Reprinted with permission from Kim, E., Kang, H., Lee, H., Lee, H.-J., M.-W., Song, J.-J., et al., 2014. Morphological brain network assessed using graph theory and network filtration in deaf adults. Hear. Res. 315, 88–98, Elsevier. Copyright 2014.*

normal-hearing adults (mean age=49.5 years) using resting-state VBM. There were no significant changes in gray matter density in A1. However, unlike postlingual, prelingual deafness showed increased bilateral temporal connectivity of the primary auditory cortex compared with the hearing adults (Fig. 7.6). In the prelingual deaf group, connectivity was stronger between the right dorsolateral superior frontal gyrus and the right median cingulate or left amygdala, between the left amygdala and bilateral supplementary motor areas, and between the left superior temporal pole and left inferior temporal gyrus (Fig. 7.6A), than in the hearing adults. In contrast, the connectivity between the right superior temporal pole and left inferior parietal gyrus, and between the right parahippocampal and left middle temporal gyrus, was weaker in the prelingual deaf group compared with the hearing adults (Fig. 7.6A). (Kim et al., 2014) "proposed that the preserved density of auditory cortex associated with increased connectivity in pre-lingual deafness, and closer coupling between certain brain areas, represent distinctive reorganization of auditory and related cortices compared with hearing or post-lingual deaf adults. The differential network reorganization in the pre-lingual deaf adults could be related to the absence of auditory speech experience."

(Kim et al., 2017) carried out a tract-based statistical analysis of diffusion tensor fractional anisotropy in adults with prelingual and postlingual deafness. Compared with hearing controls, patients with deafness showed significant decreases in diffusion anisotropy at the right internal capsule, the right thalamus, and the splenium of the corpus callosum as well as within the bilateral superior temporal gyrus (including Heschl gyrus) and right temporal white matter (cf. (Kim et al., 2009)). Compared with patients with postlingual deafness, those with prelingual deafness showed lower diffusion anisotropy in the bilateral temporal white matter, including Heschl's gyrus, which might indicate microstructural neural atrophy within the white matter bundles of auditory association areas. (Kim et al., 2017) suggested that, "in patients with deafness, reception of early stimuli before language acquisition might be more critical to white matter maturation and brain organization than the duration of disuse." Thus gray matter connectivity (Li et al., 2015) and white matter connectivity is affected predominantly in prelingual deafness.

Using resting-state fMRI, (Wolak et al., 2019) found the most profound functional connectivity differences between patients with prelingual (before language development, PRE) versus postlingual onset (after language development, POST) of sloping SNHL. An early onset was related to enhancement in long-range network connections, including the default-mode network, the dorsal-attention network and the frontoparietal network, as well as in local sensory networks, the visual and the sensorimotor. A number of multisensory brain regions in frontal and parietal cortices, as well as the cerebellum, were also more internally connected. At the same time, POST patients showed enhanced FC between the salience network and multisensory parietal areas, as well as with the hippocampus, when they were compared with those with PRE hearing loss. Signals in several visual cortex regions were also more intracorrelated in POST versus PRE patients. This outcome might point to more attention resources directed to multisensory as well as memory experience. Finally, audiological scores correlated with FC in several sensory and high-order brain regions in all patients.

7.5 Summary

Animal research showed that delta-theta activity in auditory cortex was dramatically different from activity in an adjacent nonauditory responsive region. Sound-induced gamma band oscillations are prominent and originate in auditory cortex. UHL, sudden as well as long-term, shows changes in the limbic and paralimbic systems, and in the auditory network brain areas but these are critically dependent on the side of hearing loss. Right, but not left, UHL can lead to a downward cross-modal plasticity. Compared with normal hearing controls, patients with long-term bilateral hearing loss, as in UHL, presented weaker brain region activations in A1 and the language network, and enhanced neural activities in the DMN. Inefficient modulation of the salience network in SNHL might contribute to central effects and associated cognition and emotion deficits. No significant changes in gray matter density in A1 for prelingual and postlingual deafness are found. However, unlike postlingual, prelingual deafness shows increased bilateral temporal connectivity of the A1 compared with the hearing adults.

References

Barth, D.S., MacDonald, K.D., 1996. Thalamic modulation of high-frequency oscillating potentials in auditory cortex. Nature 383, 78–81.

Bonetti, L., Haumann, N.T., Brattico, E., Kliuchko, M., Vuust, P., Särkämö, T., et al., 2018. Auditory sensory memory and working memory skills: association between frontal MMN and performance scores. Brain Res. 1700, 86–98.

Brosch, M., Budinger, E., Scheich, H., 2002. Stimulus-related gamma oscillations in primate auditory cortex. J. Neurophysiol. 87, 2715–2725.

Burton, H., Firszt, J.B., Holden, T., Agato, A., Uchansk, R.M., 2012. Activation lateralization in human core, belt, and parabelt auditory fields with unilateral deafness compared to normal hearing. Brain Res. 1454, 33–47.

Eggermont, J.J., 2015. Auditory Temporal Processing and Its Disorders. Oxford University Press, Oxford, UK, ISBN: 978-0-19-871909-0.

Eggermont, J.J., 2019. The Auditory Brain and Age-Related Hearing Impairment. Academic Press, London, UK, ISBN: 978-0-12-815304-8.

Eggermont, J.J., Moore, J.K., 2012. Morphological and functional development of the auditory nervous system. In: Werner, L.A., et al. (Eds.), Human Auditory Development. Springer Handbook of Auditory Research 42, Springer Science + Business Media, pp. 61–105, https://doi.org/10.1007/978-1-4614-1421-6_3.

Farley, B.J., Noreña, A.J., 2013. Spatiotemporal coordination of slow-wave ongoing activity across auditory cortical areas. J. Neurosci. 33 (8), 3299–3310.

Fontolan, L., Morillon, B., Liegeois-Chauvel, C., Giraud, A.-L., 2014. The contribution of frequency-specific activity to hierarchical information processing in the human auditory cortex. Nat. Commun. 5, 1–10. https://doi.org/10.1038/ncomms569.

Franowicz, M.N., Barth, D.S., 1995. Comparison of evoked potentials and high-frequency (gamma-band) oscillating potentials in rat auditory cortex. J. Neurophysiol. 74 (1), 96–112.

Giraud, A.-L., Poeppel, D., 2012. Cortical oscillations and speech processing: emerging computational principles and operations. Nat. Neurosci. 15, 511–517.

Gourévitch, B., Le Bouquin Jeannès, R., Faucon, G., Liégeois-Chauvel, C., 2008. Temporal envelope processing in the human auditory cortex: response and interconnections of auditory cortical areas. Hear. Res. 237, 1–18.

Haegens, S., Zion Golumbic, E., 2018. Rhythmic facilitation of sensory processing: a critical review. Neurosci. Biobehav. Rev. 86, 150–165.

Husain, F.T., Carpenter-Thompson, J.R., Schmidt, S.A., 2014. The effect of mild-to-moderate hearing loss on auditory and emotion processing networks. Front. Syst. Neurosci. 8 (10). https://doi.org/10.3389/fnsys.2014.00010.

Jung, M.E., Colletta, M., Coalson, R., Schlaggar, B.L., Lieu, J.E.C., 2017. Differences in interregional brain connectivity in children with unilateral hearing loss. Laryngoscope 127, 2636–2645.

Kim, J., Choi, J.Y., Eo, J., Park, H.-J., 2017. Comparative evaluation of the white matter fiber integrity in patients with prelingual and postlingual deafness. NeuroReport 28, 1103–1107.

Kim, E., Kang, H., Lee, H., Lee, H.-J., Suh, M.-W., Song, J.-J., et al., 2014. Morphological brain network assessed using graph theory and network filtration in deaf adults. Hear. Res. 315, 88–98.

Kim, D.-J., Park, S.-T., Kim, J., Lee, D.H., Park, H.-J., 2009. Alterations of white matter diffusion anisotropy in early deafness. NeuroReport 20, 1032–1036.

Li, W., Li, J., Wang, Z., Li, Y., Liu, Z., Yan, F., et al., 2015. Grey matter connectivity within and between auditory, language and visual systems in prelingually deaf adolescents. Restor. Neurol. Neurosci. 33, 279–290. https://doi.org/10.3233/RNN-140437.

Luan, Y., Wang, C., Jiao, Y., Tang, T., Zhang, J., Lu, C., et al., 2020. Abnormal functional connectivity and degree centrality in anterior cingulate cortex in patients with long-term sensorineural hearing loss. Brain Imag. Behav. 14 (3), 682–695.

Luan, Y., Wang, C., Jiao, Y., Tang, T., Zhang, J., Teng, G.-J., 2019. Prefrontal-temporal pathway mediates the cross-modal and cognitive reorganization in sensorineural hearing loss with or without tinnitus: a multimodal MRI study. Front. Neurosci. 13, 222. https://doi.org/10.3389/fnins.2019.00222.

Majerus, S., Péters, F., Bouffier, M., Cowan, N., Phillips, C., 2017. The dorsal attention network reflects both encoding load and top–down control during working memory. J. Cogn. Neurosci. 30 (2), 144–159.

O'Connell, M.N., Barczak, A., Ross, D., McGinnis, T., Schroeder, C.E., Lakatos, P., 2015. Multi-scale entrainment of coupled neuronal oscillations in primary auditory cortex. Front. Hum. Neurosci. 9, 655. https://doi.org/10.3389/fnhum.2015.00655.

Pantev, C., Makeig, S., Hoke, M., Galambos, R., Hampson, S., Gallen, C., 1991. Human auditory evoked gammaband magnetic fields. Proc. Natl. Acad. Sci. U.S.A. 88, 8996–9000.

Pefkou, M., Arnal, L.H., Fontolan, L., Giraud, A.-L., 2017. q-Band and b-band neural activity reflects independent syllable tracking and comprehension of time-compressed speech. J. Neurosci. 37 (33), 7930–7938.

Potes, C., Brunner, P., Gunduz, A., Knight, R.T., Schalk, G., 2014. Spatial and temporal relationships of electrocorticographic alpha and gamma activity during auditory processing. NeuroImage 97, 188–195.

Qiao, Y., Li, X., Shen, H., Zhang, X., Sun, Y., Hao, W., et al., 2019. Downward cross-modal plasticity in single-sided deafness. NeuroImage 197, 608–617.

Shahin, A.J., Picton, T.W., Miller, L.M., 2009. Brain oscillations during semantic evaluation of speech. Brain Cogn. 70, 259–266.

Shannon, R.V., Zeng, F.-G., Kamath, V., Wygonski, J., Ekelid, M., 1995. Speech recognition with primarily temporal cues. Science 270, 303–304.

Tseng, Y.-L., Liu, H.-H., Liou, M., Tsai, A.C., Chien, V.S.C., Shyu, S.-T., et al., 2019. Lingering sound: event-related phase-amplitude coupling and phase-locking in fronto-temporo-parietal functional networks during memory retrieval of music melodies. Front. Hum. Neurosci. 13, 150. https://doi.org/10.3389/fnhum.2019.00150.

Vander Ghinst, M., Bourguignon, M., Op de Beeck, M., Wens, V., Marty, B., Hassid, S., et al., 2016. Left superior temporal gyrus is coupled to attended speech in a cocktail-party auditory scene. J. Neurosci. 36 (5), 1596–1606.

Wolak, T., Ciesla, K., Pluta, A., Włodarczyk, E., Biswal, B., Skarzynski, H., 2019. Altered functional connectivity in patients with sloping sensorineural hearing loss. Front. Hum. Neurosci. 13, 284. https://doi.org/10.3389/fnhum.2019.00284.

Xu, H., Fan, W., Zhao, X., Li, J., Zhang, W., Lei, P., et al., 2016. Disrupted functional brain connectome in unilateral sudden sensorineural hearing loss. Hear. Res. 335, 138–148.

Xu, L.-C., Zhang, G., Zou, Y., Zhang, M.-F., Zhang, D.-S., Ma, H., et al., 2017. Abnormal neural activities of directional brain networks in patients with long-term bilateral hearing loss. Oncotarget 8 (48), 84168–84179.

Xu, X.-M., Jiao, Y., Tang, T.-Y., Lu, C.-Q., Zhang, J., Salvi, R., et al., 2019a. Altered spatial and temporal brain connectivity in the salience network of sensorineural hearing loss and tinnitus. Front. Neurosci. 13, 246. https://doi.org/10.3389/fnins.2019.00246.

Xu, X.-M, Jiao, Y., Tang, T.-Y., Zhang, J., Lu, C.-Q., Luan, Y., et al., 2019b. Dissociation between cerebellar and cerebral neural activities in humans with long-term bilateral sensorineural hearing loss. Neural Plast. 2019. https://doi.org/10.1155/2019/835484.

Yang, M., Chen, H.-J., Liu, B., Huang, Z.-C., Feng, Y., Li, J., et al., 2014. Brain structural and functional alterations in patients with unilateral hearing loss. Hear. Res. 316, 37–43.

Zhang, G.-Y., Yang, M., Liu, B., Huang, Z.-C., Chen, H., Zhang, P.-P., et al., 2015. Changes in the default mode networks of individuals with long-term unilateral sensorineural hearing loss. Neuroscience 285, 333–342.

Zhang, Y., Mao, Z., Feng, S., Liu, X., Lan, L., Zhang, J., et al., 2018a. Altered functional networks in long-term unilateral hearing loss: a connectome analysis. Brain Behav. 8. https://doi.org/10.1002/brb3.912.

Zhang, Y., Mao, Z., Feng, S., Wang, W., Zhang, J., Yu, X., 2018b. Convergent and divergent functional connectivity patterns in patients with long-term left-sided and right-sided deafness. Neurosci. Lett. 665, 74–79.

CHAPTER

8

Neuromodulation of brain rhythms and connectivity

Neuromodulation is a technique to manipulate brain rhythms and connectivity, and allows the identification of mechanisms that change cognitive processes or tasks such as motor acts. In addition, neuromodulation may demonstrate that brain rhythms are causal to these processes. Several methods are available to modulate brain oscillations (Herrmann et al., 2016). We note effects of pharmacological, rhythmic sensory stimulation, transcranial magnetic stimulation (TMS), transcranial alternating current stimulation/transcranial direct current stimulation (tACS/tDCS), and neurofeedback. Here, we will only present an overview of the effects induced by TMS, tACS, and tDCS. In addition, we will discuss recently developed neuromodulation procedures based on focused low intensity ultrasound. In Chapter 10, we illustrate the use of neurofeedback and TMS in the treatment of tinnitus.

8.1 Transcranial magnetic stimulation

TMS involves delivering a brief high-intensity magnetic pulse to the head through a figure-8-shaped coil. This pulse induces electrical currents in a focal area underneath the coil, which interacts with ongoing activity in the neural tissue (Thut and Miniussi, 2009). Repetitive TMS (rTMS), up to 100 Hz, may transiently modulate neural excitability (Walsh and Cowey, 2000), with the net effect depending on the repetition frequency. Usually, low-frequency (≤ 1 Hz) stimulation results in inhibition, whereas high-frequency stimulation can result in excitatory changes in the stimulated area (Kobayashi and Pascual-Leone, 2003). Paired stimulation of different sites may involve spike-time-dependent plasticity (STDP) and may result in long-term potentiation (LTP) or long-term depression (LTD) in these cortical areas.

https://doi.org/10.1016/B978-0-12-819818-6.00003-0

8.1.1 Single-pulse TMS

Using single-pulse TMS, Fuggetta et al. (2005) evaluated the event-related electroencephalography (EEG) power and event-related coherence transformations of alpha and beta rhythms. Event-related power reflects regional oscillatory activity, whereas event-related coherence reflects the interregional functional coupling of this oscillatory neural activity (Chapters 5 and 6). The event-related power showed that the magnetic pulse-modulated cortical oscillations within the first half second for both alpha and beta ranges. For the alpha rhythm, stimulation with threshold-TMS induced a small decrease in the amplitude of EEG oscillations over the stimulation site, while for both alpha and beta rhythm, a progressive interhemispheric synchronization was observed as the intensity of TMS was increased in both hemispheres.

8.1.2 Repetitive TMS effects on synchrony and memory

Eldaief et al. (2011) used rTMS to modulate cortical activity in the left posterior inferior parietal lobule (lpIPL) of the default mode network (DMN), and functional magnetic resonance imaging (fMRI) to assess the effect of this stimulation on functional correlations throughout the network. Two different frequencies of rTMS to this DMN node induced two topographically distinct changes in functional connectivity. High-frequency (20 Hz; Fig. 8.1C) rTMS to lpIPL (bottom right in each panel) decreased functional correlations between cortical DMN nodes, but not between these nodes and the hippocampal formation (HF) bilaterally. In contrast, low frequency (1 Hz; Fig. 8.1B) rTMS to lpIPL did not alter connectivity between DMN nodes but increased functional correlations between lpIPL and HF bilaterally. The Eldaief et al. (2011) study emphasizes the dynamic relationships between regions of intrinsic brain networks, which changing functional relationships among regions plays a role in supporting different processing demands.

Romei et al. (2016) showed that sensorimotor cortex rhythms are entrained by rTMS stimulation in the beta-band but that the elicited responses are strongest at the intrinsic individual beta-peak frequency. While these entrainment effects are of short duration, even subthreshold rTMS pulses propagate through the network and elicit significant corticospinal coupling, particularly when stimulated at the individual's beta-frequency. These results show that externally imposed rhythmicity interacts with intrinsic brain rhythms such that the individual peak frequency determines the effect of rTMS.

Auditory cortical processing is organized according to two pathways, that is, dorsal and ventral streams (Hickok and Poeppel, 2007). While the

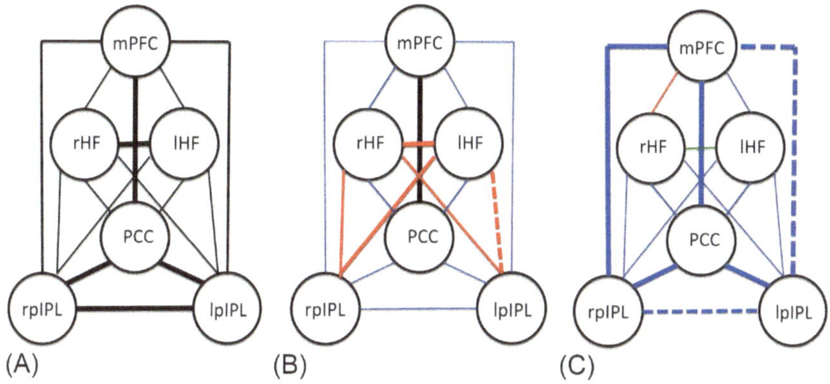

FIG. 8.1 Schematic representation of functional connectivity changes across multiple regions within the default network. The thickness of the lines connecting default network regions is proportional to the connectivity strength between these regions. Connectivity is displayed before 1- and 20-Hz rTMS (A). Changes in functional connectivity between default network regions as a result of 1-Hz (B) and 20-Hz (C) rTMS. Scaling of line thickness to changes in correlation strength is different between (B), (C), and (A). Decreases in functional connectivity are depicted with *blue lines,* and increases are depicted in *red. Large-dashed lines* signify significant changes (corrected for multiple comparisons). *Small-dashed lines* signify significant correlation changes not surviving multiple comparisons corrections. These results suggest coupling between regions in the default network is dynamic. *Redrawn from Eldaief, M.C., Halko, M.A., Buckner, R.L., Pascual-Leone, A., 2011. Transcranial magnetic stimulation modulates the brain's intrinsic activity in a frequency-dependent manner. Proc. Natl. Acad. Sci. U. S. A. 108(52), 21229–21234.*

ventral pathway supports simple memory processing [simple melodies comparison, mismatch negativity (Albouy et al., 2013)], the dorsal auditory stream is recruited when high-level processing is required, such as memory manipulation (Foster and Zatorre, 2010). Albouy et al. (2017) used simultaneous magnetoencephalography (MEG) and EEG, to investigate the causal relationship between brain oscillations and behavior by applying theta-frequency rTMS over the left intraparietal sulcus during the silent interval between the two stimuli. Theta-frequency rTMS-entrained theta oscillation and boosted participants' recall accuracy. These effects were not seen for a melody-comparison control task and were not observed for arrhythmic TMS. These data establish theta activity in the dorsal stream as causally related to memory manipulation. Kim et al. (2018) tested this putative distinction. There was no increase in anterior-temporal network activity due to rTMS, whereas there was for the posterior-medial network. The posterior-medial network is therefore a functional unit for memory processing that is distinct from the anterior-temporal network. These findings suggest that targeted stimulation can lead to network-specific increases in excitability during memory formation.

8.1.3 Paired associative stimulation

STDP is a temporally asymmetric form of Hebbian learning induced by tight temporal correlations between the spikes of presynaptic and postsynaptic neurons. As with other forms of synaptic plasticity it underlies learning and information storage in the brain (Bi and Poo, 2001). With STDP, repeated presynaptic spike arrival a few milliseconds before postsynaptic action potentials leads in many synapse types to LTP of the synapses, whereas repeated spike arrival after postsynaptic spikes leads to LTD of the same synapse (Sjöstrom and Gerstner, 2010; Fig. 8.2).

Koch et al. (2013) used a paired associative TMS protocol to repeatedly activate the short-latency connection between the posterior parietal cortex (PPC) and the primary motor cortex (M1) of the left hemisphere. They systematically varied the temporal relationships between the stimuli and the preferential activation of different M1 neuronal populations by applying TMS over M1 with different coil orientations and in different states of cortical excitability (rest vs muscular contraction). They found evidence for the existence of both Hebbian and anti-Hebbian STDP in human long-range connections. Subsequently, Santarnecchi et al. (2018) used pairs

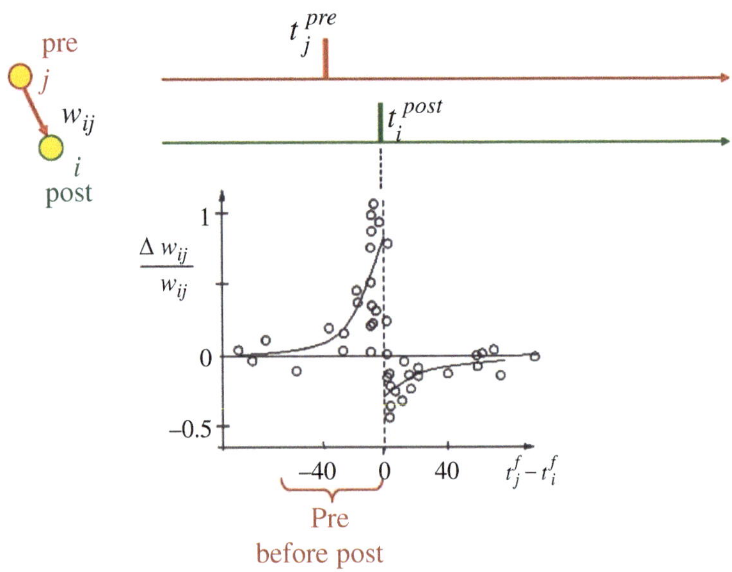

FIG. 8.2 Spike-timing-dependent plasticity (STDP) (schematic). The STDP function shows the change of synaptic connections as a function of the relative timing of presynaptic and postsynaptic spikes after 60 spike pairings. *From Sjöström, J., Gerstner, W., 2010. Spike-timing dependent plasticity. Scholarpedia 5(2), 1362. https://doi.org/10.4249/scholarpedia.1362 (Revision #184913). Open access.*

of TMS pulses at specific interstimulus intervals that were repeatedly delivered over two negatively correlated nodes of the DMN and the task-positive network defined on the basis of individual fMRI data. This induced a modulation of network-to-network connectivity, even reversing correlation from negative to slightly positive in 30% of cases. Results also suggested a baseline-dependent effect, with a greater connectivity modulation observed in participants with weaker between-networks connectivity strength right before TMS.

8.2 Transcranial direct current stimulation

Transcranial direct current stimulation uses constant, low-level direct current delivered via electrodes on the head.

8.2.1 Changes in excitability of motor cortex

Nitsche and Paulus (2000) demonstrated the possibility of a noninvasive modulation of motor cortex excitability by the application of weak tDCS (1–5 min; 0.2–1.0 mA) through the scalp. Excitability changes of up to 40%, revealed by the response to single-pulse TMS lasted for several minutes after the end of tDCS. Excitation and inhibition were achieved selectively by placing respectively the anode and the cathode above motor cortex. Nitsche and Paulus (2001) showed that 1-mA tDCS induces sustained cortical excitability elevations. As revealed by single-pulse TMS, motor cortical excitability increased approximately 150% above baseline for up to 90 min after the end of stimulation. The findings suggest that tDCS does modulate synaptic strength within the cortex. Antal et al. (2011) investigated the online effects of short periods of anodal and cathodal tDCS on brain activity of healthy subjects and associated hemodynamics by concurrent blood-oxygenation-level-dependent (BOLD) fMRI at 3 T. Using a block design, 20 s periods of 1 mA tDCS over the left M1 altered with 20 s periods without tDCS. In different fMRI runs, the effect of anodal or cathodal tDCS was assessed at rest or during finger tapping. Neither anodal nor cathodal tDCS over the primary motor cortex (M1) for 20 s—with or without finger tapping—induced a detectable BOLD signal change in supplementary motor area, premotor cortex, and M1. Thus motor-evoked potentials (MEPs) after tDCS stimulation of the M1 (Nitsche and Paulus, 2000) are not paralleled by analogous changes in regional BOLD signals. This difference implies (Antal et al., 2011) that the BOLD signal and measurements of MEPs probe diverse physiological mechanisms, that is, the MEP amplitude reflects synchronous changes in excitability of pyramidal neurons while the BOLD signal is a measure of synchronous and asynchronous synaptic activity of all cortical neurons.

8.2.2 Excitability changes in auditory cortex

Zaehle et al. (2011) applied tDCS over the left temporal and the left temporoparietal cortex and investigated effects on cortical auditory-evoked potentials (CAEPs) after anodal, cathodal, and sham stimulation. They found that anodal and cathodal tDCS modified auditory cortex reactivity. While anodal tDCS over the temporal cortex increased auditory P50 amplitudes, cathodal tDCS over the temporoparietal cortex induced larger N100 amplitudes. The same group (Heimrath et al., 2014) separately stimulated with anodal tDCS the left or right auditory cortex (ACx; cathode on vertex) while participants performed a between-channel gap detection task. Results show that stimulation of the left but not right ACx deteriorated the auditory perception of this rapidly changing acoustic information. This indicates a left hemispheric dominance for the processing of rapid temporal cues in auditory nonspeech sounds. Heimrath et al. (2015) confirmed this by assessed the influence of anodal and cathodal high-definition (HD)-tDCS delivered over the left or right ACx on auditory mismatch negativity (MMN) in response to temporal and spectral deviants. HD-tDCS applies a ring of reference electrodes (cathodal or anodal) close around the active electrode (anodal or cathodal). The results showed that MMN to temporal deviants was significantly enhanced by anodal HD-tDCS applied over the left ACx only. Kunzelmann et al. (2018) investigated the potential modulatory effect of tDCS on the excitability of left ACx in detail. They included healthy participants in a crossover design, applying sham and anodal stimulation in two measurement sessions 1 week apart. Amplitudes and latencies of P50, N100, and P200 components of the CAEP were compared between anodal and sham stimulation, and between time points before, during, and after tDCS. They found no significant differences between stimulation types or time points for any of the investigated CAEP amplitudes or latencies. Furthermore, a topographical analysis did not show any topographical differences during peak time periods of the investigated CAEP for stimulation types and time points besides a habituation effect. Clearly, results differ from those reported by the Zaehle-group (Zaehle et al., 2011; Heimrath et al., 2014, 2015), and it is noted that the effects in these studies were small, so that strong conclusions are not warranted.

Neuling et al. (2012) conducted an auditory detection experiment with humans with concurrent EEG recording, and simultaneously applied oscillating tDCS at 10 Hz (alpha-tDCS). Note that unlike for tACS (next section), the oscillation in alpha-tDCS is around the DC component, and does not change polarity. Auditory detection thresholds were dependent on the phase of the oscillation that was entrained by alpha-tDCS (Fig. 8.3). This behavioral effect was accompanied by an electrophysiological effect: alpha-power was enhanced after alpha-tDCS as compared

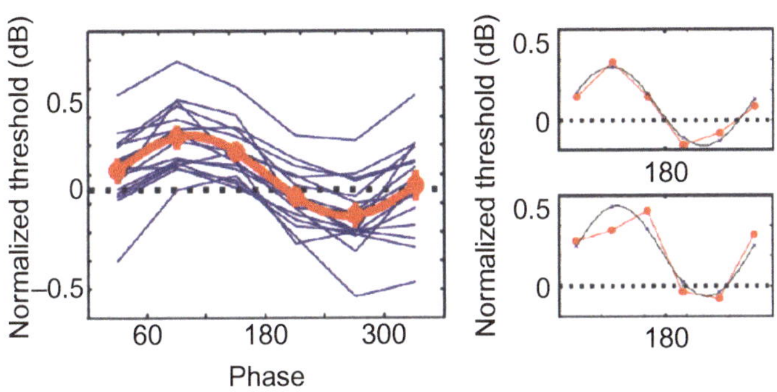

FIG. 8.3 Behavioral data. *(Left) Blue lines* represent the fits for single subject data. The *red line* shows the average fit in a higher resolution for illustration purposes. *Red dots* denote the mean of the fitted data ± SEM. *(Right)* Sine fits for two exemplary subjects. The *red line* shows the raw data and the *black line* shows the fitted sine wave. *Reprinted with permission from Neuling, T., Rich, S., Wagner, S., Wolters, C.H., Herrmann, C.S., 2012. Good vibrations: Oscillatory phase shapes perception. NeuroImage 63, 771–778, Elsevier. Copyright 2012.*

with a prestimulation period. Neuling et al. (2012) observed that manipulation of the phase resulted in different detection thresholds, which supported the notion that oscillatory processes can periodically modulate perception.

8.3 Transcranial random noise stimulation

Terney et al. (2008) introduced transcranial random noise stimulation (tRNS) whereby a random wide-band frequency electrical signal is applied over the motor cortex. tRNS induced consistent excitability increases that lasted for 60 min after stimulation. These effects were observed through both physiological measures and behavioral tasks. Higher tRNS frequencies (100–640 Hz) appeared to be responsible for generating this excitability increase. Terney et al. (2008) remarked that stochastic resonance (SR) may play a role in tRNS, specifically at the higher frequency ranges. SR is a phenomenon where a signal that is normally too weak to be detected by an electrode or magnetic sensor can be boosted by adding broad-band noise to the signal. Neural synchrony in the brain (specifically in the gamma wave frequency band) has been suggested as a possible neural mechanism for SR by researchers who have investigated the perception of "subconscious" visual sensation (Moss et al., 2004). The long-lasting effect, however, requires mechanisms different from stochastic resonance. Van Doren et al. (2014) used EEG to assess tRNS-induced changes in resting-state activity and in auditory steady state responses (ASSRs). Stimuli were 1 kHz carrier tones that were amplitude modulated at 20 and 40 Hz and applied over ACx bilaterally in randomized order. Van Doren et al. (2014) found a significant increase in the ASSR in response to 40 Hz frequency modulated tone, in line with increase of stochastic resonance.

Rufener et al. (2017) stimulated the human ACx bilaterally with tRNS while EEG was continuously measured. Modulations in the participants' temporal and spectral resolution ability were investigated by means of a gap detection task and a pitch discrimination task, respectively. Compared with sham, auditory tRNS increased the detection rate for near-threshold stimuli in the temporal domain only, while no such effect was present for the discrimination of spectral features. Behavioral findings were paralleled by reduced peak latencies of the P50 and N100 component of the CAEP, indicating an impact on early sensory processing. The facilitating effect of tRNS was limited to the processing of near-threshold stimuli while stimuli clearly below and above the individual perception threshold were not affected by tRNS. This nonlinear relationship between the signal-to-noise level of the presented stimuli and the effect of stimulation again suggests SR as the underlying mechanism of tRNS on auditory processing.

8.4 Transcranial alternating current stimulation

Instead of applying a direct electrical current, tACS applies a sinusoidal current—with mean level of zero—at a chosen frequency to interact with the brain's intrinsic cortical oscillations. When a single, consonant frequency is applied, the exogenous oscillation can entrain with the brain's endogenous frequency. As such, tACS effects are contingent on the frequency, amplitude, and phase.

8.4.1 Effects on EEG-frequency-band power

8.4.1.1 Visual system

Kanai et al. (2008) demonstrated that tACS interacts with ongoing rhythmic activities in the visual cortex in a frequency-specific fashion and induces visual sensations (phosphenes). They delivered tACS over the occipital cortex, and compared the effects of tACS under "Light" or "Dark" conditions. Stimulation over the occipital cortex induced perception of continuously flickering light most effectively when the beta frequency range was applied in an illuminated room, whereas the most effective stimulation frequency shifted to the alpha frequency range during testing in darkness (Fig. 8.4). Stimulation with theta or gamma frequencies did not produce any visual phenomena. The shift of the effective stimulation frequency indicates that the frequency dependency is caused by interactions with the dominant ongoing oscillatory activity in the stimulated cortex.

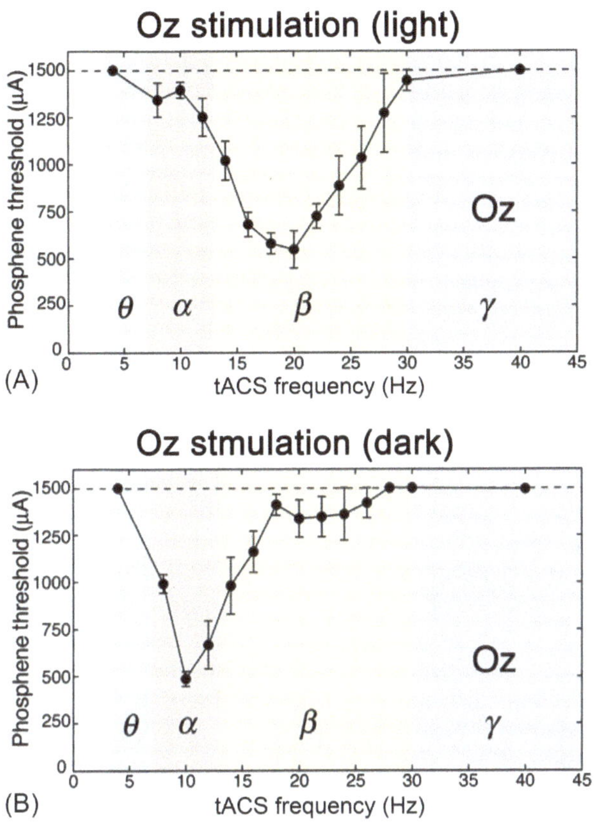

FIG. 8.4 Thresholds for phosphenes induction. (A) Mean phosphene thresholds in the Light condition. The error bars correspond to one standard error of the mean (SEM). (B) Mean phosphene thresholds in the dark condition. *From Kanai, R., Chaieb, L., Antal, A., Walsh, V., Paulus, W., 2008. Frequency-dependent electrical stimulation of the visual cortex. Curr. Biol. 18, 1839–1843. Open access.*

Poor performance on visual tasks has been associated with increased EEG alpha power (Jensen and Mazaheri, 2010). tACS over occipitoparietal cortex at 10 Hz (alpha-tACS) can selectively enhance EEG alpha power. Clayton et al. (2019) assessed the contribution of alpha oscillations to attention, by delivering alpha-tACS across four experiments while participants performed sustained visual or auditory attention tasks. In contrast to the prediction, alpha-tACS was found to prevent deteriorations in visual performance that otherwise occurred during sham- and 50-Hz tACS. Such effects were not seen in an auditory control task, indicating specificity to the visual domain.

8.4.1.2 Sensorimotor system

Antal et al. (2008) compared the effects of tDCS with tACS. Motor-evoked potentials evoked by single-pulse TMS, EEG-power, and reaction times measured in a motor implicit learning task, were analyzed to detect changes in cortical excitability after 2–10 min of tACS and sinusoidally modulated tDCS, that is, maintaining the polarity, by using 1, 10, 15, 30, and 45 Hz and sham stimulation over the primary motor cortex in healthy subjects. Only 10-Hz tACS significantly improved implicit motor learning. No significant changes were observed in any of the analyzed frequency bands of EEG and with regard to the MEP amplitudes after tACS or tDCS. Moisa et al. (2016) applied gamma-band and beta-band frequency tACS over M1 in healthy humans during a visuomotor task and concurrent fMRI. Beta tACS had no statistically significant effects, whereas gamma tACS improved both the velocity and acceleration of visually triggered movements. Gamma tACS induced motor performance enhancements correlated with changed BOLD activity in the stimulated M1. This is in clear contrast to what was found with tDCS (Antal et al., 2011; Section 8.2.1). Moreover, Moisa et al. (2016) found frequency- and task-specific neural compensatory activity modulations in the dorsomedial prefrontal cortex (dmPFC), suggesting a key regulatory role of this region in motor performance. Connectivity analyses revealed that the dmPFC interacted functionally with M1 and with regions within the executive motor system. These results suggest a role of the dmPFC for motor control and show that tACS-induced behavioral changes not only result from activity modulations underneath the stimulation electrode but also reflect compensatory modulation within connected and functionally related brain networks (Moisa et al., 2016).

8.4.1.3 Auditory system

Riecke et al. (2015) measured detection performance for near-threshold auditory stimuli (4-Hz click trains) that were presented at various moments during ongoing tACS (two synchronous 4-Hz alternating currents applied to the two cerebral hemispheres). They found that changes in the relative timing of acoustic and electric stimulation cause corresponding perceptual changes that oscillate predominantly at the 4-Hz frequency of the electric stimulation, which is consistent with results based on 10-Hz tACS (Neuling et al., 2012). Rufener et al. (2016) analyzed repetition-induced perceptual learning effects. The participants received either 6- or 40-Hz tACS over the bilateral auditory cortex, while repeatedly performing a phoneme categorization task. A 40-Hz tACS attenuated the repetition-induced improvement in phoneme categorization abilities.

Temporal resolution of cortical, auditory processing mechanisms may be linked to peak frequency of neuronal gamma oscillations in auditory cortex (individual gamma frequency, IGF). Individuals with higher IGF tend to have better temporal resolution. Behavioral measures of individual auditory temporal resolution, determined by a between-channel gap detection (leading and trailing masker of different frequency; Phillips et al., 1997), could be linked to an IGF of auditory cortex areas (Baltus and Herrmann, 2015). Baltus et al. (2018) identified the IGF and temporal resolution using an ASSR paradigm and a between-channel gap detection task in healthy participants with one stimulation electrode per hemisphere. They found that the performance of improved significantly during tACS with 4 Hz above IGF in comparison with performance during tACS with 4 Hz below IGF. They concluded that: "The possibility to improve gap detection with individualized stimulation protocols for tACS further supports the link between oscillatory activity and temporal resolution, whereby the improvement of temporal resolution is particularly relevant for the clinical aspect of auditory tACS."

8.4.2 Effects on phase locking with intrinsic oscillations

Helfrich et al. (2014) found that 10-Hz tACS increases parietooccipital alpha activity and synchronized cortical oscillators with similar intrinsic frequencies to the entrainment frequency. Additionally, they demonstrated that tACS modulates target detection performance in a phase-dependent fashion highlighting the causal role of alpha oscillations for visual perception (Neuling et al., 2013).

Polanía et al. (2012) used 6-Hz tACS simultaneously applied over left prefrontal and parietal cortices with an in-phase (synchronized condition) or counter-phase (desynchronized condition) difference to the intrinsic theta in the dorsolateral prefrontal cortex (dlFPC) or posterior parietal cortex (PPC), or a placebo stimulation condition, during which healthy subjects performed a delayed letter discrimination task. They selected EEG electrode sites approximately belonging to dlFPC and PPC a priori (event-related potentials in Fig. 8.5A) and identified a significant increase of phase synchronization at ~ 4–7 Hz occurring ~ 200–500 ms after memory probe onset (Fig. 8.5B). A representative trial band-pass filtered at 6 ± 1 Hz is shown to illustrate theta phase synchronization occurring ~ 200 ms after probe onset (Fig. 8.5C). This exogenously induced frontoparietal theta synchronization significantly improved visual memory-matching reaction times as compared with placebo stimulation. In contrast, exogenously induced frontoparietal theta desynchronization deteriorated performance. These findings provide evidence of causality of theta phase-coupling of distant cortical areas for cognitive performance in healthy humans (Polanía et al., 2012).

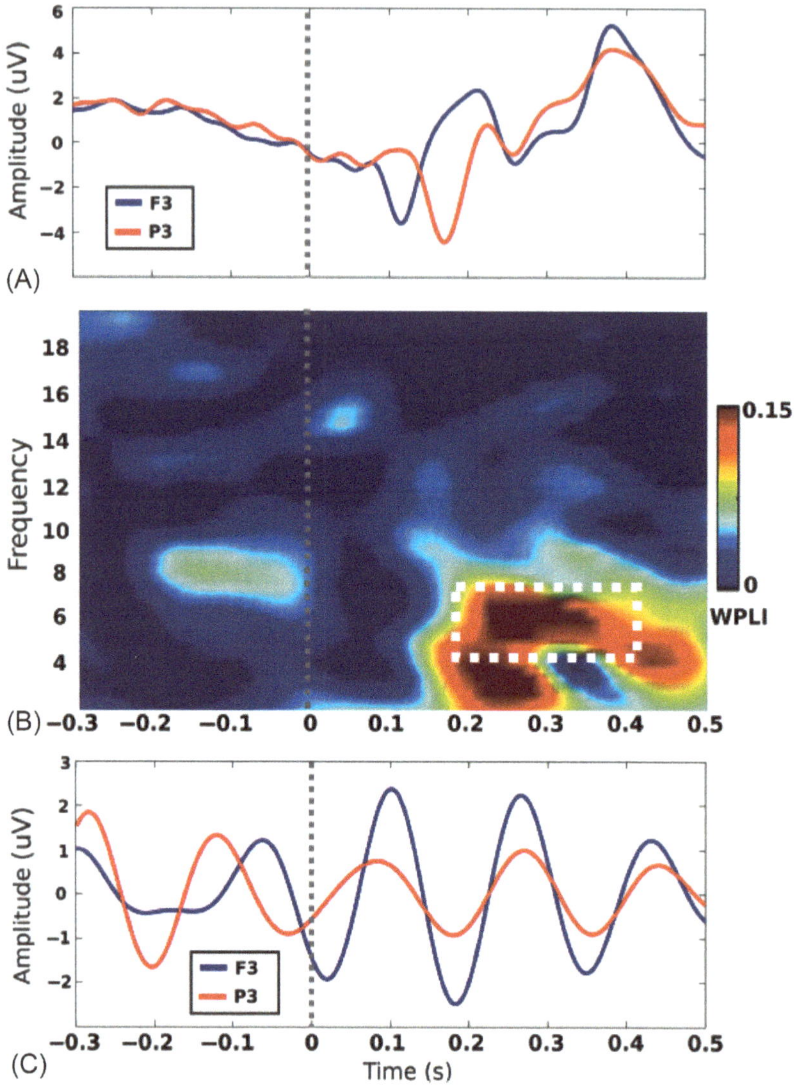

FIG. 8.5 Effects of tACS on EEG. (A) Amplitude of the event-related potentials from channels F3 and P3, and (B) their weighted phase lag index. (C) A representative trial band-pass filtered at 6 ± 1 Hz is shown to illustrate theta phase synchronization occurring −200 ms after probe onset times (x axis) in (A), (B), and (C) are shown with respect to probe onset. *From Polanía, R., Nitsche, M.A., Korman, C., Batsikadze, G., Paulus, W., 2012. The importance of timing in segregated theta phase-coupling for cognitive performance. Curr. Biol. 22, 1314–1318. Open access.*

8.5 Ultrasound

An emerging technology called transcranial-focused ultrasound (tFUS) neuromodulation makes use of low-level sound energies. Ultrasonic waves can be noninvasively steered and focused into millimeter-scale regions across the human body and brain, and their application in generating modulation of neuronal activity could therefore potentially have implications for neuroscience (Naor et al., 2016).

8.5.1 Findings in animal experiments

Daniels et al. (2018) used tFUS suppression of inferior colliculus AEPs in 22 rats and cortical AEPs in 5 pigs: Repetitive sounds were produced, and the induced AEPs were recorded before and repeatedly after tFUS treatment of the auditory pathway, that is, the inferior colliculus (rats) and auditory cortex (pigs). All treated animals exhibited a decrease in AEP amplitude posttreatment in contrast to animals undergoing the sham treatment. Suppression was weaker for lower stimulus levels. AEP amplitudes in the treated pigs decreased to 27.7% ± 5.9% of baseline. This effect lasted between 30 min and 1 month in most treated animals. No evidence of heating during treatment or later brain damage/edema was observed. Folloni et al. (2019) showed effects of tFUS in macaque cortex anterior cingulate cortex (ACC) and amygdala. They measured neuromodulatory effects by examining relationships between activity in these areas and the rest of the brain using fMRI. tFUS of each area, ACC and amygdala, had specific effects that were distinct from one another. In control conditions without tFUS, activity in a given area was related to activity in interconnected regions, but such relationships were reduced after sonication, specifically for the targeted areas. Dissociable and focal effects on neural activity could not be explained by auditory confounds because the auditory stimulation associated with the tFUS application ceased after the sonication, but the neural activity measurements were initiated 10 min later.

8.5.2 Effects in humans

Mueller et al. (2014) examined the effects on the instantaneous phase and phase rate of EEG recordings before, during, and after transmission of tFUS to human somatosensory cortex. These recordings showed that tFUS alters the phase distribution of intrinsic brain activity for beta frequencies, but not gamma. Additionally, tFUS modulated phase distributions in the beta band of early sensory-evoked activity but did not affect late sensory-evoked activity, lending support to the spatial specificity of tFUS for neuromodulation. Using EEG, Legon et al. (2014) showed that tFUS significantly attenuated the amplitudes of somatosensory-evoked potentials elicited by median nerve stimulation. They also found that tFUS significantly modulated the spectral content of sensory-evoked brain oscillations. Compared with sham, they found that tFUS significantly decreased the power of alpha-band (7–12 Hz) and beta-band (13–30 Hz) baseline activity recorded from EEG sites C3 and P3 in the 100 ms following the onset of tFUS transmission before median nerve stimulation. Legon et al. (2014) also found that tFUS produced a significant attenuation in the power of short-latency-evoked gamma-band (30–55 Hz) activity occurring within 70 ms of median nerve stimulation. As was also evident to varying degrees across the EEG channels recorded, tFUS significantly modulated the power of late-onset alpha-, beta-, and gamma-band activity occurring about 200 ms after median nerve stimulation or later. The changes produced by tFUS on sensory-evoked brain activity were abolished when the acoustic beam was focused 1-cm anterior or posterior to primary somatosensory cortex (S1). Behavioral investigations showed that tFUS targeted to S1 enhanced performance on sensory discrimination tasks without affecting task attention or response bias. Legon et al. (2014) concluded that tFUS may be used to focally modulate human cortical function.

Lee et al. (2016) reported tFUS effects on the primary visual cortex (V1) in humans, resulting in elicited activation not only from the sonicated brain area but also from the network of regions involved in visual and higher-order cognitive processes (as revealed by simultaneous acquisition of BOLD fMRI). Accompanying phosphene perception was also reported. The tFUS-evoked responses showed distinct peaks associated with the stimulation, which were roughly comparable with visual-evoked potentials—with the exception of the P250.

8.6 Summary

Here, we presented an overview of the effects induced by TMS, tACS, and tDCS. In addition, we discuss recently developed neuromodulation procedures based on focused low-intensity ultrasound. TMS may induce dynamic relationships between regions of intrinsic brain networks. Theta-TMS induces activity in the dorsal auditory stream, causally related to memory manipulation. Paired associative TMS can induce both Hebbian and anti-Hebbian STDP in human long-range connections. tDCS does modulate synaptic strength within motor cortex, as visualized in the MEP but absent in the BOLD. tDCS effects in auditory cortex were small. However, oscillating tDCS modulated auditory detection thresholds in sinusoidal manner. tRNS works by inducing stochastic resonance and enhances threshold detection in phase-dependent fashion. tACS at 4 and 10 Hz also resulted in periodic modulation of auditory detection thresholds, and enhancement in auditory temporal processing. tFUS produced a significant attenuation in the power of short-latency-evoked gamma-band (30–55 Hz) activity occurring within 70 ms of median nerve stimulation. tFUS alters the phase distribution of intrinsic brain activity for beta frequencies but not gamma.

References

Albouy, P., Mattout, J., Bouet, R., Maby, E., Sanchez, G., Aguera, P.E., et al., 2013. Impaired pitch perception and memory in congenital amusia: The deficit starts in the auditory cortex. Brain 136, 1639–1661.

Albouy, P., Weiss, A., Baillet, S., Zatorre, R.J., 2017. Selective entrainment of theta oscillations in the dorsal stream causally enhances auditory working memory performance. Neuron 94, 193–206.

Antal, A., Boros, K., Poreisz, C., Chaieb, L., Terney, D., Paulus, W., 2008. Comparatively weak after-effects of transcranial alternating current stimulation (tACS) on cortical excitability in humans. Brain Stimul. 1, 97–105.

Antal, A., Polania, R., Schmidt-Samoa, C., Dechent, P., Paulus, W., 2011. Transcranial direct current stimulation over the primary motor cortex during fMRI. Neuroimage 55, 590–596.

Baltus, A., Herrmann, C.S., 2015. Auditory temporal resolution is linked to resonance frequency of the auditory cortex. Int. J. Psychophysiol. 98, 1–7.

Baltus, A., Wagner, S., Wolters, C.H., Herrmann, C.S., 2018. Optimized auditory transcranial alternating current stimulation improves individual auditory temporal resolution. Brain Stimul. 11, 118–124.

Bi, G.-Q., Poo, M.-M., 2001. Synaptic modification by correlated activity: Hebb's postulate revisited. Annu. Rev. Neurosci. 24, 139–166.

Clayton, M.S., Yeung, N., Cohen Kadosh, R., 2019. Electrical stimulation of alpha oscillations stabilizes performance on visual attention tasks. J. Exp. Psychol. Gen. 148 (2), 203–220.

Daniels, D., Sharabi, S., Last, D., Guez, D., Salomon, S., Zivli, Z., et al., 2018. Focused ultrasound-induced suppression of auditory evoked potentials in vivo. Ultrasound Med. Biol. 44 (5), 1022–1030.

Eldaief, M.C., Halko, M.A., Buckner, R.L., Pascual-Leone, A., 2011. Transcranial magnetic stimulation modulates the brain's intrinsic activity in a frequency-dependent manner. Proc. Natl. Acad. Sci. U. S. A. 108 (52), 21229–21234.

Folloni, D., Verhagen, L., Mars, R.B., Fouragnan, E., Constans, C., Aubry, J.-F., et al., 2019. Manipulation of subcortical and deep cortical activity in the primate brain using transcranial focused ultrasound stimulation. Neuron 101, 1–8.

Foster, N.E., Zatorre, R.J., 2010. A role for the intraparietal sulcus in transforming musical pitch information. Cereb. Cortex 20, 1350–1359.

Fuggetta, G., Fiaschi, A., Manganotti, P., 2005. Modulation of cortical oscillatory activities induced by varying single-pulse transcranial magnetic stimulation intensity over the left primary motor area: a combined EEG and TMS study. Neuroimage 27, 896–908.

Heimrath, K., Kuehne, M., Heinze, H.-J., Zaehle, T., 2014. Transcranial direct current stimulation (tDCS) traces the predominance of the left auditory cortex for processing of rapidly changing acoustic information. Neuroscience 261, 68–73.

Heimrath, K., Breitling, C., Krauel, K., Heinze, H.-J., Zaehle, T., 2015. Modulation of pre-attentive spectro-temporal feature processing in the human auditory system by HD-tDCS. Eur. J. Neurosci. 41, 1580–1586.

Helfrich, R.F., Schneider, T.R., Rach, S., Trautmann-Lengsfeld, S.A., Engel, A.K., Herrmann, C.S., 2014. Entrainment of brain oscillations by transcranial alternating current stimulation. Curr. Biol. 24, 333–339.

Herrmann, C.S., Strüber, D., Helfrich, R.F., Engel, A.K., 2016. EEG oscillations: from correlation to causality. Int. J. Psychophysiol. 103, 12–21.

Hickok, G., Poeppel, D., 2007. The cortical organization of speech processing. Nat. Rev. Neurosci. 8, 393–402.

Jensen, O., Mazaheri, A., 2010. Shaping functional architecture by oscillatory alpha activity: gating by inhibition. Front. Hum. Neurosci. 4, 186. https://doi.org/10.3389/fnhum.2010.00186.

Kanai, R., Chaieb, L., Antal, A., Walsh, V., Paulus, W., 2008. Frequency-dependent electrical stimulation of the visual cortex. Curr. Biol. 18, 1839–1843.

Kim, S., Nilakantan, A.S., Hermiller, M.S., Palumbo, R.T., VanHaerents, S., Voss, J.L., 2018. Selective and coherent activity increases due to stimulation indicate functional distinctions between episodic memory networks. Sci. Adv. 4, eaar2768.

Kobayashi, M., Pascual-Leone, A., 2003. Transcranial magnetic stimulation in neurology. Lancet Neurol. 2, 145–156.

Koch, G., Ponzo, V., Di Lorenzo, F., Caltagirone, C., Veniero, D., 2013. Hebbian and anti-Hebbian spike-timing-dependent plasticity of human cortico-cortical connections. J. Neurosci. 33 (23), 9725–9733.

Kunzelmann, K., Meier, L., Grieder, M., Morishima, Y., Dierks, T., 2018. No effect of transcranial direct current stimulation of the auditory cortex on auditory-evoked potentials. Front. Neurosci. 12, 880. https://doi.org/10.3389/fnins.2018.00880.

Lee, W., Kim, H.-C., Jung, Y., Chung, Y.A., Song, I.U., Lee, J.-H., et al., 2016. Transcranial focused ultrasound stimulation of human primary visual cortex. Sci. Rep. 6, 34026. https://doi.org/10.1038/srep34026.

Legon, W., Sato, T.F., Opitz, A., Mueller, J., Barbour, A., Williams, A., et al., 2014. Transcranial focused ultrasound modulates the activity of primary somatosensory cortex in humans. Nat. Neurosci. 17 (2), 322–329.

Moisa, M., Polania, R., Grueschow, M., Ruff, C.R., 2016. Brain network mechanisms underlying motor enhancement by transcranial entrainment of gamma oscillations. J. Neurosci. 36 (47), 12053–12065.

Moss, F., Ward, L.M., Sannita, W.G., 2004. Stochastic resonance and sensory information processing: a tutorial and review of application. Clin. Neurophysiol. 115, 267–281.

Mueller, J., Legon, W., Opitz, A., Sato, T.F., Tyler, W.J., 2014. Transcranial focused ultrasound modulates intrinsic and evoked EEG dynamics. Brain Stimul. 7, 900–908.

Naor, O., Krupa, S., Shoham, S., 2016. Ultrasonic neuromodulation. J. Neural Eng. 13. https://doi.org/10.1088/1741-2560/13/3/031003.

Neuling, T., Rach, S., Wagner, S., Wolters, C.H., Herrmann, C.S., 2012. Good vibrations: oscillatory phase shapes perception. Neuroimage 63, 771–778.

Neuling, T., Rach, S., Herrmann, C.S., 2013. Orchestrating neuronal networks: sustained after-effects of transcranial alternating current stimulation depend upon brain states. Front. Hum. Neurosci. 7, 161. https://doi.org/10.3389/fnhum.2013.00161.

Nitsche, M.A., Paulus, W., 2000. Excitability changes induced in the human motor cortex by weak transcranial direct current stimulation. J. Physiol. (Lond.) 527 (3), 633–639.

Nitsche, M.A., Paulus, W., 2001. Sustained excitability elevations induced by transcranial DC motor cortex stimulation in humans. Neurology 57, 1899–1901.

Phillips, D.P., Taylor, T.L., Hall, S.E., Carr, M.M., Mossop, J.E., 1997. Detection of silent intervals between noises activating different perceptual channels: some properties of "central" auditory gap detection. J. Acoust. Soc. Am. 101, 3694–3705.

Polanía, R., Nitsche, M.A., Korman, C., Batsikadze, G., Paulus, W., 2012. The importance of timing in segregated theta phase-coupling for cognitive performance. Curr. Biol. 22, 1314–1318.

Riecke, L., Formisano, E., Herrmann, C.S., Sack, A.T., 2015. 4-Hz transcranial alternating current stimulation phase modulates hearing. Brain Stimul. 8, 777–783.

Romei, V., Bauer, M., Brooks, J.L., Economides, M., Penny, W., Thut, G., et al., 2016. Causal evidence that intrinsic beta-frequency is relevant for enhanced signal propagation in the motor system as shown through rhythmic TMS. Neuroimage 126, 120–130.

Rufener, K.S., Zaehle, T., Oechslin, M.S., Meyer, M., 2016. 40 Hz-transcranial alternating current stimulation (tACS) selectively modulates speech perception. Int. J. Psychophysiol. 101, 18–24.

Rufener, K.S., Ruhnau, P., Heinze, H.-J., Zaehle, T., 2017. Transcranial random noise stimulation (tRNS) shapes the processing of rapidly changing auditory information. Front. Cell. Neurosci. 11, 162. https://doi.org/10.3389/fncel.2017.00162.

Santarnecchi, E., Momi, D., Sprugnoli, G., Neri, F., Pascual-Leone, A., Rossi, A., et al., 2018. Modulation of network-to-network connectivity via spike-timing-dependent noninvasive brain stimulation. Hum. Brain Mapp. 39, 4870–4883.

Sjöström, J., Gerstner, W., 2010. Spike-timing dependent plasticity. Scholarpedia 5 (2), 1362. https://doi.org/10.4249/scholarpedia.1362 (Revision #184913).

Terney, D., Chaieb, L., Moliadze, V., Antal, A., Paulus, W., 2008. Increasing human brain excitability by transcranial high-frequency random noise stimulation. J. Neurosci. 28 (52), 14147–14155.

Thut, G., Miniussi, C., 2009. New insights into rhythmic brain activity from TMS–EEG studies. Trends Cogn. Sci. 13 (4), 182–189.

Van Doren, J., Langguth, B., Schecklmann, M., 2014. Electroencephalographic effects of transcranial random noise stimulation in the auditory cortex. Brain Stimul. 7, 807–812.

Walsh, V., Cowey, A., 2000. Transcranial magnetic stimulation and cognitive neuroscience. Nat. Rev. Neurosci. 1, 73–79.

Zaehle, T., Beretta, M., Jäncke, L., Herrmann, C.S., Sandmann, P., 2011. Excitability changes induced in the human auditory cortex by transcranial direct current stimulation: direct electrophysiological evidence. Exp. Brain Res. 215, 135–140.

CHAPTER

9

Lifespan changes in brain rhythms and networks

9.1 Introduction

In this chapter, we explore the lifespan changes in brain rhythms and whole-brain network connectivity, and their implications for cognitive functioning. This will be based on data from structural magnetic resonance imaging (MRI), diffusion tensor imaging (DTI), resting-state functional imaging (rs-fMRI), electroencephalography (EEG), and magnetoencephalography (MEG).

One of the more dramatic changes in the brain—synaptic pruning—takes place throughout the first 20 years of life (Huttenlocher and Dabholkar, 1997), and particularly in late childhood and adolescence, and may result in decreased local connectivity measures. Conversely, increased myelination throughout childhood and adolescence facilitates increased long-range functional correlations (Brody et al., 1987; Paus et al., 2001). There are significant but not perfect correlations between the resulting adult structural and functional connectivity (Honey et al., 2007), as functional connectivity between regions can also result from their common inputs (Power et al., 2010). Increases in functional connectivity during development, particularly in task control network (TCN) and default mode network (DMN) tends to take place between portions of the brain that are functionally related in adults (Fair et al., 2008). A significant finding that emerged from recent graph-theory-based studies (Chapter 6) is that brain networks throughout development possess a common structure and are organized in a way that facilitates efficiency on local and global network scales through the lifespan (Fair et al., 2009).

The balance between short- and long-range connections contributes to the integration and segregation of networks during development. Disturbance of the balance of these connections is associated with

https://doi.org/10.1016/B978-0-12-819818-6.00013-3

mental disorders such as autism and schizophrenia (Ouyang et al., 2017; Chapters 11 and 12). Short-range connectivity in particular is characterized by unique and spatiotemporally heterogeneous dynamics from infancy to adulthood. The development of short-range functional and structural connections at the macro level shows an inverted U-shaped pattern of maturation progressing from primary- to higher-order brain regions (Fig. 9.1). Based on the underlying maturation time scales, we divide this chapter in sections on preterm and term babies, infants (< 2 years), children (2–10 years), adolescents (< 15 years), and remaining lifespan including old age.

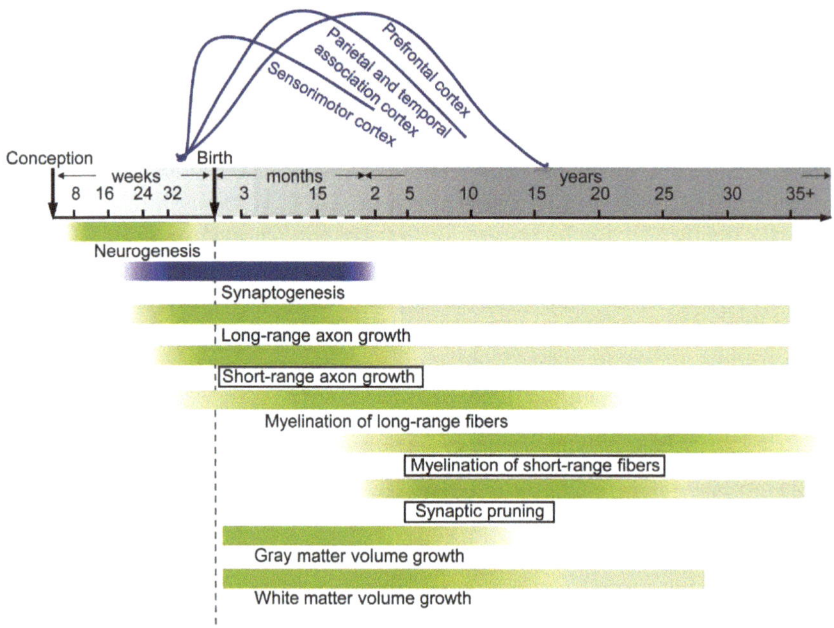

FIG. 9.1 Timeline of spatiotemporally distinctive human brain maturational processes. These processes include neurogenesis, synaptogenesis, long-range and short-range axon growth, myelination of long-range and short-range fibers, synaptic pruning, and gray and white matter volume growth. Time axis is in postconceptional weeks (before birth), postnatal months (until 24 months), and postnatal years (after 2 years). Processes related to maturation of short-range connections are highlighted with *black boxes*. The color intensity in each bar corresponds to the rate of developmental changes. The spatial progression is illustrated using synaptogenesis as an example. Besides temporal information, the spatial progression from primary sensorimotor cortex to higher-order prefrontal cortex of synaptogenesis (*blue bar*) is illustrated by the curves in corresponding color above the time axis. *Reprinted with permission from Ouyang, M., Kang, H., Detre, J.A., Roberts, T.P.L., Huang, H., 2017. Short-range connections in the developmental connectome during typical and atypical brain maturation. Neurosci. Biobehav. Rev. 83, 109–122. Copyright 2017 Elsevier.*

9.2 Preterm and term

9.2.1 Preterm

9.2.1.1 Longitudinal network development

Using rs-fMRI, Doria et al. (2010) mapped the development of different resting-state networks (RSNs) in infants born between 29 and 43 weeks of postmenstrual age (PMA). PMA is most frequently applied for the perinatal period beginning after the day of birth. At the day of birth, PMA is equal to gestational age (GA). Doria et al. (2010) described the development of connectivity in RSNs from being often fragmentary at 30 weeks PMA to adult-like patterns at term. Auditory, somatosensory, visual, motor, DMN, frontoparietal (FPN), and executive control networks (ECN) develop at different rates. However, at term all networks are present, several of which are integrated with thalamic activity. The auditory network shows significant interhemispheric connectivity from the earliest age, and thereafter changes little. This is consistent with the significant interhemispheric coherence of the auditory cortices of the early preterm group (Fig. 9.2A). Except for the auditory network, the results of Doria et al. (2010) place the emergence of RSNs largely during the period of rapid neural growth in the third trimester of gestation (Fig. 9.2), suggesting that they are formed before the acquisition of cognitive competencies in later childhood.

In a longitudinal study, Ball et al. (2014) demonstrated that a rich club of interconnected cortical hubs (cf. Fig. 6.3; Chapter 6) is already present by 30-week gestation. Subsequently, until the time of normal birth, the principal development is a proliferation of connections between core hubs and the rest of the brain. In the adult human brain, the rich club is formed by dense, long-distance connections linking cortical hub regions of functional specialization. Rich-club organization is a property common to complex networks across many domains and is hypothesized as a basis for efficient global information transfer in the brain (van den Heuvel et al., 2012). The rich-club network configuration increases metabolic demands compared with surrounding nonrich-club regions (Collin et al., 2014), a high cost that may be offset by the functional benefits of efficient information flow (Bullmore and Sporns, 2012).

Using high resolution DTI in preterm-born and full-term neonates scanned at 32–42 weeks PMA, Zhao et al. (2019) performed graph-theoretic analyses. Faster increases of nodal efficiency were mainly located at the brain hubs distributed in primary sensorimotor regions, superior-middle frontal, and precuneus regions (Fig. 9.3A). The hub regions were then categorized as connector hubs, which occupied high intermodule connections, and provincial hubs, which occupied high intramodule connections. Higher rates of edge strength increases were found in the rich-club and within-module connections (Fig. 9.3B), compared with other connections. The edge strength of

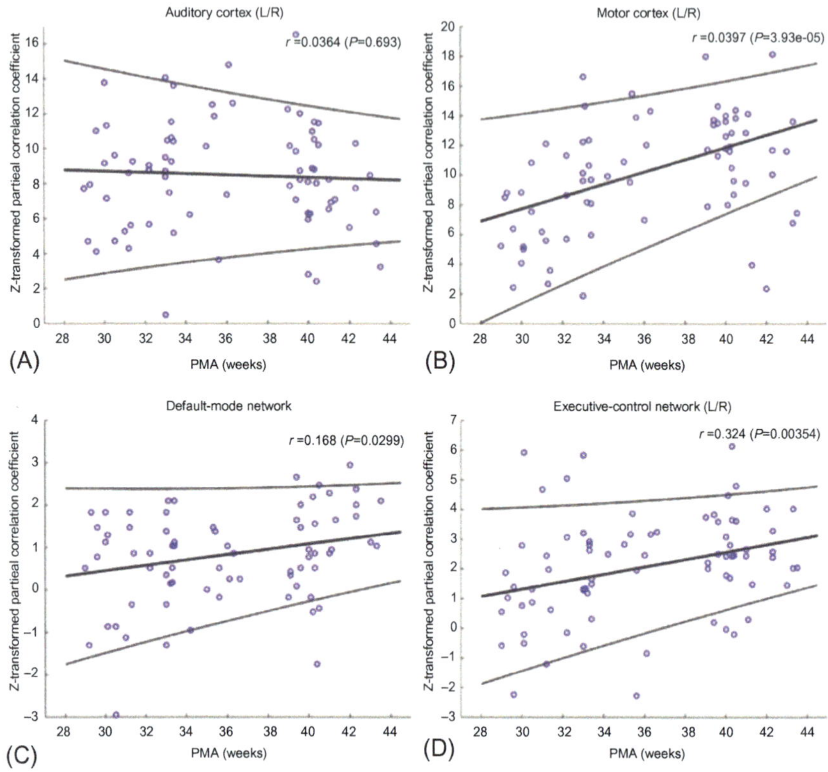

FIG. 9.2 Quantification of developmental changes of connectivity in the different networks. For the auditory (A) and motor (B) the connectivity is expressed as the z-transformed partial correlation coefficient between the right and left cortices. For the complex networks DMX (C) and executive control (D) the connectivity is expressed as the average of the partial correlation coefficients computed between the different pair of nodes within each network and z-transformed. The *line plots* show the robust regression line with prediction error bounds. *From Doria, V., Beckmann, C.F., Arichi, T., Merchant, N., Groppo, M., Turkheimer, F.E., et al., 2010. Emergence of resting state networks in the preterm human brain. Proc. Natl. Acad. Sci. USA 107(46), 20015–20020. Open access.*

short-range connections increased faster than that of long-range connections (Fig. 9.3C). Nodal efficiencies of the hubs predicted individual PMA more accurately than those of nonhubs. These nodal efficiencies jointly underlie network segregation and differentiated emergence of brain functions.

Using rs-fMRI and voxel-based graph theoretical analysis, Cao et al. (2017) investigated the topological organization of whole-brain networks in preterm and full-term babies of 31–42 weeks PMA. They found disproportional increases of functional connectivity strength mainly in the primary sensorimotor, visual, and language regions, but much less in high-order DMN and ECN regions. These largely resulted from enhancement of short-to-middle range connections. While small-world, modular, and

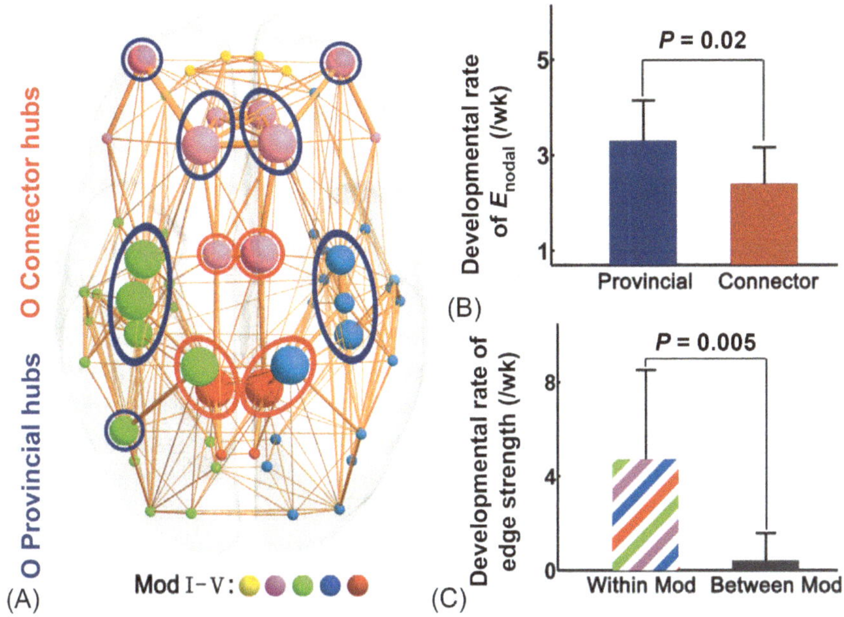

FIG. 9.3 Higher developmental rates of edge strength, of provincial hubs compared with connector hubs, and higher developmental rates of within-module connections compared with between-module connections. (A) A 3D representation of the modular structure of the group-averaged backbone network with nodes in colors corresponding to different modules and size encoding average regional efficiency. Module I was composed of bilateral orbital-frontal regions (*yellow*). Module II was composed of bilateral prefrontal regions and dorsal cingulate gyrus (*purple*). Module III mainly consisted of left pre/postcentral gyrus, temporal, and superior parietal regions (*green*). Module IV mainly consisted of right pre/postcentral gyrus, temporal, and parietal regions (*blue*). Module V consisted of bilateral posterior parietal regions (*red*). The provincial hubs and connector hubs were marked with *blue* and *red circles*, respectively. (B) The bar plot showing higher developmental rate of provincial hubs than that of connector hubs. (C) The bar plot showing higher developmental rate of within-module (Within Mod) connections compared with that of between-module (Between Mod) connections. *Reprinted with permission from Zhao, T., Mishra, V., Jeon, T, Ouyang, M., Peng, Q. Chalak, L, et al., 2019. Structural network maturation of the preterm human brain. NeuroImage 185, 699–710. Copyright 2019, Elsevier.*

rich-club structures were present at 31.3 weeks, local clustering and the shortest path length increased gradually and rich-club structure expanded from 31.3 to 41.7 weeks PMA, resulting in the architectural transformation from a relatively random to more organized network configurations. Cao et al. (2017) found that for all networks, gradually enhanced functional network segregation appeared in the third trimester, which was primarily driven by the rapid increases of functional connectivity of the primary regions. Using fMRI in the fetal brain, Turk et al. (2019) found that key features of the functional connectome are already present in the second and third trimesters of pregnancy.

9.2.1.2 Effects of extrauterine sensory stimulation

Karolis et al. (2016) used structural MRI and diffusion tensor imaging (DTI) data obtained at five developmental phases—preterm and term, infancy (1 year), young and middle childhood, and adolescence—to reconstruct the connectome in very preterm-born at a mean age of 29.26 ± 2.13 years in comparison with an age-matched (term born) group. They assessed its rich-club organization and modularity of the RSNs (Chapters 1 and 6), which reflect the capacity to support global and local information exchange, respectively. Karolis et al. (2016) found that the very-preterm brain exhibited stronger rich-club architecture than the control brain, despite possessing a relative paucity of white-matter resources. Premature exposure to the extrauterine environment, as in preterm babies, is associated with altered network architecture and reduced network capacity, which may in part account for the high prevalence of cognitive problems in preterm infants (Karolis et al., 2016).

In preterm infants between 26 and 42 weeks PMA and a control group of term-born infants, Monson et al. (2018) used DTI to characterize the maturation of primary auditory cortex (A1) and nonprimary auditory cortex (A2). They could distinguish between A1 and A2 as early as 28 weeks PMA, a time at which the sulcal boundaries of Heschl's gyrus are just beginning to appear. Changes in cortical microstructure during maturation had largely finished in A1 by 28 weeks PMA (cf. Fig. 9.2); however, rapid changes were still taking place in A2 between 26 and 42 weeks PMA. At term equivalent PMA, diffusion parameters for auditory cortex were different between preterm infants and term control infants, reflecting either delayed maturation or injury. No effect of acoustic environment in the NICU was observed. For the preterm group, disturbed maturation of A2 (but not A1) was associated with poorer language performance at 2 years of age.

Environmental enrichment with music for preterm newborns might be useful to guide the development of the brain. Using rs-fMRI, Lordier et al. (2019) found that preterm infants exposed to music in the neonatal intensive care units have significantly increased coupling between brain networks that were decreased in premature infants. These are the salience network with the superior frontal, auditory, and sensorimotor networks, and the salience network with the thalamus and precuneus networks. Therefore music exposure leads to functional brain architectures that are more similar to those of full-term newborns, providing evidence for a beneficial effect of music on the preterm brain. This contrasts the findings of Monson et al. (2018) who found no effect of acoustic environment (but did not test for music).

9.2.2 Term birth

Fransson et al. (2011) reported that in the term brain (mean age = 39-week GA) large parts of the auditory cortex were found to be bilaterally functional (cf. Fig. 9.2). He and Parikh (2016) also found strong interhemispheric connectivity in the auditory network from early gestational age on with minimally changes until term. However, at 52 weeks the connections were significantly stronger than at 39 weeks corrected age. This early maturation may explain why topographical functional connectivity exists in the majority of congenitally, prelingually deaf (Striem-Amit et al., 2016).

Most structural changes in the brain occur during the first postnatal year (Dubois et al., 2016a; Fig. 9.4A and B). During early infancy, all major

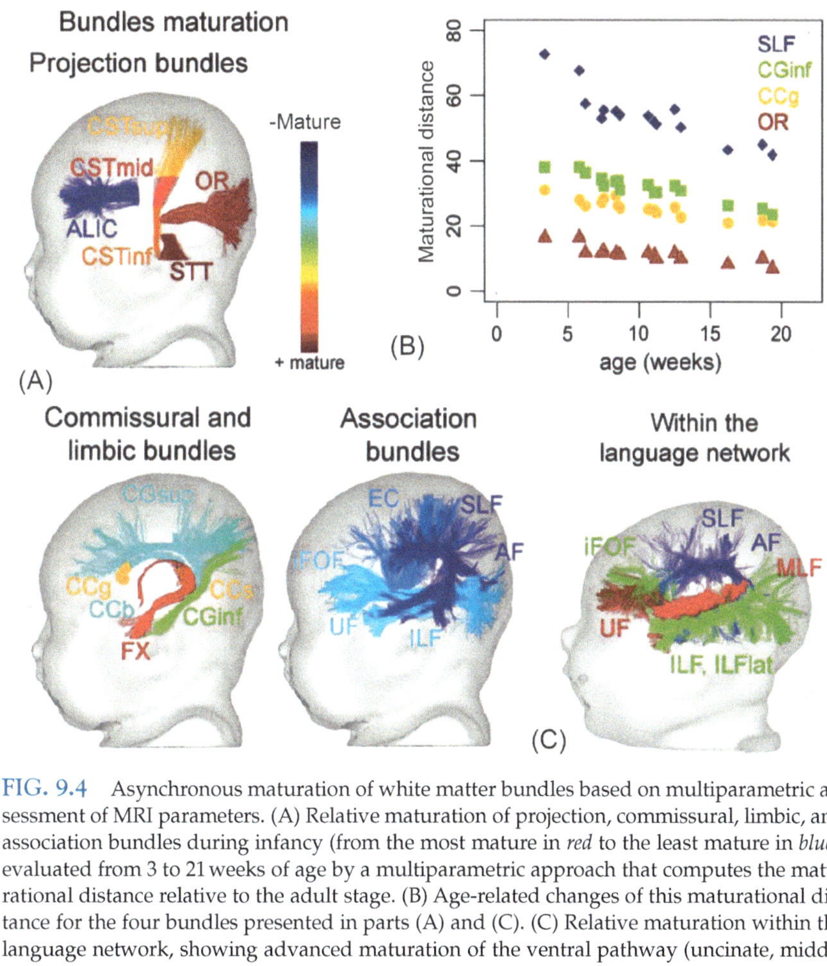

FIG. 9.4 Asynchronous maturation of white matter bundles based on multiparametric assessment of MRI parameters. (A) Relative maturation of projection, commissural, limbic, and association bundles during infancy (from the most mature in *red* to the least mature in *blue*), evaluated from 3 to 21 weeks of age by a multiparametric approach that computes the maturational distance relative to the adult stage. (B) Age-related changes of this maturational distance for the four bundles presented in parts (A) and (C). (C) Relative maturation within the language network, showing advanced maturation of the ventral pathway (uncinate, middle and inferior longitudinal, inferior frontooccipital fascicles; compared with the dorsal pathway (arcuate and superior longitudinal fascicles) during infancy. The color scales of maturation are not comparable in (A) and (C). *Abbreviations:* Projection bundles: *ALIC*, anterior limb of the internal capsule; *CST*, corticospinal tract (inf/mid/sup, inferior/middle/superior portions); *OR*, optic radiations; *STT*, spinothalamic tract. Callosal bundles: *CC*, corpus callosum (g/b/s, genu/body/splenium). Limbic bundles: *CG*, cingulum (inf/sup, inferior/superior parts); *FX*, fornix. Association bundles: *AF*, arcuate fasciculus; *EC*, external capsule; *iFOF*, inferior fronto-occipital fasciculus; *ILF*, inferior longitudinal fasciculus; *SLF*, superior longitudinal fasciculus; *UF*, uncinate fasciculus. *From Dubois, J., Adibpour, P., Poupon, C., Hertz-Pannier, L., Dehaene-Lambertz, G., 2016a. MRI and M/EEG studies of the white matter development in human fetuses and infants: review and opinion. Brain Plast. 2, 49–69. doi:10.3233/BPL-160031. Open access.*

white matter bundles, albeit weakly myelinated, were identified with DTI and tractography (Kulikova et al., 2015; Dubois et al., 2006). These include commissural tracts of the corpus callosum, projection tracts, limbic tracts, and associative tracts. The trajectory and morphology of these bundles remain stable, as revealed by a longitudinal study, between birth and 2 years of age (Geng et al., 2012). Short-range connections can also be visualized in the infant brain by MRI (Dubois et al., 2016b). At the whole-brain level, an increase in global efficiency and integration, and a decrease in segregation are observed in the structural networks during the first two postnatal years. Dubois et al. (2016a) also emphasized that the auditory network is already functional in utero but further development is more protracted throughout infancy and toddlerhood than the visual and somatosensory modalities (cf. Fig. 9.2). In contrast, the visual network develops mainly after birth.

Zhang et al. (2017) used EEG power spectrum analysis and auditory event-related potentials to identify the characteristics of auditory development in normal full-term neonates. They found that delta power—the main brain activity rhythm in the neonatal period—gradually increased from 1 to 10 days of age, peaked at 11 to 20 days, and gradually decreased from 21 to 28 days. In analogy with adult performance (Chapters 3 and 4), Zhang et al. (2017) concluded that: "in the first several days after birth, the EEG delta power manifested the characteristic performance of identifying task information" (Chapter 3).

9.3 Infants

9.3.1 Structural and functional MRI

Using rs-fMRI in normal developing 3-week to 2-year-old infants, Gao et al. (2011) investigated the temporal and spatial evolution of brain network topology. They found that the brain possesses the small-world topology immediately after term birth (as in preterm babies, Section 9.2.1), followed by a remarkable improvement in whole brain wiring efficiency in 1-year olds, which becomes more stable in 2-year olds. Neonates show two apparent clusters of hubs, anterior and posterior clusters, suggesting that major functional developments may have already occurred in both the frontal and occipital lobes (Tóth et al., 2017). In addition, most of the connections between hubs are interhemispheric with no clear connections between the anterior and posterior clusters. In 1-year olds, the hubs are well connected and more anatomically centered with fewer hubs located in the anterior and posterior areas of the brain than seen in neonates. Finally the distribution of hubs becomes more spatially uniform in 2-year olds. For spatial distribution of hubs, clearly there is a major focus on primary functions during early infancy: For motor-related functions, bilateral caudate (associated with voluntary movement control) emerge as hubs in neonates; bilateral supplementary motor areas (SMA) are among the hubs in 1-year olds; For visual functions, bilateral middle occipital gyri appear in neonates; bilateral inferior temporal gyrus and fusiform regions are among the hubs in both 1- and 2-year olds; besides, left middle occipital gyrus also emerges in 2-year olds (Fransson et al., 2011). The bilateral insula areas (cf. Fig. 7.2) are consistently observed as hubs for all three age groups.

Based on a large sample with longitudinal rs-fMRI scans and cognitive data collected during the first 2 years of life, Alcauter et al. (2014) examined the functional relationship between the thalamus and nine cortical functional networks and determined the correlation between the thalamocortical connectivity and cognitive performance at ages 1 and 2 years. Alcauter et al. (2014) found that the thalamus-sensorimotor and thalamus-salience connectivity networks were already present in neonates, whereas the thalamus-medial visual and thalamus-DMN connectivity emerged later, at 1 year of age. More importantly, brain-behavior analyses based on the Mullen Early Learning Composite Score and visual-spatial working memory performance measured at 1 and 2 years of age showed significant correlations with the thalamus-salience network connectivity.

Damaraju et al. (2014) found that the functional network-connectivity map was largely similar at both 4 and 9 months. However, at 9 months the connectivity strength decreased within local networks and increased between more distant networks. The connectivity within the DMN, which contains both local and more distant nodes, also increased in strength with age. The low-frequency (~ 0.02 Hz) power spectrum increases with age only in the posterior DMN and the posterior cingulate cortex (PCC), which is a central node in the DMN. Components of the DMN putatively support self-reflection and social cognition, with the PCC particularly important for emotional processing of faces. The developmental change occurring between 4 and 9 months reflect the strengthening of working memory and object permanence, that is, the ability to maintain a mental representation of an object even when it is not present (references in Damaraju et al., 2014).

9.3.2 Electrophysiology

In a longitudinal study from infancy to early childhood, Marshall et al. (2002) observed a distinct brain rhythm over central sites at 10, 14, 24, and 51 months of age during periods of quiet attention to a visual stimulus. No similar spectral peak was visible at 5 months. The rhythm had a peak frequency of 7–8 Hz in infant, increasing to 9 Hz in early childhood. Saby and Marshall (2012) noted a finding by Lindsley (1938) that visual stimulation reduced the infant posterior low-frequency EEG rhythm amplitude, suggesting a functional similarity to the adult alpha rhythm (Chapter 3). More recent work (Stroganova et al., 1999) has corroborated this suggestion. Thorpe et al. (2016) analyzed event-related desynchronization, that is, power reduction, of the alpha rhythm in 12-month-old infants, 4-year-old children, and adults. They showed that the frequencies at which power spectral density (PSD) peaks occurred across samples were concentrated in two distinct clusters corresponding to the peaks in the mean spectrum. These clusters had mean frequencies of about 4.5 and 7.4 Hz, respectively. For 4-year olds a similar pair of peaks is evident at slightly higher frequencies. For adult subjects, the PSD spectrum did not show the same clear double-peaked structure as 12-month and 4-year subjects. Rather, a broad peak spanning the classic adult alpha range, with a mean peak frequency of ~ 10.5 Hz was present.

Benasich et al. (2008) studied resting-state EEG-gamma in children aged 16, 24, and 36 months. Individual differences in the distribution of frontal gamma power during rest were highly correlated with concurrent language and cognitive skills at all ages. Gamma power was also associated with attention measures; children who were observed as having better inhibitory control and more mature attention shifting abilities had higher gamma power density functions.

9.4 Children

9.4.1 Structural and functional MRI

Fair et al. (2008) used rs-fMRI to characterize the development of the brain's default mode network. As in adults, interhemispheric functional connections between homotopic regions in the contralateral hemisphere were relatively strong in 7- to 9-year-old children; yet, overall the DMN was still only sparsely connected. With age the DMN became significantly more integrated (Fig. 9.5).

What does the continuing development of correlated intrinsic activity within the default network mean? Zhong et al. (2014) examined 6- to 10-year-old children's structural MRI to identify areas with cortical thinning, rs-fMRI to identify functional organization in parallel to cortical development, and working memory/response inhibition tasks to assess executive functioning. They found that neuroanatomical changes in cortical thinning over bilateral frontal, parietal, and occipital regions. These regions were engaged in three functional networks: sensorimotor and auditory, ECN, and DMN. Working memory and response inhibition only associated with regional functional connectivity but not with local and global efficiency of information transfer of these functional networks. Functional connections associated with "bottom-up" as opposed

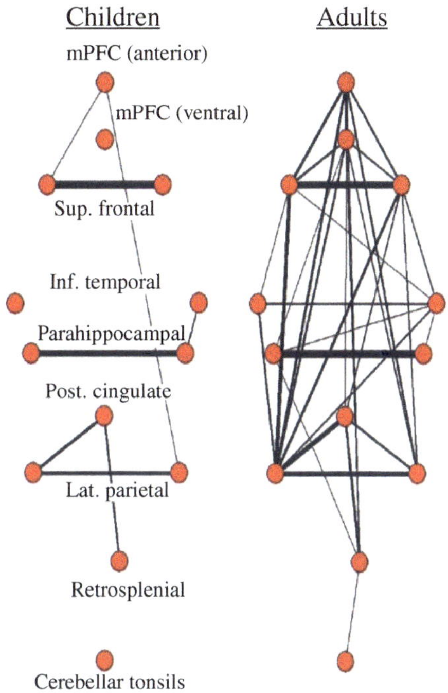

FIG. 9.5 Graph visualization of correlation matrices of the DMN represented in a pseudo-anatomical organization. Default regions are only very sparsely connected in children aged 7–9 years. In adults aged 21–31 years, the default regions are highly integrated (strongly functionally connected). Only connections with $r > 0.15$ are shown (line width is proportional to the connection strength). *From Fair, D.A., Cohen, A.L., Dosenbach, N.U.F., Church, J.A., Miezin, F.M., Barch, D.M., et al., 2008. The maturing architecture of the brain's default network. Proc. Natl. Acad. Sci. USA 105(10), 4028–4032. Open access.*

to "top-down" processing were more clearly related to children's performance on working memory and response inhibition, implying an important role for brain systems involved in late childhood.

Using DTI, Krogsrud et al. (2016) scanned children between 4 and 11 years of age twice; at the first testing the age range was 4.2–9.3 years; at the second testing, it was 5.8–11 years with an intertest interval of 436–819 days. Krogsrud et al. (2016) found a significant linear increase in global fractional anisotropy and decrease in mean and radial diffusivity over time. For mean axial diffusivity, a weak decrease was observed. The developmental changes in specific white matter tracts showed regional differences. Overall, a greater rate of change was found in the left hemisphere. There was a posterior-to-anterior gradient with greater change in frontal regions for all metrics. Sex did not affect change in any DTI metric.

9.4.2 Electrophysiology

Bathelt et al. (2013) applied source reconstruction with age-matched templates to resting-state high-density EEG recordings in typically developing children between 2 and 6 years of age. Graph theory was then used to assess the association strengths of 68 cortical regions of interest. They found that the modules resembled MRI-based hub networks, and that connection strength within functional modules (cf. Fig. 6.2) did not change with age. Network parameters were similar for networks based on different EEG-frequency bands. Miskovic et al. (2015) examined EEG maturation across 1-year windows in children aged between 7 and 11 years. Overall, they found major decreases in absolute EEG spectral density (particularly in the slow wave range) across cortical lobes as a function of age. The peak of the alpha frequency increased with chronological age and there was a redistribution of relative spectral density toward the higher frequency ranges, consistent with earlier findings (Yordanova and Kolev, 1997). There were age differences in long-range functional brain connectivity, particularly in the alpha frequency band, culminating in the densest and spatially variable networks in the oldest children. Miskovic et al. (2015) also found age-related reductions in characteristic path lengths, modularity and homogeneity of alpha-band cortical networks from early to late childhood. Meyer et al. (2019) investigated whether EEG theta-power in 4-year-olds is sensitive to task engagement and to different cognitive task demands. In a within-subject design, children had three different instructions before watching videos including either no demands (No Task), language-related (Color-naming Task), or action-related (Imitation Task) demands. They found more frontomedial theta-power when children were engaged in a task, which increased during their cognitive engagement. Theta-power was stronger over left frontotemporal sites for language-related demands compared with action-related demands.

9.5 Adolescence

Adolescence is a period that coincides with heightened vulnerability to mental illness and behavioral problems that may correlate with distinct changes in brain networks.

9.5.1 Structural and functional MRI

Using longitudinal DTI studies, Baker et al. (2015) characterized changes in connectivity between 80 cortical and subcortical anatomical regions over a 2-year period in adolescents between the ages of 15 and 19 years. Only a small subset of connections showed significant developmental changes over the study period, with 8% of connections demonstrating decreased (subcortical hubs) and 6% increased (frontal-subcortical and frontal-parietal hub-hub) structural connectivity. This suggests that late adolescence is characterized by selective, yet significant remodeling of hub-hub connectivity, with the topological organization of hubs shifting emphasis from subcortical (including thalamus, putamen, globus pallidus, and hippocampus) hubs in favor of an increasingly prominent role for frontal hub regions. Andre et al. (2016) conducted a systematic review and coordinate based metaanalyses of all the available primary fMRI studies that mapped white matter function in healthy adolescents (10–17years) and young adults (18–30years). They found that activation during adolescence increased with age principally in higher-order cortices, part of the core working memory network, while reductions were detected in more diffuse and potentially more immature neural networks. In a structural MRI study, Narvacan et al. (2017) reported a within-subject analysis in which 65%–80% of 13–17year olds underwent a longitudinal decrease in volume between scans (~ 4years apart) for the putamen, globus pallidus, and hippocampus, suggesting unique developmental processes during adolescence. Zhong et al. (2017) used diffusion and structural MRI to construct hemispheric white matter networks, and applied graph-theory to quantify topological parameters of the hemispheric networks. In both adolescents and young adults, rightward asymmetry in both global and local network efficiencies was consistently observed between the two hemispheres. At the nodal level, the young adults exhibited less rightward asymmetry of nodal efficiency mainly around the parasylvian area, posterior tempoparietal cortex, and fusiform gyrus.

9.5.2 Histology and electrophysiology

Huttenlocher and Dabholkar (1997) reported a dramatic change in the mean synaptic density of auditory cortex between 3.5years and 12years of age. On average, the synaptic density at 12years is about 60% of that at 3.5years. In a review, Blakemore (2012) noted that in a study in postmortem human brains aged 1week to 91years, "Petanjek et al. (2011) demonstrated that prefrontal dendritic spine density increases in childhood, resulting in numbers that exceed adult levels two- or threefold by puberty, and then decreases gradually after puberty. The elimination of synaptic spines continued beyond adolescence throughout the third decade of life, providing evidence for astonishingly protracted dendritic reorganization in the human prefrontal cortex."

Tierney et al. (2013) defined the developmental trajectory of resting gamma throughout childhood and into adulthood by collecting EEG data from subjects aged 3–38years while they were awake but not engaged in a task. Subjects were divided among preschoolers (3–5years old), school-aged children (6–13years), adolescents (14–17years), and adults (18–38years). Older subjects showed significantly less resting gamma than younger subjects, in which the distribution of gamma power across EEG electrodes was strikingly similar across the age groups, suggesting that topography of resting gamma power did not vary with age. Cho et al. (2015) studied a large sample aged 8–22 years with EEG recording of the auditory steady-state response (ASSR) for 20-, 30-, and 40-Hz click trains. Both 40-Hz-evoked power and phase-locking factor (PLF) increased monotonically from 8 to 16 years, and subsequently decreased toward ages 20–22 years. Cross-frequency coupling (CFC) followed a similar pattern, with strongest age-related modulation of 40-Hz amplitude by the phase of delta oscillations (Fig. 9.6; Chapter 3). In contrast, the evoked power, PLF and CFC for the 20- and 30-Hz stimulation were distinct from the 40-Hz condition, with flat or decreasing profiles from childhood to early adulthood. The inverted U-shaped developmental trajectory of gamma oscillations may be consistent with interacting maturational processes—such as increasing fast GABA inhibition that enhances gamma activity and synaptic pruning that decreases gamma activity—that may continue from childhood through to adulthood.

Rodrıíguez-Martínez et al. (2015) examined a possible co-maturation pattern between different frequency ranges on 3 min of resting-state EEG during development from 6 to 26 years old. They found a power spectral density decrease in all frequency-band ranges between 1 and 46 Hz. Furthermore, there was a high correlation during development between theta and beta ranges, and between high-beta and gamma frequency ranges. The correlation of theta and beta was particularly high between the anterior and posterior electrode sites. The same group (Gómez et al.,

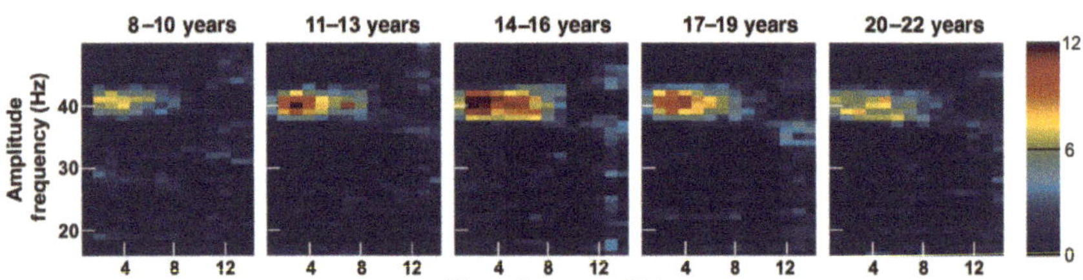

FIG. 9.6 Age-related changes in phase-amplitude cross-frequency coupling (CFC) in response to 40-Hz click trains. Shown are t-value maps of CFC compared with prestimulus baseline. The x-axis shows the lower-modulating frequency used for phase estimation whereas the y-axis shows the higher-modulated frequency used for amplitude estimation. The amplitude of the 40-Hz response is coupled to the phase of frequencies between 2 and 6 Hz. The frequency specificity of coupling appears consistent across development but the strength of coupling is greatest at 14–16 years. *From Cho, R.Y., Walker, C.P., Polizzotto, N.R., Wozny, T.A., Fissell, C., Chen C-M. A., et al., 2015. Development of sensory gamma oscillations and cross-frequency coupling from childhood to early adulthood. Cereb. Cortex 25, 1509–1518. By permission of Oxford University Press.*

2017) recorded spontaneous MEG activity in three age groups: children/adolescents (7–14 years), young adults (17–20 years), and adults (21–26 years). Results showed not only a broadband PSD decrease in frontal areas, which suggests the late maturation of this region, but also a mild increase in high frequency PSD with age in posterior areas. These findings suggest that the intensity of the neural sources during spontaneous brain activity decreases with age, which may be related to synaptic pruning.

9.6 General lifespan

Changes in the connectome in older age are extensively discussed in Eggermont (2019). Here, I review additional structural studies and especially changes in brain rhythms across the lifespan.

9.6.1 Structural and functional MRI

Narvacan et al. (2017) examined T1-weighted images of three healthy cohorts: (1) a cross-sectional cohort aged 5–83 years; (2) a longitudinal neurodevelopment cohort scanned twice approximately 4 years apart, aged 5–27 years at first scan; and (3) a longitudinal aging cohort scanned twice approximately 3 years apart, aged 46–83 years at first scan. Cross-sectional cohort trajectories are shown in Fig. 9.7 and reflect the overall decrease in gray matter, the initial increase in change of white matter volume peaking at 30–49 years of age and followed by a gradual decrease with advancing age. Overall brain volume peaks in adolescence.

Zhao et al. (2015) examined a cohort of healthy individuals (ages 9–85 years) with both diffusion and structural MRI. For each participant, the high-resolution white matter structural networks were constructed by tractography that were quantified with graph theoretical analyses. The global network properties followed an inverted U-shaped trajectory with a peak age around the third decade. The brain areas with the most significantly nonlinear changes were located in the prefrontal and temporal cortices. The posterior cingulate and lateral temporal cortices displayed prolonged maturation/degeneration compared with the prefrontal cortices. Rich-club organization was evident across the lifespan, whereas hub integration decreased linearly with age, especially accompanied by the loss of frontal hubs and their connections (Fig. 9.8). As shown, for the 9–20 years and the 30–40 years. Groups, the hub regions were mainly distributed in the medial frontal, parietal, and occipital cortices. In contrast, reduced frontal regions and their connections in the rich-club component were observed for the elderly group (> 70 years old) compared with the younger groups.

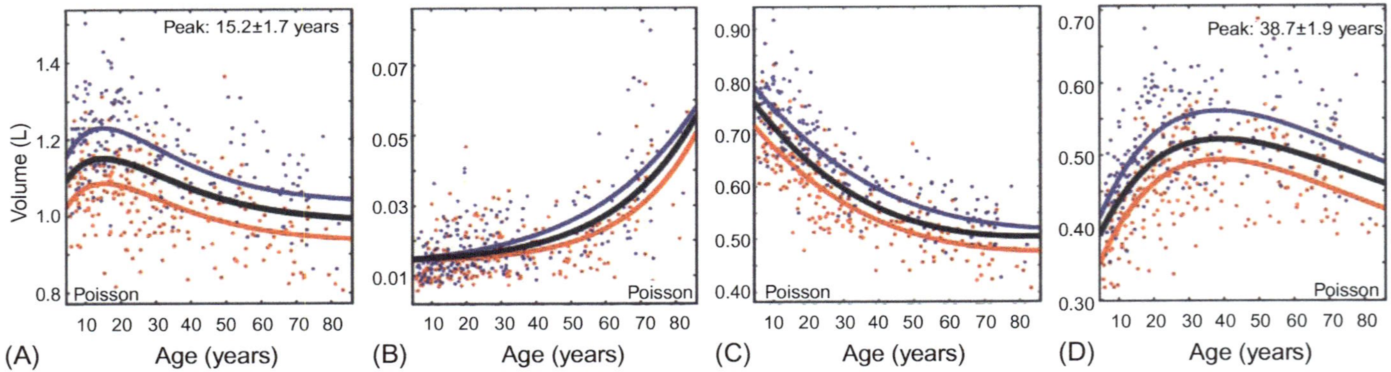

FIG. 9.7 Cross-sectional cohort trajectories. (A) Total brain volume, (B) ventricles, (C) cortical gray matter, and (D) white matter volume change with age ($n = 406$). Males (*blue lines*) on average show greater volumes over all ages than females (*red lines*), but both sexes show very similar trajectories, with the exception of the ventricles with an age-by-sex interaction indicating earlier increases of volume in male than females. *From Narvacan, K., Treit, S., Carnicioli, R., Martin, W., Beaulieu, C., 2017. Evolution of deep gray matter volume across the human lifespan. Hum. Brain Mapp. 38, 3771–3790. Copyright 2017 Wiley Periodicals, Inc.*

Petrican et al. (2017) used rs-fMRI and behavioral data from a large lifespan sample and tested the interactions among 10 intrinsic neural systems. Their results revealed three distinguishable profiles, which characterized developmental differences in connectivity within the 10 systems, between networks thought to underlie cognitive control and noncontrol systems, and among the noncontrol networks (Fig. 9.9). The within-network connectivity profile (along the main diagonal, *dashed*) showed decreased connectivity within two external processing networks (auditory/ language, AUD, and ventral attention, VAN) and increase for the visual system (VIS). The noncontrol-to-noncontrol connectivity profile featured increased separation between networks involved in external processing, including language (DAN and AUD) and those linked to internally generated cognitions and category learning (DMN, subcortical, and SUB).

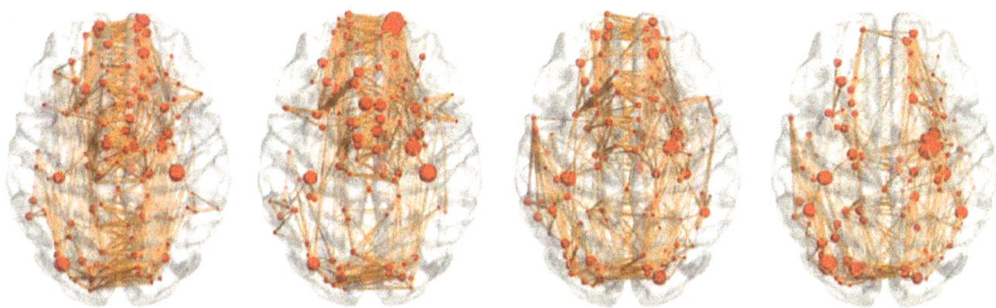

FIG. 9.8 Rich-club organization of different age subgroups. The *red nodes* represent the hub regions with mean *z*-scores of the nodal strength as nodal size. The *orange lines* represent the WM connections between huh regions identified from the average matrix of different subgroups. *From Zhao, T., Cao, M., Niu, H., Zuo, X-N., Evans, A., He, Y., et al., 2015. Age-related changes in the topological organization of the white matter structural connectome across the human lifespan. Hum. Brain Mapp. 36, 3777–3792. Copyright 2015 Wiley Periodicals, Inc.*

	AUD	VIS	SAL	CON	SM	FPC	SUB	VAN	DMN	DAN
AUD	⇓		⇑	⇑			⇓			⇑
VIS		⇑	⇑							
SAL	⇑	⇑		⇑	⇑			⇑		
CON	⇑		⇑			⇑		⇑		
SM			⇑							
FPC				⇑			⇑			
SUB	⇓					⇑			⇑	
VAN			⇑	⇑				⇓		
DMN							⇑			⇓
DAN	⇑								⇓	

FIG. 9.9 Connectivity changes over the lifespan. *AUD*, Auditory; *CON*, cinguloopercular; *DAN*, dorsal attention; *DMN*, default mode; *FPC*, frontoparietal; *SAL*, salience; *SM*, somatomotor; *SUB*, subcortical; *VAN*, ventral attention; *VIS*, visual.

Finally, the third connectivity profile was characterized by increased coupling of the three control networks—frontoparietal (FPC), salience (SAL), cinguloopercular (CON)—with one another and with the remaining systems, particularly the SUB and AUD and VAN showing declining segregation with age. All three profiles showed significant associations with behavior during young adulthood, and degraded during late middle age and older adulthood. Nonetheless, significant lifespan reductions in segregation, that is, increased connectivity, were noticed only in two systems linked to processing external information, specifically AUD, which is relevant to speech perception and higher-order language processes and the VAN, which is relevant to bottom-up processing of task-related distracters and unexpected task event (Fig. 9.9). These findings are consistent with previous reports on the declining functional relevance of bottom-up processes in cognition from childhood onwards (Zhong et al., 2014), while they also provide greater specificity with regards to the systems most vulnerable to reduced segregation across the lifespan.

Vij et al. (2018) assessed age-related changes in spatial and temporal features of independent components in a cross-sectional lifespan sample (ages 6–85 years). While large-scale network configurations remained consistent throughout the lifespan, changes persisted in both local and global organization of these networks. They showed that the spatial extent of the majority of functional networks exhibited linear decreases (auditory, visual, and SAL) and both positive and negative quadratic trajectories (DMN, ECN, basal ganglia, and sensorimotor network) across the lifespan. Kundu et al. (2018) used multiecho (ME)-fMRI to characterize functional brain changes with age. From resting-state ME-fMRI in persons between 8.3 and 46.2 years of age, they found a highly significant exponential decrease in the number of BOLD components with age, particularly dorsolateral prefrontal cortices, parietal cortex, and cerebellum. Note that: Historically, the cerebellum has been viewed mostly as a motor control region, but more recently there is increased recognition of its role in the regulation of autonomic function and cognition (Singer et al., 2004) and in affective processing (Schmahmann and Caplan, 2006). Note also the emerging connectivity of the cerebellum with the DMN with age (Fig. 9.5).

9.6.2 Electrophysiology

Babiloni et al. (2006) analyzed the correlation between cortical EEG rhythms and age. Resting-state eyes-closed EEG data were recorded in young (range: 18–50 years) and old (range: 51–85 years) subjects. Delta sources in the occipital area had significantly less magnitude in old compared with young subjects. Similarly, alpha1 (8–10.5 Hz) and alpha2

(10.5–13 Hz) sources in the parietal, occipital, temporal, and limbic areas had significantly less magnitude in old compared with young subjects. They found a significant negative linear correlation of age with the magnitude of delta sources in the occipital area and of alpha1 sources in occipital and limbic areas. In addition, there was a negative linear correlation with age for alpha2 sources in the parietal, occipital, temporal, and limbic areas. Caplan et al. (2015) recorded EEG during rest, alternating eyes open and closed, in healthy younger (18–25 years) and older (60–74 years) participants. In contrast to Babiloni et al. (2006), they did not observe a decrease of low-frequency oscillations in healthy aging. Rather, low-frequency rhythmic activity, in both the delta and alpha bands, was present during rest and remained stable between the two healthy age groups. The clearest age effect was increased oscillations in the beta band in healthy older compared with healthy young participants. Knyazeva et al., 2018 found a multicomponent structure of alpha in 78% of 20- to 81-year-old healthy adults. The spatial distribution of the alpha2 occipitoparietal component broadened with age, with its maximum moving from secondary/associative visual cortex to fusiform gyrus. The alpha1 component originating from the occipitotemporal regions in young adults also moves anteriorly with age, while maintaining its maximum within the fusiform gyrus. Both components slowed down by 1 Hz over the adult lifespan. The multicomponent alpha rhythm appears more common in younger subjects and is transformed into a single-component alpha rhythm is featured in older subjects.

9.7 Summary

Rich-club organization is evident across the lifespan, whereas hub integration decreased linearly with age, especially accompanied by the loss of frontal hubs and their connections. The very preterm brain exhibited stronger rich-club architecture than the control brain, despite possessing a relative paucity of white-matter resources. Both resting-state and task-related gamma oscillations emerge during early childhood, and precise temporal coordination through neural synchrony continues to mature until early adulthood. The DMN structure in children deviates significantly from that in the adult. Over age, however, the default network becomes significantly more integrated. Connection strength within functional modules did not change with age, and the modules resembled hub networks previously described for MRI. Late adolescence is characterized by selective, yet significant remodeling of hub-hub connectivity, with the topological organization of hubs shifting emphasis from subcortical hubs in favor of an increasingly prominent role for frontal hub regions.

References

Alcauter, S., Lin, W., Smith, J.K., Short, S.J., Goldman, B.D., Reznick, J.S., et al., 2014. Development of thalamocortical connectivity during infancy and its cognitive correlations. J. Neurosci. 34 (27), 9067–9075.

Andre, J., Picchioni, M., Zhang, R., Toulopoulou, T., 2016. Working memory circuit as a function of increasing age in healthy adolescence: a systematic review and meta-analyses. NeuroImage Clin. 12, 940–948.

Babiloni, C., Binetti, G., Cassarino, A., Dal Forno, G., Del Percio, C., Ferreri, F., et al., 2006. Sources of cortical rhythms in adults during physiological aging: a multicentric EEG study. Hum. Brain Mapp. 27, 162–172.

Baker, S.T.E., Lubman, D.I., Yücel, M., Allen, N.B., Whittle, S., Fulcher, B.D., et al., 2015. Developmental changes in brain network hub connectivity in late adolescence. J. Neurosci. 35 (24), 9078–9087.

Ball, G., Aljabar, P., Zebari, S., Tusor, N., Arichi, T., Merchant, N., et al., 2014. Rich-club organization of the newborn human brain. Proc. Natl. Acad. Sci. USA 111 (20), 7456–7461.

Bathelt, J., O'Reilly, H., Clayden, J.D., Cross, J.H., de Haan, M., 2013. Functional brain network organisation of children between 2 and 5 years derived from reconstructed activity of cortical sources of high-density EEG recordings. NeuroImage 82, 595–604.

Benasich, A.A., Gou, Z., Choudhury, N., Harris, K.O., 2008. Early cognitive and language skills are linked to resting frontal gamma power across the first 3 years. Behav. Brain Res. 195, 215–222.

Blakemore, S.-J., 2012. Imaging brain development: the adolescent brain. NeuroImage 61, 397–406.

Brody, B.A., Kinney, H.C., Kloman, A.S., Gilles, F.H., 1987. Sequence of central nervous system myelination in human infancy. I. An autopsy study of myelination. J. Neuropathol. Exp. Neurol. 46, 283–301.

Bullmore, E., Sporns, O., 2012. The economy of brain network organization. Nat. Rev. Neurosci. 13 (5), 336–349.

Cao, M., He, Y., Dai, Z., Liao, X., Jeon, T., Ouyang, M., et al., 2017. Early development of functional network segregation revealed by connectomic analysis of the preterm human brain. Cereb. Cortex 27, 1949–1963.

Caplan, J.B., Bottomley, M., Kang, P., Dixon, R.A., 2015. Distinguishing rhythmic from non-rhythmic brain activity during rest in healthy neurocognitive aging. NeuroImage 112, 341–352.

Cho, R.Y., Walker, C.P., Polizzotto, N.R., Wozny, T.A., Fissell, C., Chen, C.-M.A., et al., 2015. Development of sensory gamma oscillations and cross-frequency coupling from childhood to early adulthood. Cereb. Cortex 25, 1509–1518.

Collin, G., Sporns, O., Mandl, R.C.W., van den Heuvel, M.P., 2014. Structural and functional aspects relating to cost and benefit of rich club organization in the human cerebral cortex. Cereb. Cortex 24 (9), 2258–2267.

Damaraju, E., Caprihan, A., Lowe, J.R., Allen, E.A., Calhoun, V.D., Phillips, J.P., 2014. Functional connectivity in the developing brain: a longitudinal study from 4 to 9 months of age. NeuroImage 84, 169–180.

Doria, V., Beckmann, C.F., Arichi, T., Merchant, N., Groppo, M., Turkheimer, F.E., et al., 2010. Emergence of resting state networks in the preterm human brain. Proc. Natl. Acad. Sci. USA 107 (46), 20015–20020.

Dubois, J., Hertz-Pannier, L., Dehaene-Lambertz, G., Cointepas, Y., Le Bihan, D., 2006. Assessment of the early organization and maturation of infants' cerebral white matter fiber bundles: a feasibility study using quantitative diffusion tensor imaging and tractography. NeuroImage 30 (4), 1121–1132.

Dubois, J., Adibpour, P., Poupon, C., Hertz-Pannier, L., Dehaene-Lambertz, G., 2016a. MRI and M/EEG studies of the white matter development in human fetuses and infants: review and opinion. Brain Plast. 2, 49–69. https://doi.org/10.3233/BPL-160031.

Dubois, J., Poupon, C., Thirion, B., Simonnet, H., Kulikova, S., Leroy, F., et al., 2016b. Exploring the early organization and maturation of linguistic pathways in the human infant brain. Cereb. Cortex 26, 2283–2298.

Eggermont, J.J., 2019. The Auditory Brain and Age-Related Hearing Impairment. Academic Press, London, UK.

Fair, D.A., Cohen, A.L., Dosenbach, N.U.F., Church, J.A., Miezin, F.M., Barch, D.M., et al., 2008. The maturing architecture of the brain's default network. Proc. Natl. Acad. Sci. USA 105 (10), 4028–4032.

Fair, D.A., Cohen, A.L., Power, J.D., Dosenbach, N.U.F., Church, J.A., Miezin, F.M., et al., 2009. Functional brain networks develop from a "local to distributed" organization. PLoS Comput. Biol. 5 (5), e1000381. https://doi.org/10.1371/journal.pcbi.1000381.

Fransson, P., Åden, U., Blennow, M., Lagercrantz, H., 2011. The functional architecture of the infant brain as revealed by resting-state fMRI. Cereb. Cortex 21, 145–154.

Gao, W., Gilmore, J.H., Giovanello, K.S., Smith, J.K., Shen, D., Zhu, H., 2011. Temporal and spatial evolution of brain network topology during the first two years of life. PLoS One 6 (9), e25278. https://doi.org/10.1371/journal.pone.0025278.

Geng, X., Gouttard, S., Sharma, A., Gu, H., Styner, M., Lin, W., et al., 2012. Quantitative tract-based white matter development from birth to age 2 years. NeuroImage 61 (3), 542–557.

Gómez, C.M., Rodríguez-Martínez, E.I., Fernández, A., Maestú, F., Poza, J., Gómez, C., 2017. Absolute power spectral density changes in the magnetoencephalographic activity during the transition from childhood to adulthood. Brain Topogr. 30, 87–97.

He, L., Parikh, N.A., 2016. Brain functional network connectivity development in very preterm infants: the first six months. Early Hum. Dev. 98, 29–35.

Honey, C., Kotter, R., Breakspear, M., Sporns, O., 2007. Network structure of cerebral cortex shapes functional connectivity on multiple time scales. Proc. Natl. Acad. Sci. USA 104, 10240–10245.

Huttenlocher, P.R., Dabholkar, A.S., 1997. Regional differences in synaptogenesis in human cerebral cortex. J. Comp. Neurol. 387, 167–178.

Karolis, V.R., Froudist-Walsh, S., Brittain, P.J., Kroll, J., Ball, G., Edwards, A.D., et al., 2016. Reinforcement of the brain's rich-club architecture following early neurodevelopmental disruption caused by very preterm birth. Cereb. Cortex 26, 1322–1335.

Knyazeva, M.G., Barzegaran, E., Vildavski, V.Y., Demonet, J.-F., 2018. Aging of human alpha rhythm. Neurobiol. Aging 69, 261–273.

Krogsrud, S.K., Fjell, A.M., Tamnes, C.K., Grydeland, H., Mork, L., Due-Tønnessen, P., et al., 2016. Changes in white matter microstructure in the developing brain—a longitudinal diffusion tensor imaging study of children from 4 to 11 years of age. NeuroImage 124, 473–486.

Kulikova, S., Hertz-Pannier, L., Dehaene-Lambertz, G., Buzmakov, A., Poupon, C., Dubois, J., 2015. Multi-parametric evaluation of the white matter maturation. Brain Struct. Funct. 220 (6), 3657–3672.

Kundu, P., Benson, B.E., Rosen, D., Frangou, S., Leibenluft, E., Luh, W.-M., et al., 2018. The integration of functional brain activity from adolescence to adulthood. J. Neurosci. 38 (14), 3559–3570.

Lindsley, D.B., 1938. Electrical potentials of the brain in children and adults. J. Genet. Psychol. 19, 285–306.

Lordier, L., Meskaldji, D.-E., Grouiller, F., Pittet, M.P., Vollenweider, A., Vasung, L., et al., 2019. Music in premature infants enhances high-level cognitive brain networks. Proc. Natl. Acad. Sci. USA 16 (24), 12103–12108.

Marshall, P.J., Bar-Haim, Y., Fox, N.A., 2002. Development of the EEG from 5 months to 4 years of age. Clin. Neurophysiol. 113, 1199–1208.

Meyer, M., Endedijk, H.M., van Ede, F., Hunnius, S., 2019. Theta oscillations in 4-year-olds are sensitive to task engagement and task demands. Sci. Rep. 9, 6049. https://doi.org/10.1038/s41598-019-42615-x.

Miskovic, V., Ma, X., Chou, C.-A., Fan, M., Owens, M., Sayama, H., et al., 2015. Developmental changes in spontaneous electrocortical activity and network organization from early to late childhood. NeuroImage 118, 237–247.

Monson, B.B., Eaton-Rosen, Z., Kapur, K., Liebenthal, E., Brownell, A., Smyser, C.D., et al., 2018. Differential rates of perinatal maturation of human primary and nonprimary auditory cortex. eNeuro 5 (1). https://doi.org/10.1523/ENEURO.0380-17.2017.

Narvacan, K., Treit, S., Camicioli, R., Martin, W., Beaulieu, C., 2017. Evolution of deep gray matter volume across the human lifespan. Hum. Brain Mapp. 38, 3771–3790.

Ouyang, M., Kang, H., Detre, J.A., Roberts, T.P.L., Huang, H., 2017. Short-range connections in the developmental connectome during typical and atypical brain maturation. Neurosci. Biobehav. Rev. 83, 109–122.

Paus, T., Collins, D.L., Evans, A.C., Leonard, G., Pike, B., Zijdenbos, A., 2001. Maturation of white matter in the human brain: a review of magnetic resonance studies. Brain Res. Bull. 54, 255–266.

Petanjek, Z., Judaš, M., Šimic, G., Rasin, M.R., Uylings, H.B., Rakic, P., et al., 2011. Extraordinary neoteny of synaptic spines in the human prefrontal cortex. Proc. Natl. Acad. Sci. USA 108 (32), 13281–13286.

Petrican, R., Taylor, M.J., Grady, C.L., 2017. Trajectories of brain system maturation from childhood to older adulthood: implications for lifespan cognitive functioning. NeuroImage 163, 125–149.

Power, J.D., Fair, D.A., Schlaggar, B.L., Petersen, S.E., 2010. The development of human functional brain networks. Neuron 67, 735–748.

Rodríguez-Martínez, E.I., Barriga-Paulino, C.I., Rojas-Benjumea, M.A., Gómez, C.M., 2015. Co-maturation of theta and low-beta rhythms during child development. Brain Topogr. 28, 250–260.

Saby, J.N., Marshall, P.J., 2012. The utility of EEG band power analysis in the study of infancy and early childhood. Dev. Neuropsychol. 37 (3), 253–273.

Schmahmann, J.D., Caplan, D., 2006. Cognition, emotion and the cerebellum. Brain 129, 290–292.

Singer, T., Seymour, B., O'Doherty, J., Kaube, H., Dolan, R.J., Frith, C.D., 2004. Empathy for pain involves the affective but not sensory components of pain. Science 303, 1157–1162.

Striem-Amit, E., Almeida, J., Belledonne, M., Chen, Q., Fang, Y., Han, Z., et al., 2016. Topographical functional connectivity patterns exist in the congenitally, prelingually deaf. Sci. Rep. 6, 29375. https://doi.org/10.1038/srep29375.

Stroganova, T.A., Orekhova, E.V., Posikera, I.N., 1999. EEG alpha rhythm in infants. Clin. Neurophysiol. 110, 997–1012.

Thorpe, S.G., Cannon, E.N., Fox, N.A., 2016. Spectral and source structural development of mu and alpha rhythms from infancy through adulthood. Clin. Neurophysiol. 127, 254–269.

Tierney, A., Strait, D.L., O'Connell, S., Kraus, N., 2013. Developmental changes in resting gamma power from age three to adulthood. Clin. Neurophysiol. 124, 1037–1042.

Tóth, B., Urbán, G., Háden, G.P., Márk, M., Török, M., Stam, C.J., et al., 2017. Large-scale network organization of EEG functional connectivity in newborn infants. Hum. Brain Mapp. 38, 4019–4033.

Turk, E., van den Heuvel, M.I., Benders, M.J., de Heus, R., Franx, A., Manning, J.H., et al., 2019. Functional connectome of the fetal brain. J. Neurosci. 39 (49), 9716–9724.

van den Heuvel, M.P., Kahn, R.S., Goñi, J., Sporns, O., 2012. High-cost, high-capacity backbone for global brain communication. Proc. Natl. Acad. Sci. USA 109 (28), 11372–11377.

Vij, S.G., Nomi, J.S., Dajani, D.R., Uddin, L.Q., 2018. Evolution of spatial and temporal features of functional brain networks across the lifespan. NeuroImage 173, 498–508.

Yordanova, J., Kolev, V., 1997. Alpha response system in children: changes with age. Int. J. Psychophysiol. 26, 411–430.

Zhang, Q., Cheng, Q., Li, H., Dong, X., Tu, W., 2017. Evaluation of auditory perception development in neonates by quantitative electroencephalography and auditory event-related potentials. PLoS One 12 (9), e0183728. https://doi.org/10.1371/journal.pone.0183728.

Zhao, T., Cao, M., Niu, H., Zuo, X.-N., Evans, A., He, Y., et al., 2015. Age-related changes in the topological organization of the white matter structural connectome across the human lifespan. Hum. Brain Mapp. 36, 3777–3792.

Zhao, T., Mishra, V., Jeon, T., Ouyang, M., Peng, Q., Chalak, L., et al., 2019. Structural network maturation of the preterm human brain. NeuroImage 185, 699–710.

Zhong, J., Rifkin-Graboi, A., Ta, A.T., Yap, K.L., Chuang, K., Meaney, M.J., et al., 2014. Functional networks in parallel with cortical development associate with executive functions in children. Cereb. Cortex 24, 1937–1947.

Zhong, S., He, Y., Shu, H., Gong, G., 2017. Developmental changes in topological asymmetry between hemispheric brain white matter networks from adolescence to young adulthood. Cereb. Cortex 27, 2560–2570.

CHAPTER

10

Brain rhythms and connectivity changes in tinnitus patients

Tinnitus is the conscious percept of sound heard in the absence of physical sound sources external or internal to the body, and reflected in aberrant neural synchrony of spontaneous or resting-state brain activity. About 20% of tinnitus occurrence has been conclusively attributed to noise-induced hearing loss (Eggermont, 2012, 2016). Tinnitus is an enigma; we do not really know what causes it; we know the necessary bottom-up conditions but not the sufficient top-down ones, and we have no idea how to cure it. What we do now is the plethora of potential etiologies; as a consequence, tinnitus is not a disease but a symptom. The tinnitus percept can be located in either one or both ears or in the head. One of the proposed neural substrates is increased spontaneous firing rate (SFR) or increased burst firing in central auditory structures, another independent one is increased neural firing synchrony.

There is evidence from animal studies that initially the cause may be increased neural synchrony (Noreña and Eggermont, 2003) and central gain resulting for example, from deafferentation by hearing loss, which amplifies residual spontaneous firings in auditory nerve fibers and brainstem neurons. However, after about 2 months post noise trauma, midbrain auditory structures show intrinsic increased SFR, independent of cochlear afferent input (Robertson et al., 2013). Thus there are differences in the electrophysiology of recent-onset and long-existing tinnitus.

Qu et al. (2019) studied noise-induced temporary threshold shift in mice with behavioral signs of tinnitus—gap-prepulse inhibition of the acoustic startle reflex. The mice showed reduced auditory brainstem response (ABR) wave I amplitude, and an enhanced wave IV amplitude and wave IV/I amplitude ratio, as compared with control and nontinnitus mice (cf. Kujawa and Liberman, 2009). Resting-state functional magnetic resonance imaging (rs-fMRI) was used to identify abnormal connectivities and to determine the relationship with tinnitus characteristics. They found increased amplitude of low-frequency (~0.1 Hz) fluctuation (ALFF) in the

https://doi.org/10.1016/B978-0-12-819818-6.00010-8

auditory cortex and decreased ALFF in the amygdala and hippocampus at day 1 postexposure. However, at day 28 after noise exposure in tinnitus mice, there was decreased ALFF in the auditory cortex and increased ALFF in the amygdala. This suggests maladaptive functional connectivity between the central auditory pathway and limbic structures, such as the amygdala and hippocampus (Qu et al., 2019).

In humans, the described changes in the ABR, however, are not consistently present in people with tinnitus and clinical normal hearing (Guest et al., 2017). Furthermore, tinnitus in humans is typically not an isolated symptom but has comorbidities in hearing loss, hyperacusis, distress, or depression, to name a few. Each of those has been shown to affect brain rhythms, networks, and their connectivity, few studies agree on changes caused by tinnitus in isolation. Structural and functional MRI, electroencephalography (EEG) and magnetoencephalography (MEG) have been used to separate tinnitus effects from comorbid influences on oscillations and networks in the brain.

10.1 Structural and functional MRI networks

10.1.1 Structural changes

To separate anatomical substrates of tinnitus from the effects of comorbid hearing loss, Husain et al. (2011) used MRI and voxel-based morphometry (VBM) to identify cortical gray matter changes, and diffusion tensor imaging (DTI) to delineate changes in white matter tracts. The VBM data showed that hearing loss individuals without tinnitus had gray matter (GM) decreases in the anterior cingulate cortex (ACC) and in the superior and medial frontal gyri compared with those with hearing loss and tinnitus. Compared with normal hearing controls, participants with only hearing loss showed gray matter decreases in superior and medial frontal gyri. In comparison with normal hearing controls, DTI analysis in the hearing loss group showed decreases in fractional anisotropy values in the right superior and inferior longitudinal fasciculi, corticospinal tract, inferior frontooccipital tract, superior occipital fasciculus, and anterior thalamic radiation (for a visualization of these nerve tracts see Fig. 111.1). Husain et al. (2011) concluded that hearing loss rather than tinnitus had the greatest influence on gray and white matter alterations. Consequently, the interaction between hearing loss and tinnitus on structural changes in cortex, potentially affecting the ACC, is more complicated than an additive effect of tinnitus on a baseline hearing impaired condition.

To compare structural MRIs of tinnitus subjects and nontinnitus controls, Melcher et al. (2013) also used VBM. The audiograms of all participants were very well matched and normal or near normal at standard audiometric frequencies (≤8kHz). Mean thresholds up to 14kHz, age, sex, and handedness were also matched between groups. In agreement with Boyen et al. (2013), there were no significant differences between the tinnitus and control groups in volume or density maps of gray matter. In the study by Melcher et al. (2013), increased hearing thresholds at frequencies above 8kHz were correlated with decreases in GM volume in the ventral posterior cingulate cortex, the dorsomedial prefrontal cortex (DMPFC), and a subcallosal region that included the ventromedial prefrontal cortex (VMPFC). This supports the proposal of Leaver et al. (2011) that decreasing activity in the VMPFC could disinhibit the action of the thalamic reticular nucleus (TRN) on the ventral part of the medial geniculate body, thereby allowing spontaneous subcortical activity more access to the auditory cortex (ACx). The results of Melcher et al. (2013) and Boyen et al. (2013) also support the notion that structural changes in the brain are not specific to tinnitus but rather result from auditory deafferentation (Adjamian et al., 2014).

Allan et al. (2016) examined changes in brain anatomy associated with tinnitus from 128 participants with tinnitus and hearing loss, tinnitus with clinically (≤8kHz) normal hearing, and nontinnitus controls with clinically normal hearing. The groups were matched for hearing loss, age, and gender. They used voxel- and surface-based morphometry to investigate gray and white matter volume and cortical thickness. They found a decrease in cortical volume with hearing loss in left Heschl's gyrus (HG). However, the main significant tinnitus-related findings were found outside sensory auditory structures, and include a decrease in cortical thickness for the tinnitus group compared with controls in the left superior frontal gyrus. Allan et al. (2016) found no changes in the subcallosal region.

The general conclusion is that only sporadic changes in gray and white matter were found that could be specifically attributed to tinnitus. Most changes, including those outside the auditory areas, have to be attributed to hearing loss.

10.1.2 Functional network changes

Several intrinsic resting-state networks (RSNs) have been implicated in tinnitus; these include the default mode network (DMN), auditory, visual, and dorsal attention networks (DANs). The DMN (cf. Fig. 1.1) has gained prominence as the canonical RSN; it typically comprises nodes (subnetworks) in the posterior cingulate/precuneus, bilateral superior frontal gyrus, medial frontal gyrus, and angular gyrus (Mantini et al., 2007; Chapter 6). The DMN is the most active at rest and shows reduced activity when a subject enters a task-based state involving attention or goal-directed behavior (Chapter 1). An opposite pattern is seen with other

RSNs, which show heightened correlated activity in the task-based state but retain connectivity (although with reduced activity) during rest. Some detail on the network changes observed in tinnitus patients is provided in the following.

Studies reviewed in Husain and Schmidt (2014) suggest that tinnitus causes consistent modifications to brain networks, including greater connectivity between limbic areas and cortical networks not traditionally involved with emotion processing and increased connectivity between attention and auditory processing brain regions. For example, Maudoux et al. (2012a,b) found that the tinnitus and control groups showed different connectivity patterns. In the control group, the connectivity graph could be divided into two distinct anticorrelated networks. The first one comprised the ACx and the insula, which does also respond to sound (cf. Fig. 7.2). The second one consisted of the frontoparietal cortex and ACC, brainstem, amygdala, basal ganglia/nucleus accumbens (NAc), and parahippocampal regions. In the tinnitus group, only the ACx-insula network was present, and showed increased functional connectivity between ACx and the left parahippocampal region, which was part of the second network in controls. Connectivity in nonauditory regions such as the basal ganglia/NAc, cerebellum and parahippocampal, right prefrontal, parietal and sensorimotor areas was increased in tinnitus patients.

Schmidt et al. (2017) examined two sources of variability in tinnitus subgroups: tinnitus severity and the length of time a person has had chronic tinnitus. Decreased correlations between two seed regions in the DMN—mPFC and PCC—and the precuneus were consistent across individuals who have had tinnitus for longer than 1 year, with more bothersome tinnitus demonstrating stronger decreases (Fig. 10.1). Schmidt et al. (2017) proposed DMN-precuneus decoupling as a potential marker of long-term tinnitus. Patients with moderately severe tinnitus showed increased correlations between seeds in the DAN—FEF and pIPS—and the precuneus. The precuneus is related to conscious and internal awareness, consequently coupling of the precuneus and DAN at rest is associated with bothersome tinnitus and is not always observed in patients with mild tinnitus and tinnitus duration between 6 and 12 months. Thus mild, recent-onset tinnitus may form a subgroup of tinnitus patients where rs-FC patterns have not yet changed from those seen in controls.

The same group (Zimmerman et al., 2018) studied patterns of leader-follower relationships between two signals, to rs-fMRI data of brain regions acquired from subjects with and without tinnitus. ... They found that the pairs of brain regions that contributed most to the dissociation between tinnitus and controls were those connected to the amygdala. These comprised parts of the DMN and the DAN (Fig. 10.2). Surprisingly the dominant role of the precuneus—decreased connectivity to the DMN and increased connectivity to the DAN in tinnitus (Schmidt et al., 2017)—is not

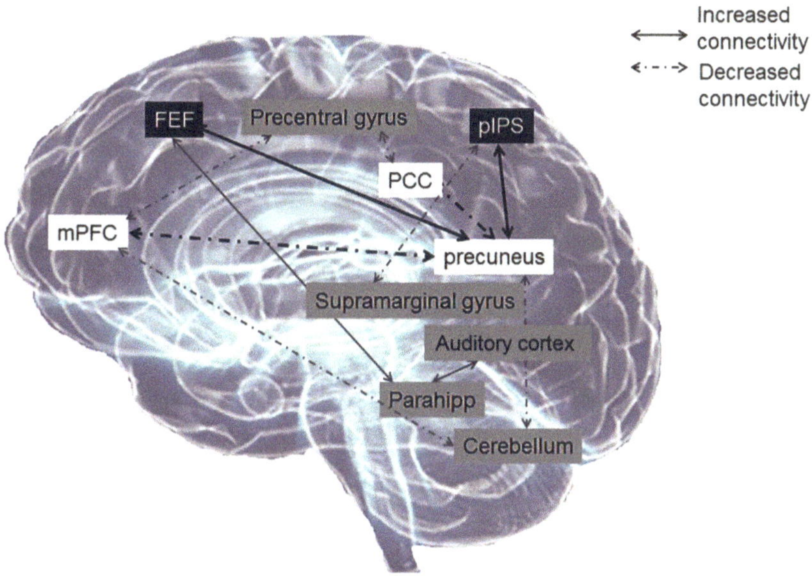

FIG. 10.1 An illustration of the effect of tinnitus on resting state functional connectivity, compared with controls. The *black hoses* correspond to regions of the dorsal attention network (DAN), while the *white boxes* are regions associated with the default mode network. In tinnitus patients, connectivity between the precuneus and the other regions of the default mode is decreased. This effect is more pronounced as tinnitus severity increases. Connectivity between the precuneus and regions in the DAN are increased in patients with bothersome tinnitus, but the effect is inconsistent in patients with mild tinnitus. The *gray boxes* and *lines* indicate findings that were noted previously (Husain and Schmidt, 2014) but were not replicated in the current study. *FEF*, frontal eye fields; *mPFC*, medial prefrontal cortex; *PCC*, posterior cingulate cortex; *pIPS*, posterior intraparietal sulci; *parahipp*, parahippocampus. *From Schmidt, S.A., Carpenter-Thompson, J., Husain, F.T., 2017. Connectivity of precuneus to the default mode and dorsal attention networks: a possible invariant marker of long-term tinnitus. Neuroimage Clin. 16, 196–204. Open access.*

obvious in this presentation. In the controls, there were consistent temporal patterns across frontal, parietal, and limbic regions and amygdalar activity, whereas in tinnitus subjects, this pattern was much more variable.

Leaver et al. (2016) demonstrated tinnitus-related reductions in resting-state functional connectivity (rs-FC) between specific brain regions and networks. They identified one RSN in tinnitus patients that was not apparent in controls. This tinnitus RSN included auditory regions—inferior colliculus (IC) and medial Heschl's gyrus (mHG)—as well as classically nonauditory regions such as mediodorsal nucleus of the thalamus (MDN), striatum, lateral prefrontal cortex (PFC), and orbitofrontal cortex (OFC). Tinnitus loudness was positively correlated with rs-FC in the MDN. Functional connectivity between the caudal and rostral mHG was stronger in tinnitus patients than in controls. Also, connections between

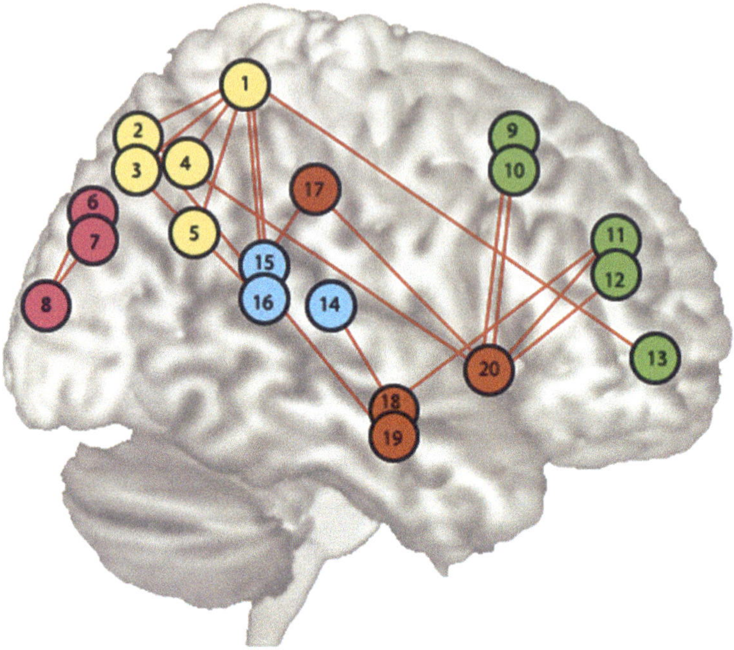

FIG. 10.2 Graphical representation of the 20 most discriminating ROI pairs that help distinguish tinnitus from normal hearing controls. *Green* is the dorsal attention network; *red* is the limbic network; *yellow* is the default mode network; *blue* is the auditory network; and *purple* is the visual network. 1, Precuneus; 2, L posterior intraparietal sulcus; 3, R posterior intraparietal sulcus; 4, R ventral intraparietal sulcus; 5, L inferior parietal lobe; 6, L superior occipital lobe; 7, R superior occipital lobe; 8, primary visual cortex; 9, L frontal eye field; 10, R frontal eye field; 11, L midfrontal gyrus; 12, R mid frontal gyrus; 13, medial prefrontal cortex; 14, L primary auditory cortex; 15, L superior temporal junction; 16, R superior temporal junction; 17, posterior cingulate cortex; 18, L parahippocampus; 19, R parahippocampus; 20, L Amygdala. *From Zimmerman, B.J., Abraham, I., Schmidt, S.A., Baryshnikov, Y., Husain, F.T., 2018. Dissociating tinnitus patients from healthy controls using resting-state cyclicity analysis and clustering. Network Neurosci. 3(1), 67–89. https://doi.org/10.1162/netn_a_00053. Open access.*

the left OFC with the right lateral PFC, the caudal mHG, and the rostral mHG were stronger in controls than in tinnitus patients. Leaver et al.'s (2016) data support the idea that tinnitus involves network dysfunction and suggest the importance of communication between auditory and frontostriatal circuits in tinnitus pathophysiology.

Using rs-fMRI in tinnitus patients and hearing loss-matched controls, Chen et al. (2015, 2016) analyzed the interhemispheric functional coordination between homotopic points of the brain. They found that compared with healthy controls chronic tinnitus patients had disrupted rs-FC strength and Granger-causal connectivity mostly in nonauditory regions, especially the PFC. In addition, chronic tinnitus patients had significantly enhanced rs-FC between the ACC and left precuneus, and disrupted rs-FC patterns within the DMN regions, which were correlated with tinnitus distress.

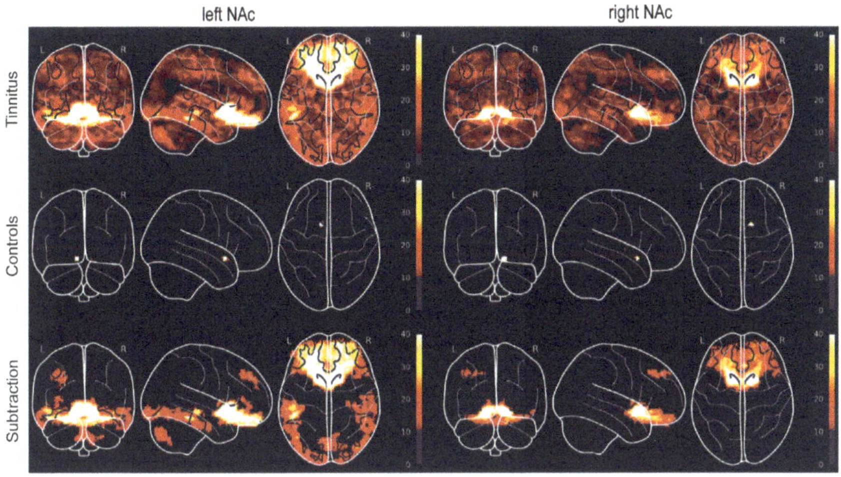

FIG. 10.3 Glass-brain visualization; of voxel wise rs-FC with left/right NAc seed regions. All analyses include age and whole-brain rs-FC as covariates. The within-group results for the tinnitus and control subjects (top, middle) are presented at height thresholds of $P<0.05$ and cluster thresholds of $P<0.05$, both including the FDR correction for multiple comparisons. The subtraction analysis (bottom) is presented at a height threshold of $P=0.0001$ (uncorrected) and a cluster threshold of 200 voxels. These height thresholds are reflected by the grayed-out regions of each color bar. The color-bar range over all subfigures has also been restricted to a maximum height of 40 (F-statistic) both for the purposes of direct visual comparison and also to better display the overall variance in the data; the true maxima of the subfigures are much higher, owing to the high correlations of the seed regions with the voxels in and around them. *From Hullfish, J., Abenes, I., Yoo, H.B., De Ridder, D., Vanneste, S., 2019. Frontostriatal network dysfunction as a domain-general mechanism underlying phantom perception. Hum. Brain Mapp. 40, 2241–2251.*

Hullfish et al. (2019) showed that a large sample of patients with chronic tinnitus had extensive rs-FC with the seeded NAc that was largely absent in nontinnitus controls (Fig. 10.3). The greatest level of rs-FC in the tinnitus subjects with the NAc involved VMPFC, OFC, and subgenual ACC (sgACC). They observed significant differences between tinnitus and controls in connectivity between left NAc and (1) the inferior temporal gyrus, extending posteriorly into the occipital lobe, (2) the left superior frontal gyrus, and (3) the cerebellum. Differences in right NAc-related connectivity beyond ventral/orbital frontal cortex were limited to the left superior frontal gyrus. NAc-vmPFC connections correlated negatively with age, mean hearing loss, and tinnitus percept loudness. The latter one disappeared when controlling for the effects of age and hearing loss, suggesting a strong effect of hearing loss—including age-related—on the findings. The strongest correlation, which did survive after controlling for the effects of age and mean hearing loss, was between the connection strength between NAc and the parahippocampal cortex and tinnitus loudness (Hullfish et al., 2019). This loudness correlation is consistent with existing

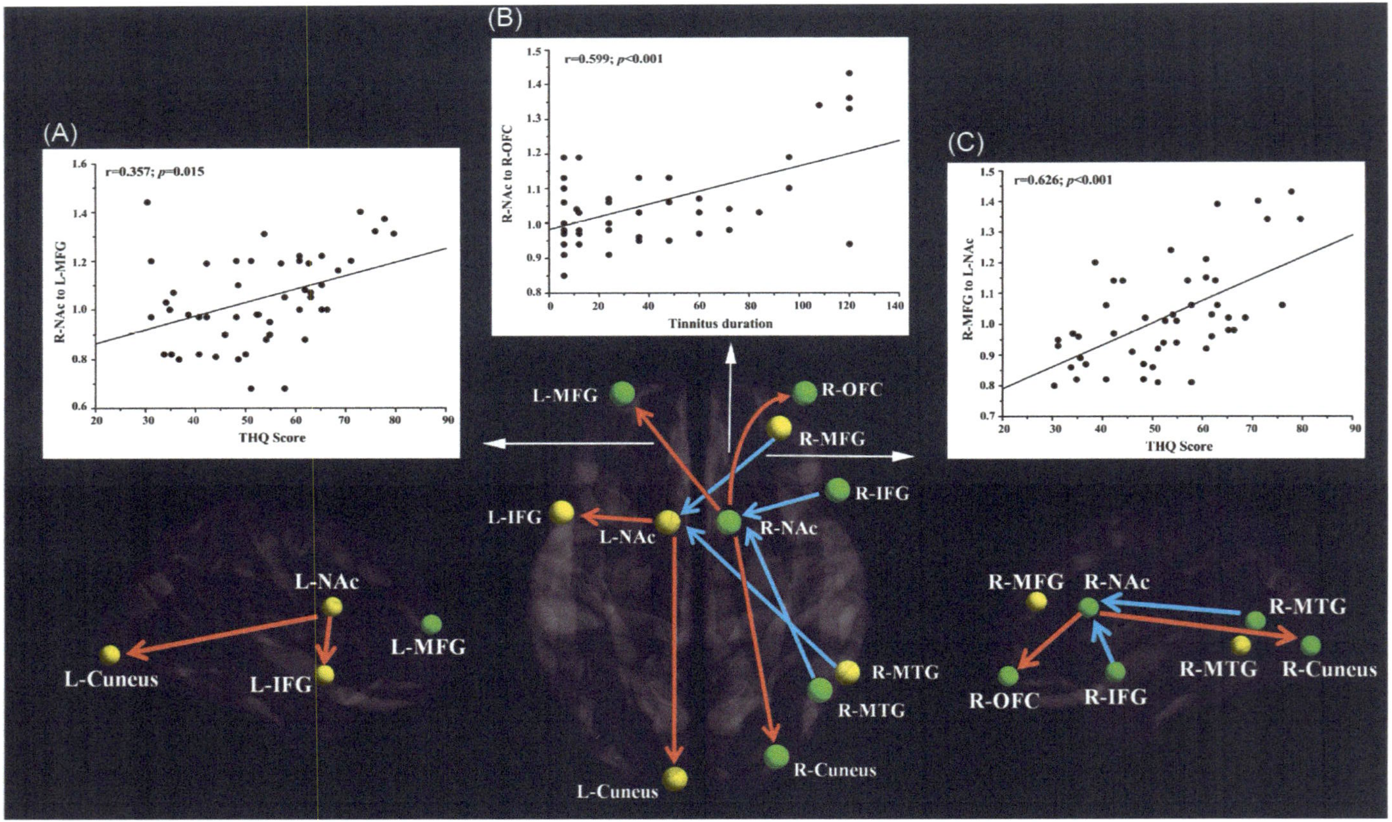

FIG. 10.4 The directional functional connectivity networks, of the bilateral NAc. The *red line* represents the directional functional connectivity from the bilateral NAc to the other brain regions; the *blue line* represents the directional functional connectivity from the other brain regions to the bilateral NAc. (A) Positive correlation between the increased directional connectivity from the right NAc to left MFG and the THQ score ($r=0.357$, $P=0.015$). (B) Positive correlation between the increased directional connectivity from the right NAc to right OFC and the tinnitus duration ($r=0.599$, $P<0.001$). (C) Positive correlation between the enhanced directional connectivity from the right MFG to left NAc and the THQ score ($r=0.626$, $P<0.001$). *From Xu, J.-J., Cui, J., Feng, Y., Yong, W., Chen, H., Chen, Y.-C., et al., 2019. Chronic tinnitus exhibits bidirectional functional dysconnectivity in frontostriatal circuit. Front. Neurosci. 13, 1299. https://doi.org/10.3389/fnins.2019.01299. Open access.*

literature (De Ridder et al., 2014; Schmidt et al., 2017) identifying the parahippocampus as a tinnitus generator or alternatively in combination with the PFC maintaining tinnitus presence (Araneda et al., 2018).

Hullfish et al. (2019) did not find direct evidence supporting a prediction of the frontostriatal gating theory, for example, the decreased connectivity. The frontostriatal gating theory (Rauschecker et al., 2010) assumes that dysfunction in the circuit comprising the VMPFC and sgACC may affect perception. Damage to the descending projection from VMPFC/sgACC by the nucleus accumbens (NAc) to the TRN was suggested to prevent the suppression of irrelevant sensory signals, such as those that result in tinnitus, and could increase the intensity thereof. Given the increased connectivity between VMPFC and the NAc, this could still increase the activity of the TRN, whereas the decreased connection between VMPFC and TRN (Leaver et al., 2011) may in some way counteract that.

On the basis of the NAc as a seed region, Xu et al. (2019) conducted a Granger causality analysis to investigate the directional connectivity and the relationship with tinnitus duration or distress. Tinnitus patients exhibited abnormal directional connectivity between the NAc and the prefrontal cortex, principally the middle frontal gyrus (MFG), OFC, and inferior frontal gyrus (IFG). Additionally, they found positive correlations between tinnitus handicap questionnaire (THQ; higher values indicate increased tinnitus problems) scores and increased directional connectivity from the right NAc to the left MFG and from the right MFG to the left NAc (in concurrence with Hullfish et al., 2019). Enhanced directional connectivity from the right NAc to the right OFC was positively associated with the duration of tinnitus (Fig. 10.4). Note that Xu et al. (2019) did not detect any significant differences of GM volume between tinnitus patients and controls. This is in contrast to previous studies (Leaver et al., 2011; Boyen et al., 2013; Adjamian et al., 2014) that found reduced GM volumes in the NAc, vmPFC, superior frontal gyrus, superior temporal gyrus, cingulate cortex, hippocampus, and occipital lobe in tinnitus patients.

10.1.3 Neurochemistry

Animal models (Wang et al., 2011) suggest that tinnitus results from a maladaptive plastic net down-regulation of inhibitory amino acid neurotransmission in the central auditory pathway. This loss of inhibition may be a compensatory response to loss of afferent input such as that caused by acoustic insult and/or age-related hearing loss, the most common causes of tinnitus. To examine the roles of GABA and glutamate in tinnitus, Brozoski et al. (2012) used high-resolution point-resolved proton magnetic resonance spectroscopy (^{1}H-MRS) in rats with noise-induced tinnitus to quantify their levels in the dorsal cochlear nucleus (DCN), inferior colliculus (IC), medial geniculate body (MGB), and primary auditory cortex (A1) of rats. They found decreased GABA levels only in the MGB especially in the contralateral MGB. There was a suggestion of minor reduction of GABA levels in A1. Glutamate was moderately and bilaterally elevated both in the DCN and in A1. In the MGB, glutamate was increased ipsilaterally but decreased contralaterally.

In, up to now, the only human study tightly controlled for hearing loss and hyperacusis, Sedley et al. (2015b) used magnetic resonance spectroscopy (MRS; Chapter 2) to noninvasively quantify GABA in the left and right ACx of a group of patients with lateralized tinnitus and controls matched for age, sex, and hearing. They also explored the potential relationships with other brain metabolites (i.e., choline, *N*-acetylaspartate, and creatine). The presence of tinnitus was associated with a reduction in auditory cortex GABA concentration. Regardless of tinnitus laterality, post hoc testing indicated reductions that were significant in right ACx and nonsignificant in left ACx. Tinnitus severity and hearing loss were correlated positively with right ACx choline but not GABA. Note that only the right ACx has significantly reduced choline and GABA regardless of tinnitus laterality.

10.2 EEG/MEG-based network connectivity changes in tinnitus

10.2.1 Undirected network connections

Schlee et al. (2009) found a significant decrease of interareal coupling in the alpha-band (9–12 Hz) and an increase of inter-areal coupling in the 48–54 Hz gamma-frequency range in tinnitus patients relative to the control group. Audiograms were not obtained. They noted an effect of tinnitus duration on the distribution of the gamma network: In patients with a tinnitus duration of less than 4 years, the left temporal cortex was predominant in the gamma-network whereas in patients with tinnitus duration of more than 4 years, the gamma-network was more widely distributed including more frontal and parietal regions. Vanneste et al. (2011) analyzed the source-localized EEG recordings of a very homogenous group of patients, but with mean hearing loss at the tinnitus frequency of 27 dB HL (recent onset group; <4 years) and 35 dB HL (chronic group; 4–10 years) with left-sided narrow-band-noise like tinnitus. Results indicated that the generators involved in tinnitus of recent onset seem to change over time with increased activity in several brain areas—auditory cortex, supplementary motor area and dorsal ACC plus insula—associated with a decrease in connectivity between the different auditory and nonauditory brain structures. These changes in ACx could have been driven by a subcortical, for example, the dorsal cochlear nucleus, tinnitus generator (Shore and Wu, 2019). In contrast, Vanneste et al. (2011) found an increase in gamma-band connectivity between the left primary and secondary ACx and the left insula, and also between the ACx and the right DLPFC. These networks are both connected to the left parahippocampal area. Thus acute and chronic tinnitus is related to differential activity and connectivity in a network comprising the ACx, insula, dorsal ACC, and premotor cortex. The above studies suggest that tinnitus perception relies on the interplay between auditory and nonauditory brain areas. According to the Global Brain Model (Schlee et al., 2011), a network of global fronto-parietal-cingulate areas is important in the generation and maintenance of the conscious perception of tinnitus. Thus the distress experienced by many individuals with tinnitus is related to the top—down influence of this global network on auditory areas and predicts a decrease in alpha activity in the auditory cortices (see also Section 10.3.3.2).

Based on EEG and MEG data, De Ridder et al. (2014) considered the tinnitus percept as an emergent property of multiple, parallel, dynamically changing, and partially overlapping subnetworks, each with a specific spontaneous oscillatory pattern and functional connectivity. De Ridder et al. (2014) suggested that the gamma network is principally related to the auditory system and could therefore be largely reflecting the hearing loss and central gain. This could thus also underlie hyperacusis and tinnitus loudness. The lower frequency rhythms were suggested to reflect distress and mood, and may therefore not be unique to tinnitus either (cf. Balkenhol et al., 2013; Meyer et al., 2014, 2017), The connection between gamma and the lower-frequency brain rhythms is suggested to take place in the parahippocampus.

Kandeepan et al. (2019) used fMRI to investigate correlations between behavioral tinnitus scores and functional brain connectivity of five RSNs comprising the auditory, the DMN, the executive control network (ECN) left and right, and the salience network (SAL). They found that functional connectivity changes were not limited to one single network. In particular, tinnitus distress showed a strong correlation with the connectivity pattern within and between the right ECN and the other four RSNs, indicating that it is probably the consequence of a hyperactive attention condition. In this tinnitus population, Kandeepan et al. (2019) observed that some auditory-related brain regions had resting BOLD activity that was more correlated with the time course of the DMN, left and right ECN, and SAL than the control population. They suggested that tinnitus is associated with a reorganization of functional connectivity between these networks. However, tinnitus objective loudness, that is, the strength of the auditory network response, was not correlated with any behavioral scores.

10.2.2 Directed network connections

Ahn et al. (2017) used cross-frequency coupling of lower-frequency (1–13 Hz) rhythms with high-gamma (Chapters 2 and 6), and Granger causality analysis to evaluate resting-state EEG data in healthy participants and patients with tinnitus. They found abnormally pronounced distributions of low-frequency activity, particularly delta activity, in the tinnitus group compared with the control group. Patients with tinnitus lacked robust frontal delta-phase/central high-gamma amplitude coupling that was clearly observed in healthy participants. Resting-state Granger causal connectivity was centered on the ACC and DLPFC in all frequency bands in both the control and tinnitus groups. This also showed prominent bidirectional connectivity of delta and theta between frontal top-down areas and primary auditory bottom-up areas in the tinnitus group compared with controls. Taken together, Ahn et al.'s (2017) findings suggest that maladaptive neuroplasticity or abnormal reorganization occurs between the auditory cortex and the DMN of patients with tinnitus. I only show here the delta and high-gamma connectivity (Fig. 10.5) from their study. Note the loss of directed high-gamma activity from primary auditory cortex to left ventral ACC and dorsal ACC in the tinnitus group.

Paraskevopoulos et al. (2019) used resting state MEG whole-head recordings to identify how chronic tonal tinnitus relates to altered functional connectivity of brain's intrinsic cortical networks. Using phase transfer entropy (Chapter 6), they estimated directed functional connectivity of the resting brain, at a whole cortex level. They found that tinnitus patients showed increased connectivity from right dorsal prefrontal to right medial temporal areas. This suggested to them that the role of the left parahippocampal area is dictated by a modulation from DMPFC; a region that is part of the dorsal attention network (DAN), as well as implicated in the regulation of emotional processing. Additionally, this whole-cortex analysis showed a crucial role of the left inferior parietal cortex, which modulated the activity of the right superior temporal gyrus. One has to wonder about the role of hearing loss (Section 10.1.1) in these findings: they wrote that "tinnitus patients did not exhibit hearing loss above 70 dB HL in the frequency ranges of one half octave above and below the tinnitus frequency."

It is clear that the dominant factors in changing cortical networks in tinnitus patients are foremost the degree and type of hearing loss, and comorbidities such as distress and mood. So far, no definite changes have been attributed to tinnitus proper, albeit that changes in connectivity between the dorsal attention network and the parahippocampal area appears to be a potential candidate.

10.3 Functional electrophysiology

10.3.1 Brain rhythms

Using MEG, Weisz et al. (2005) were the first to show that the resting-state neuronal activity of a group of individuals with tinnitus is characterized by a marked reduction in alpha (8–12 Hz) power together with a,

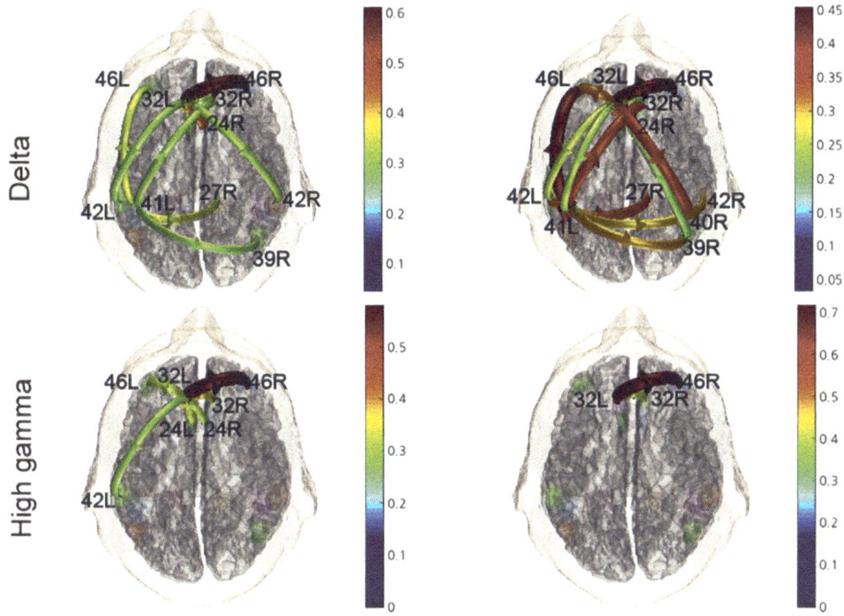

FIG. 10.5 Resting-state Granger causal connectivity of grand-averaged delta (1–4 Hz), and high-gamma (50–230 Hz) activity in healthy participants and patients with tinnitus. *Color-scaled directional arrows* link two causally connected ROIs across the 16 ROIs [Brodmann areas (BAs) 24 L/R, 27 L/R, 32 L/R, 39 L/R, 40 L/R, 41 L/R, 42 L/R, and 46 L/R] when Granger causality was statistically significant. The numbers of BAs demonstrating pronounced connectivity are marked at their positions on the topography. BA 24 is the ventral anterior cingulate cortex (ACC) and BA 32 is the dorsal ACC. BA 27 is the hippocampus/parahippocampal region. BA 39 (the angular gyrus) and BA 40 (the supramarginal gyrus) are parts of Wernicke's area involved in the comprehensive processing of presented auditory stimuli. BA 41 and 42 closely correspond to the primary auditory cortex, which is also involved in the processing of presented auditory stimuli. BA 46 is the part of the dorsolateral prefrontal cortex (DLPFC), *Color-scaled arrows* in the cortical connectivity image represent the degree of causal connectivity (ranging from 0 to 1). The topographical view is from the vertex, with the nose at the top of the image. *From Ahn, M.-H., Hong, S.K., Min, B.-K., 2017. The absence of resting-state high-gamma cross-frequency coupling in patients with tinnitus. Hear. Res. 356, 63–73. Open access.*

smaller, enhancement in delta (1.5–4 Hz) as compared with a normal hearing control group. This pattern was especially pronounced over temporal regions. Moreover, correlations with tinnitus-related distress revealed strong associations with this abnormal spontaneous activity pattern, particularly in right temporal and left frontal areas. Since no audiograms were provided for the tinnitus group, one cannot ignore hearing loss as a potential confound in these conclusions (cf. Section 10.1.1). Subsequently, the same group (Lorenz et al., 2009; Schlee et al., 2009) reported an inverse relationship between auditory gamma and alpha activity in tinnitus and control subjects (see Chapter 2). Extending these studies, Schlee et al. (2014) analyzed MEG resting-state recordings of tinnitus subjects and healthy controls with respect to the mean and the variability of spectral power in the alpha-frequency band over temporal areas. A significant decrease of auditory cortex activity was detected for the low alpha-band (8–10 Hz) but not for the upper alpha-band (10–12 Hz). Again, no audiograms were provided.

Pierzycki et al. (2016) investigated relationships between scalp-recorded EEG spectral power and behavioral aspects of tinnitus. Their patients had pure tone audiometric average < 60 dB (0.5, 1, 2, and 4 kHz) in the ear where tinnitus is perceived and the ability to hear all stimulation tones presented by the sound therapy device. They first made sure that test-retest correlation for both EEG band power measures and tinnitus variables were high. Yet Pierzycki et al. (2016) found no relationship between whole scalp EEG band power and psychoacoustic or psychosocial variables. Pierzycki et al. (2016) concluded that: "resting state whole scalp EEG should not be used as a biomarker for tinnitus." Amplifying this, Lau et al. (2018) measured oscillatory brain activity of 88 participants (46 with a chronic tinnitus percept, 42 without) in resting-state MEG. In a whole group analysis in sensor space, increased activity in the delta frequency band was found in tinnitus patients. However, Lau et al. (2018) observed no differences in oscillatory activity between tinnitus and control groups once subjects with mental health comorbidity were excluded. Again, no audiograms were provided. These findings counter the results of Weisz et al. (2005) and many subsequent ones, however, confounds of hearing loss abound (Chapter 7).

10.3.2 Effects of residual inhibition

If suprathreshold masking sounds are presented continuously for ~30–60 s and then switched off, the tinnitus sensation often disappears and then gradually recovers over a time period ranging from seconds to minutes. This phenomenon of "residual inhibition" (RI) has been reported to be more robust when masking sounds are in the tinnitus frequency region, although not all subjects show this frequency dependence (Roberts, 2007). Applying this technique, Kahlbrock and Weisz (2008) found a significant reduction of power in the delta (1.3–4.0 Hz) frequency band in temporal regions under residual inhibition. This implies that RI effects might reflect the transient reestablishment of balance between excitatory and inhibitory neuronal assemblies, by reafferentation, that have been perturbed (in most tinnitus individuals) by hearing loss. In this context, from animal experiments Galazyuk et al. (2019) found that the key mechanisms that govern neural suppression of spontaneous activity, that is, adaptation as in forward masking (Chapter 6 in Eggermont, 2015) in animals closely resemble clinical psychoacoustic findings of RI observed in tinnitus patients. This suppression is mediated by metabotropic glutamate receptors (mGluRs). They found that Groups I and II, but not Group III mGluRs play a role in modulation of spontaneous and sound-evoked firing in the inferior colliculus. They also observed that sound-induced suppression of spontaneous firing was reversibly blocked by Group I mGluR antagonists.

Sedley et al. (2015a) made extensive intracranial recordings from an awake tinnitus patient during short-term modifications in perceived tinnitus loudness by residual inhibition. As expected, they observed tinnitus-linked delta oscillations, thought to be triggered by low-frequency bursting in the thalamus, as found in animals with behavioral signs of tinnitus (Kalappa et al., 2014). Contrary to expectation, these delta changes extended far beyond circumscribed auditory cortical regions to encompass almost all of auditory cortex, plus large parts of temporal, parietal, sensorimotor, and limbic cortex. In discrete auditory, parahippocampal, and inferior parietal "hub" regions, these delta oscillations interacted with alpha, beta, and gamma activity, resulting in a coherent system of tightly coupled oscillations putatively associated with high-level functions including memory and perception (Chapters 2–4). The time course of oscillatory power changes before, during, and after the presentation of maskers that did and did not result in sustained RI, which shows that these oscillatory power changes occurred rapidly after masker offset, and the difference between RI and non-RI trials was that they lasted longer in RI trials.

10.3.3 Effect of tinnitus co-morbidities

10.3.3.1 Hearing loss and hyperacusis

Song et al. (2014) compared using resting-state EEG source-localized activity in tinnitus participants with and without hyperacusis. Those with hyperacusis showed increased beta power in the dorsal ACC and OFC, and increased alpha power in the right ACx. The hyperacusis questionnaire score was positively correlated with beta power in the OFC, alpha1 power in the right ACx, and beta1 power in the dorsal ACC. Song et al. (2014) suggested that beta power increase in the OFC/dorsal ACC may indicate increased resting-state vigilance in tinnitus patients with hyperacusis. The same group (Vanneste and De Ridder, 2016) analyzed resting-state EEG-source recordings in tinnitus patients, correlated to the mean hearing loss, the range of the hearing loss, and the hearing loss at the tinnitus frequency. They found that hearing loss correlated with theta and alpha1 activity changes over the parahippocampal area. They linked tinnitus to two very different mechanisms. In patients with little or no hearing loss, the tinnitus seems to be more related to ACx activity, but not to (para)hippocampal memory related activity. In contrast, tinnitus in patients with more severe hearing loss appears to be related to (para)hippocampal mechanisms. Furthermore, hearing loss seems to drive the communication between the auditory cortex and the parahippocampus, as measured by functional and effective connectivity (cf. Fig. 10.1). Vanneste et al. (2019) then provided evidence for a top-down type of tinnitus associated with a deficient noise-canceling mechanism (DLFPC, NAc). Top-down-related tinnitus was related to a change in the pregenual ACC that corresponded to increased activity in the ACx. Bottom-up tinnitus instead related to changes in the parahippocampal cortex. All patients had ~40 dB loss at 4–8 kHz, regardless of bottom-up or top-down origins of their tinnitus, which seems to contradict their earlier suggestion that it was related to the amount of hearing loss (Vanneste and De Ridder, 2016).

10.3.3.2 Distress

Distress appears to be nearly universal in patients with bothersome tinnitus. Joos et al. (2012) found distress mainly correlated with alpha2, beta1 and beta2 activity of the right frontopolar cortex and OFC in combination with beta2 activation of the ACC. These specific neural circuits are embedded with the parahippocampal region, which forms a crucial linkage between both tinnitus-related frequency-specific pathways. Meyer et al. (2014) investigated individuals with chronic tinnitus and nonaffected controls. They found two distinct independent components, distress and tinnitus presence, characterizing the individual tinnitus experience. Tinnitus presence was defined as the perceived loudness that increases with tinnitus duration but without being bothersome. In agreement with Joos et al. (2012), they reported that chronic tinnitus patients with high distress have increased EEG activity in the beta2 band dominantly over frontal recording sites. High-presence tinnitus showed enhanced EEG signal strength in the delta-, alpha-, and lower gamma-bands over bilateral temporal and left perisylvian electrodes.

Mohan et al. (2019) identified the core network of the domain-general distress component in tinnitus by comparing whole brain Granger causality-directed functional networks calculated from 5 min of resting state EEG data collected from tinnitus patients and nontinnitus controls. In the tinnitus network, Mohan et al. (2019) observed a reorganization of the provincial and connector hubs (cf. Fig. 6.3) in both the alpha frequency bands. The hubs that were distinctly present in the tinnitus network contain regions from the frontal, temporal, medial temporal, and cingulate cortices, which interact with one another to form the *core tinnitus network*. A disconnection among connections, going from the left hippocampus/parahippocampus to the sgACC, was observed to correlate with distress. Mohan et al. (2019) proposed this subset of regions to form the *core distress network*, and distress as the consequence of the absence of modulation of the phantom sound.

10.4 Is tinnitus caused by thalamocortical dysrhythmia?

The basic hypothesis of thalamocortical dysrhythmia (TCD), as illustrated in Fig. 10.6, is that protracted hyperpolarization—by deafferentation—of a medial thalamic subnucleus will result in low-frequency oscillations at the theta-frequency band (Llinás et al., 1999). Such oscillations, by activating return corticothalamic pathways, will entrain, through the reticular thalamic nucleus (RTN) and through direct thalamic activation, the nonspecific thalamocortical system. At the cortical level, low-frequency activation of cortical inhibitory interneurons promotes coherent gamma-frequency oscillation and thus positive symptoms. Llinás et al. (1999) recorded resting-state MEG activity in awake, healthy human controls and in patients suffering from neurogenic pain, tinnitus, Parkinson's disease, or depression. Compared with controls, all patients showed increased theta rhythmicity, in conjunction with a widespread and marked increase of coherence among high- and low-frequency oscillations. This indicated the presence of a TCD, responsible for all the earlier-mentioned conditions.

Moazami-Goudarzi et al. (2010) compared spectral parameters of resting EEG of male tinnitus patients with those of age-matched healthy males. Using LORETA source analysis, they found that the generators of these delta-, theta-, alpha-, and beta-power increases were localized dominantly to left auditory areas, temporoparietal, posterior insular, anterior cingulate, and parahippocampal cortical areas. Moazami-Goudarzi et al. (2010) also found that this overproduction of resting state delta-, theta-, and beta-band brain activities, providing support for a TCD process at the source of tinnitus (Fig. 10.6). Note that here the edge effect is visible in the beta band. It is important to note the lack of evidence that the patients were also matched for hearing loss—the degree of deafferentation is important. Only one tinnitus patient had normal hearing, and one was total deaf on one side. Consequently, the EEG power increases could have resulted from increased central gain. Note also that this does not explain tinnitus with normal hearing, that is, without deafferentation.

Adjamian et al. (2012) rigorously tested several predictions made by the TCD model in four groups of participants (tinnitus with hearing loss, tinnitus with clinically normal hearing, no tinnitus with hearing loss and no tinnitus with clinically normal hearing). MEG was used to measure oscillatory brain activity within different frequency bands in a "resting" state and during presentation of a masking noise. Results revealed that low-frequency activity in the delta-band and theta-band was significantly higher in the "tinnitus with hearing loss" group compared with the "no tinnitus with normal hearing" group. They found that this effect was unlikely to be driven by the hearing loss alone. High-frequency activity in the gamma-band (25–80 Hz) was not correlated with tinnitus (or hearing loss), contrasting the findings of Balkenhol et al. (2013) and De Ridder et al. (2014).

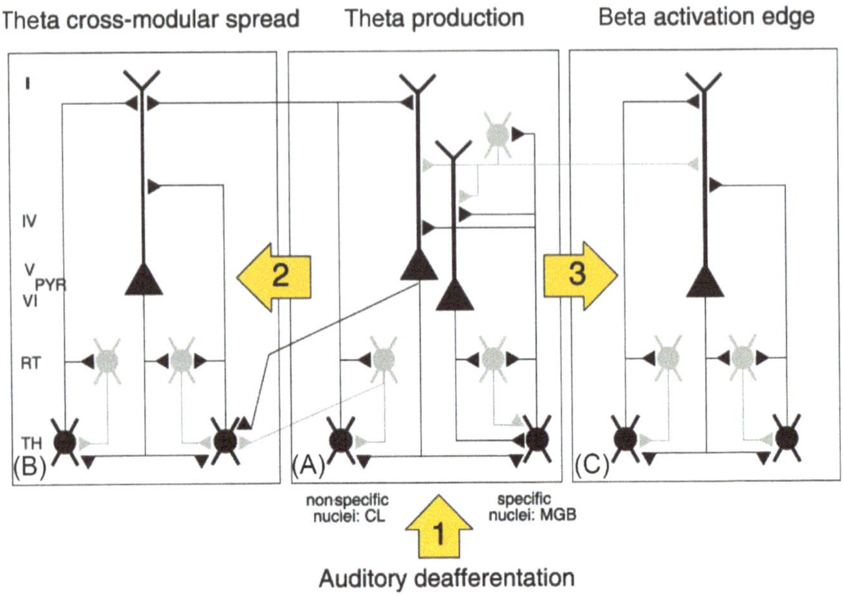

FIG. 10.6 Diagram of thalamocortical (TC) circuits. Panels (A), (B), and (C) show three TC modules (basically identical) with module A depicted in more detail. One module includes, first, cortical layers (I, IV, V, and VI) with pyramidal cells and one *gray* GABAergic inhibitory cell; second, nucleus reticularis (RT, *gray* GABAergic inhibitory cells); and third, thalamus (TH), represented by one specific cell in medial geniculate body (MGB) and one nonspecific cell (e.g., in central lateral nucleus, CL). The specific thalamic cell projects to the apical dendrites of both Layers V and VI pyramidal cells and collaterals sustain reticular feedback and cortical feed-forward inhibitions. The nonspecific thalamic cell projects to RT and to the Layer V pyramidal neuron and has a divergent connection onto the neighboring module. The corticothalamic feedback connection is depicted as intramodular onto RT and its thalamic relay cell, and divergent intra- and cross-modular onto three thalamic relay cells. There are also divergent cross-modular reticulothalamic projections. *Arrows* 1, 2, and 3 indicate the sequence of (1) auditory deafferentation, (2) theta cross-modular spread, and (3) beta activation edge. *From Moazami-Goudarzi, M., Michels, L., Weisz, N., Jeanmonod, D., 2010. Temporo-insular enhancement of EEG low and high frequencies in patients with chronic tinnitus. QEEG study of chronic tinnitus patients. BMC Neurosci. 11, 40. http://www.biomedcentral.com/1. Open access.*

It is noteworthy that the tinnitus patients in the latter study did not suffer from hyperacusis, in contrast to those tested by Adjamian et al. (2012). The findings partly support the thalamocortical dysrhythmia model and suggest that slow-wave (delta-band) activity may be a more reliable correlate of tinnitus than high-frequency activity. Zobay and Adjamian (2015) recalled that the TCD asserts that thalamocortical oscillations lock a region in the auditory cortex into an ectopic slow-wave theta-rhythm. The cortical area surrounding this region (edge effect) is hypothesized to generate abnormal gamma-oscillations giving rise to the tinnitus percept. Consequently the model predicts enhanced cross-frequency coherence in a broad range between theta- and gamma-bands. Zobay and Adjamian (2015) investigated this prediction in an MEG study involving tinnitus and control participants. Contrary to the TCD predictions, however, they did not find any indication of statistical differences in cross-frequency amplitude-amplitude coupling between tinnitus and controls and thus no evidence for the predicted enhancement of theta-gamma (Llinás et al., 1999) or theta-beta Moazami-Goudarzi et al. (2010) cross-frequency coupling in tinnitus. In a MEG study, Zobay et al. (2015) compared resting-state oscillatory activity of tinnitus participants and normal-hearing controls and found increased functional connectivity within the auditory cortices in the alpha-band. Since Zobay et al. (2015) also showed an increase in theta-connectivity they provide some support for the TCD model. However, the primary prediction of the TCD model is an increase in theta power, which is absent in these data. Overall, Zobay et al. (2015) results are not in contradiction with the theoretical TCD model of tinnitus, but they also provide only very limited support (cf. De Ridder et al., 2015). Using high resolution (7 Tesla) fMRI, Berlot et al. (2020) found reduced auditory thalamocortical connectivity in tinnitus patients. This is in accordance with a role for the auditory thalamus in tinnitus, but does not favor the TCD hypothesis, and neither the noise-cancelation hypothesis (Rauschecker et al., 2010).

10.5 Effects of neuromodulation

There have been numerous studies using neuromodulation techniques (Chapter 8) to suppress or alleviate tinnitus (Langguth, 2015; Langguth et al., 2013). Here I only describe studies that combine these interventions with MRI or EEG/MEG measures of their effects.

10.5.1 Neurofeedback

In contrast to a sham group trained with a frequency discrimination task, neurofeedback in tinnitus patients produced significant changes in EEG frequency bands and showed greater relief of tinnitus (Dohrmann et al., 2007). Furthermore, simultaneous alteration of delta- and alpha-frequency bands was strongly related to changes in tinnitus loudness matched before and after the intervention. In two patients with the greatest modulatory success, the tinnitus sensation resided completely in response to the treatment. Hartmann et al. (2014) assessed the ability of neurofeedback to focally increase alpha-power in auditory areas in comparison with the more commonly used repetitive transcranial magnetic stimulation (rTMS; Chapter 8). Resting-state MEG was measured before and after neurofeedback and rTMS intervention. Note that the patients groups were different, so conclusions are limited. Source level power and functional connectivity were analyzed with a focus on the alpha-band. Only neurofeedback produced a significant decrease in tinnitus symptoms and—more important for the context of the study—a spatially circumscribed increase in alpha-power in right auditory regions. Connectivity analysis revealed higher outgoing connectivity in a region ultimately neighboring the area in which power increases were observed. Güntensperger et al. (2019) evaluated an individually adjusted alpha/delta neurofeedback protocol. The delta band fixed between 3 and 4 Hz was chosen as the frequency for inhibition. The frequency range for the rewarded alpha band was individually determined according to each patient's intrinsic alpha peak frequency. Chronic tinnitus patients participated in 15 weekly neurofeedback training sessions and extensive pre- and posttests, and follow-up testing 3 and 6 months after training. They reported that neurofeedback led to a significant reduction of tinnitus-related distress and tinnitus loudness. While distress remained on a low level even 6 months after the completion of training, tinnitus loudness returned to baseline levels in the follow-up period. Resting-state EEG activity showed an increase in the trained alpha/delta ratio over the course of the training. This ratio increase was related to training-induced changes of tinnitus-related distress as measured with the Tinnitus Questionnaire, mainly due to power increase in the alpha frequency range. All studies mentioned here show that alpha power modulation has the largest effect.

10.5.2 Transcranial magnetic and direct-current stimulation

Faber et al. (2012) conducted a double-blind, placebo-controlled crossover study in which patients with tinnitus were randomly assigned to receive active and sham anodal transcranial direct current stimulation (tDCS) over left or right DLPFC for six sessions in a counterbalanced order; the cathode electrode was placed in the contralateral DLPFC. They found that both active conditions—irrespective of the anodal position—can decrease tinnitus annoyance but it was not associated with improvements in tinnitus loudness when comparing pre-tDCS versus post-tDCS and comparing sham-tDCS versus real tDCS.

Müller et al. (2013) examined the effects of five different stimulation protocols (including sham) by measuring tinnitus loudness and tinnitus-related brain activity with MEG before and after rTMS. Changes in oscillatory activity were analyzed for the stimulated auditory cortex and for the entire brain in the delta-, theta-, alpha-, and gamma-bands. The effect of rTMS on tinnitus loudness varied strongly across patients. Importantly, strong reductions in tinnitus loudness were associated with increases in alpha-power in the stimulated auditory cortex, while an unspecific decrease in gamma- and alpha-power, particularly in left frontal regions, was linked to an increase in tinnitus loudness.

Schecklmann et al. (2015) probed the capacity of single rTMS sessions to interfere with tinnitus-specific cortical plasticity as reflected in oscillatory power. Using EEG, they recorded from patients with bilateral chronic tinnitus and healthy controls comparable for age, sex, handedness, and hearing loss. ... Left and right temporal and left and right prefrontal cortices were each stimulated with 200 pulses at 1 Hz—assumed to be inhibitory (Chapter 8)—and with an intensity of 60% of stimulator output. Before and after each rTMS protocol 5 min of resting-state EEG were recorded. ... In tinnitus patients, left temporal rTMS decreased frontal theta and delta and increased beta2 (18.5–21 Hz) power, whereas right frontal rTMS decreased right temporal beta3 (21.5–30 Hz) and gamma-power. No changes or increases were observed in the control group. Only nonsystematic changes in tinnitus loudness were induced by rTMS.

10.5.3 Vagus nerve stimulation

Transcutaneous vagus nerve stimulation (tVNS) paired with tones or with rehabilitative training can help patients to achieve reductions in tinnitus perception (Tyler et al., 2017). It is known that tVNS paired with experience may drive neural plasticity in a controlled and therapeutic direction. A few studies investigating the neurophysiological effects, but without a change in tinnitus parameters, follow here:

Hyvärinen et al. (2015) paired auditory stimulation with tVNS in seven tinnitus patients and eight normal-hearing control subjects. They found tVNS-induced changes in synchrony of MEG brain activity in response to the auditory stimulation, as reflected in the sensor level signals. Whole-head beta band, whole-head gamma-band, and frontal gamma-band synchrony was strongly and negatively correlated with tinnitus handicap inventory scores. Hyvärinen et al. (2015) noted that no significant changes in power were found in the controls and only a nonsignificant increase in gamma power in the tinnitus group for the tVNS-on condition. Since previous studies observed that gamma activity in the auditory cortex is correlated with tinnitus loudness, Vanneste et al. (2017) assessed resting-state source-localized EEG before and after 1 to 3 months of tVNS-tone pairing in chronic tinnitus patients. tVNS-tone pairing reduced gamma band activity in left auditory cortex. tVNS-tone pairing also reduced the phase coherence between the auditory cortex and areas associated with tinnitus distress, including the cingulate cortex. These results support the hypothesis that tVNS-tone pairing can direct therapeutic neural plasticity. For both the loudness (measured with the visual analog scale) and emotional (measured with the tinnitus handicap index) components of the tinnitus percept, they observed a significant reduction after VNS treatment in comparison with before VNS treatment. After VNS treatment, there was no longer a significant correlation between loudness and the emotional component. Vanneste et al. (2017) note that the results are consistent with previous reports that reduction of tinnitus loudness is associated with desynchronization of hypersynchronous activity in the auditory cortex at the gamma frequency band. Their results are also consistent with the findings in animals that VNS-tone pairing reverses the increase in cortical synchronization observed in noise-exposed rats (Engineer et al., 2011).

10.6 Summary

Tinnitus in humans is typically not an isolated symptom but has comorbidities in hearing loss, hyperacusis, distress or depression, to name a few. Each of those has been shown to affect brain rhythms, networks and their connectivity, few studies agree on changes caused by tinnitus in isolation. Only sporadic changes in gray and white matter were found that could be specifically attributed to tinnitus. Most structural changes, including those outside the auditory areas, have to be attributed to hearing loss. Considering changes in neural networks, DMN-precuneus decoupling and increased DAN-precuneus coupling is a potential marker of long-term tinnitus. The strongest correlation of network changes and tinnitus, after controlling for the effects of age and mean hearing loss, was between the connection strength between NAc and the parahippocampal cortex and tinnitus loudness. In patients with little or no hearing loss, the tinnitus seems to be more related to ACx activity but not to (para)hippocampal memory-related activity. In contrast, tinnitus in patients with more severe hearing loss appears to be related to (para) hippocampal mechanisms.

References

Adjamian, P., Sereda, M., Zobay, O., Hall, D.A., Palmer, A.R., 2012. Neuromagnetic indicators of tinnitus and tinnitus masking in patients with and without hearing loss. J. Assoc. Res. Otolaryngol. 13 (5), 715–731.

Adjamian, P., Hall, D.A., Palmer, A.R., Allan, T.W., Langers, D.R., 2014. Neuroanatomical abnormalities in chronic tinnitus in the human brain. Neurosci. Biobehav. Rev. 45, 119–133.

Ahn, M.-H., Hong, S.K., Min, B.-K., 2017. The absence of resting-state high-gamma cross-frequency coupling in patients with tinnitus. Hear. Res. 356, 63–73.

Allan, T.W., Besle, J., Langers, D.R.M., Davies, J., Hall, D.A., Palmer, A.R., et al., 2016. Neuroanatomical alterations in tinnitus assessed with magnetic resonance imaging. Front. Aging Neurosci. 8, 221. https://doi.org/10.3389/fnagi.2016.00221.

Araneda, R., Renier, L., Dricot, L., Decat, M., Ebner-Karestinos, D., Deggouj, N., et al., 2018. A key role of the prefrontal cortex in the maintenance of chronic tinnitus: an fMRI study using a Stroop task. Neuroimage Clin. 17, 325–334.

Balkenhol, T., Wallhäusser-Franke, E., Delb, W., 2013. Psychoacoustic tinnitus loudness and tinnitus-related distress show different associations with oscillatory brain activity. PLoS One 8 (1), e53180. https://doi.org/10.1371/journal.pone.0053180.

Berlot, E., Arts, R., Smit, J., George, E., Gulban, O.F., Moerel, M., et al., 2020. A 7 Tesla fMRI investigation of human tinnitus percept in cortical and subcortical auditory areas. Neuroimage Clin. 25, 102166.

Boyen, K., Langers, D.R.M., de Kleine, E., van Dijk, P., 2013. Gray matter in the brain: differences associated with tinnitus and hearing loss. Hear. Res. 295, 67–78.

Brozoski, T., Odintsov, B., Bauer, C., 2012. Gamma-aminobutyric acid and glutamic acid levels in the auditory pathway of rats with chronic tinnitus: a direct determination using high resolution point-resolved proton magnetic resonance spectroscopy (^{1}H-MRS). Front. Syst. Neurosci. 6, 9. https://doi.org/10.3389/fnsys.2012.00009.

Chen, Y.-C., Xia, W., Feng, Y., Li, X., Zhang, J., Feng, X., et al., 2015. Altered interhemispheric functional coordination in chronic tinnitus patients. BioMed Res. Int. https://doi.org/10.1155/2015/345647.

Chen, Y.-C., Feng, Y., Xu, J.-J., Mao, C.-N., Xia, W., Ren, J., et al., 2016. Disrupted brain functional network architecture in chronic tinnitus patients. Front. Aging Neurosci. 8, 174. https://doi.org/10.3389/fnagi.2016.00174.

De Ridder, D., Vanneste, S., Weisz, N., Londero, A., Schlee, W., Elgoyhen, A.B., et al., 2014. An integrative model of auditory phantom perception: tinnitus as a unified percept of interacting separable subnetworks. Neurosci. Biobehav. Rev. 44, 16–32.

De Ridder, D., Vanneste, S., Langguth, B., Llinas, R., 2015. Thalamocortical dysrhythmia: a theoretical update in tinnitus. Front. Neurol. 6, 124. https://doi.org/10.3389/fneur.2015.00124.

Dohrmann, K., Elbert, T., Schlee, W., Weisz, N., 2007. Tuning the tinnitus percept by modification of synchronous brain activity. Restor. Neurol. Neurosci. 25, 371–378.

Eggermont, J.J., 2012. The Neuroscience of Tinnitus. Oxford University Press, Oxford, UK.

Eggermont, J.J., 2015. Auditory Temporal Processing and Its Disorders. Oxford University Press, Oxford, UK.

Eggermont, J.J., 2016. Can animal models contribute to understanding tinnitus heterogeneity in humans? Front. Aging Neurosci. 8, 265. https://doi.org/10.3389/fnagi.2016.00265.

Engineer, N.D., Riley, J.R., Seale, J.D., Vrana, W.A., Shetake, J.A., Sudanagunta, S.P., et al., 2011. Reversing pathological neural activity using targeted plasticity. Nature 470 (7332), 101–104.

Faber, M., Vanneste, S., Fregni, F., De Ridder, D., 2012. Top down prefrontal affective modulation of tinnitus with multiple sessions of tDCS of dorsolateral prefrontal cortex. Brain Stim. 5, 492–498.

Galazyuk, A.V., Longenecker, R.J., Voytenko, S.V., Kristaponyte, I., Nelson, G.L., 2019. Residual inhibition: from the putative mechanisms to potential tinnitus treatment. Hear. Res. 375, 1–13.

Guest, H., Munro, K.J., Prendergast, G., Simon Howe, S., Plack, C.J., 2017. Tinnitus with a normal audiogram: relation to noise exposure but no evidence for cochlear synaptopathy. Hear. Res. 344, 265–274.

Güntensperger, D., Thüring, C., Kleinjung, T., Neff, P., Meyer, M., 2019. Investigating the efficacy of an individualized alpha/delta neurofeedback protocol in the treatment of chronic tinnitus. Neural Plast. https://doi.org/10.1155/2019/354089.

Hartmann, T., Lorenz, I., Müller, N., Langguth, B., Weisz, N., 2014. The effects of neurofeedback on oscillatory processes related to tinnitus. Brain Topogr. 27, 149–157.

Hullfish, J., Abenes, I., Yoo, H.B., De Ridder, D., Vanneste, S., 2019. Frontostriatal network dysfunction as a domain-general mechanism underlying phantom perception. Hum. Brain Mapp. 40, 2241–2251.

Husain, F.T., Schmidt, S.A., 2014. Using resting state functional connectivity to unravel networks of tinnitus. Hear. Res. 307, 153–162.
Husain, F.T., Medina, R.E., Davis, C.W., Szymko-Bennett, Y., Simonyan, K., Pajor, N.M., et al., 2011. Neuroanatomical changes due to hearing loss and chronic tinnitus: a combined VBM and DTI study. Brain Res. 1369, 74–88.
Hyvärinen, P., Yrttiaho, S., Lehtimäki, J., Ilmoniemi, R.J., Mäkitie, A., Ylikoski, J., et al., 2015. Transcutaneous vagus nerve stimulation modulates tinnitus-related beta- and gamma-band activity. Ear Hear. 36, e76–e85.
Joos, K., Vanneste, S., De Ridder, D., 2012. Disentangling depression and distress networks in the tinnitus brain. PLoS One 7 (7), e40544. https://doi.org/10.1371/journal.pone.0040544.
Kahlbrock, N., Weisz, N., 2008. Transient reduction of tinnitus intensity is marked by concomitant reductions of delta band power. BMC Biol. 6, 4. https://doi.org/10.1186/1741-7007-6-4.
Kalappa, B.I., Brozoski, T.J., Turner, J.G., Caspari, D.M., 2014. Single unit hyperactivity and bursting in the auditory thalamus of awake rats directly correlates with behavioural evidence of tinnitus. J. Physiol. 592, 5065–5078.
Kandeepan, S., Maudoux, A., Ribeiro de Paula, D., Zheng, J.Y., Cabay, J.E., Gómez, F., et al., 2019. Tinnitus distress: a paradoxical attention to the sound? J. Neurol. 266 (9), 2197–2207. https://doi.org/10.1007/s00415-019-09390-1.
Kujawa, S.G., Liberman, M.C., 2009. Adding insult to injury: cochlear nerve degeneration after "temporary" noise-induced hearing loss. J. Neurosci. 29 (45), 14077–14085.
Langguth, B., 2015. Treatment of tinnitus. Curr. Opin. Otolaryngol. Head Neck Surg. 23, 361–368. https://doi.org/10.1097/MOO.0000000000000185.
Langguth, B., Kreuzer, P.M., Kleinjung, T., De Ridder, D., 2013. Tinnitus: causes and clinical management. Lancet Neurol. 12, 920–930.
Lau, P., Wollbrink, A., Wunderlich, R., Engell, A., Löhe, A., Junghöfer, M., et al., 2018. Targeting heterogeneous findings in neuronal oscillations in tinnitus: analyzing MEG novices and mental health comorbidities. Front. Psychol. 9, 235. https://doi.org/10.3389/fpsyg.2018.00235.
Leaver, A.M., Renier, L., Chevillet, M.A., Morgan, S., Kim, H.J., Rauschecker, J.P., 2011. Dysregulation of limbic and auditory networks in tinnitus. Neuron 69, 33–43.
Leaver, A.M., Turesky, T.K., Seydell-Greenwald, A., Morgan, S., Kim, H.J., Rauschecker, J.P., 2016. Intrinsic network activity in tinnitus investigated using functional MRI. Hum. Brain Mapp. 37, 2717–2735.
Llinás, R.R., Ribary, U., Jeanmonod, D., Kronberg, E., Mitra, P.P., 1999. Thalamocortical dysrhythmia: a neurological and neuropsychiatric syndrome characterized by magnetoencephalography. Proc. Natl. Acad. Sci. USA 96 (26), 15222–15227.
Lorenz, I., Müller, N., Schlee, W., Hartmann, T., Weisz, N., 2009. Loss of alpha power is related to increased gamma synchronization—a marker of reduced inhibition in tinnitus? Neurosci. Lett. 453, 225–228.
Mantini, D., Perrucci, M.G., Del Gratta, C., Romani, G.L., Corbetta, M., 2007. Electrophysiological signatures of resting state networks in the human brain. Proc. Natl. Acad. Sci. USA 104 (32), 13170–13175.
Maudoux, A., Lefebvre, P., Cabay, J.E., Demertzi, A., Vanhaudenhuyse, A., Laureys, S., et al., 2012a. Auditory resting-state network connectivity in tinnitus: a functional MRI study. PLoS One 7, e36222.
Maudoux, A., Lefebvre, P., Cabay, J.E., Demertzi, A., Vanhaudenhuyse, A., Laureys, S., et al., 2012b. Connectivity graph analysis of the auditory resting state network in tinnitus. Brain Res. 1485, 10–21.
Melcher, J.R., Knudson, I.M., Levine, R.A., 2013. Subcallosal brain structure: correlation with hearing threshold at supra-clinical frequencies (>8 kHz), but not with tinnitus. Hear. Res. 295, 79–86.

Meyer, M., Luethi, M.S., Neff, P., Langer, N., Büchi, S., 2014. Disentangling tinnitus distress and tinnitus presence by means of EEG power analysis. Neural Plast. https://doi.org/10.1155/2014/468546.
Meyer, M., Neff, P., Grest, A., Hemsley, C., Weidtm, S., Kleinjung, T., 2017. EEG oscillatory power dissociates between distress- and depression-related psychopathology in subjective tinnitus. Brain Res. 1663, 194–204.
Moazami-Goudarzi, M., Michels, L., Weisz, N., Jeanmonod, D., 2010. Temporo-insular enhancement of EEG low and high frequencies in patients with chronic tinnitus. QEEG study of chronic tinnitus patients. BMC Neurosci. 11, 40. http://www.biomedcentral.com/1471-2202/11/40.
Mohan, A., Davidson, C., De Ridder, D., Vanneste, S., 2019. Effective connectivity analysis of inter- and intramodular hubs in phantom sound perception—identifying the core distress network. Brain Imaging Behav. https://doi.org/10.1007/s11682-018-9989-7.
Müller, N., Lorenz, I., Langguth, B., Weisz, N., 2013. rTMS induced tinnitus relief is related to an increase in auditory cortical alpha activity. PLoS One 8, e55557. https://doi.org/10.1371/journal.pone.0055557.
Noreña, A.J., Eggermont, J.J., 2003. Changes in spontaneous neural activity immediately after an acoustic trauma: implications for neural correlates of tinnitus. Hear. Res. 183, 137–153.
Paraskevopoulos, E., Dobel, C., Wollbrink, A., Salvari, V., Bamidis, P.D., Pantev, C., 2019. Maladaptive alterations of resting state cortical network in tinnitus: a directed functional connectivity analysis of a larger MEG data set. Sci. Rep. 9, 15452. https://doi.org/10.1038/s41598-019-51747-z.
Pierzycki, R.H., McNamara, A.J., Hoare, D.J., Hall, D.A., 2016. Whole scalp resting state EEG of oscillatory brain activity shows no parametric relationship with psychoacoustic and psychosocial assessment of tinnitus: a repeated measures study. Hear. Res. 331, 101–108.
Qu, T., Qi, Y., Yu, S., Du, Z., Wei, W., Cai, A., et al., 2019. Dynamic changes of functional neuronal activities between the auditory pathway and limbic systems contribute to noise-induced tinnitus with a normal audiogram. Neuroscience 408, 31–45.
Rauschecker, J.P., Leaver, A.M., Mühlau, M., 2010. Tuning out the noise: limbic-auditory interactions in tinnitus. Neuron 66, 819–826.
Roberts, L.E., 2007. Residual inhibition. Prog. Brain Res. 166, 487–495.
Robertson, D., Bester, C., Vogler, D., Mulders, W.H., 2013. Spontaneous hyperactivity in the auditory midbrain: relationship to afferent input. Hear. Res. 295, 124–129.
Schecklmann, M., Lehner, A., Gollmitzer, J., Schmidt, E., Schlee, W., Langguth, B., 2015. Repetitive transcranial magnetic stimulation induces oscillatory power changes in chronic tinnitus. Front. Cell. Neurosci. 9, 421. https://doi.org/10.3389/fncel.2015.00421.
Schlee, W., Hartmann, T., Langguth, B., Weisz, H., 2009. Abnormal resting-state cortical coupling in chronic tinnitus. BMC Neurosci. 10, 11. https://doi.org/10.1186/1471-2202-10-11.
Schlee, W., Lorenz, I., Hartmann, T., Müller, N., Schulz, H., Weisz, N., 2011. A global brain model of tinnitus. In: Møller, A.R., Langguth, B., De Ridder, D., Kleinjung, T. (Eds.), Textbook of Tinnitus. Springer, New York, Dordrecht, Heidelberg, London, pp. 161–169.
Schlee, W., Schecklmann, M., Lehner, A., Kreuzer, P.M., Vielsmeier, V., Poeppl, T.B., et al., 2014. Reduced variability of auditory alpha activity in chronic tinnitus. Neural Plast. https://doi.org/10.1155/2014/436146.
Schmidt, S.A., Carpenter-Thompson, J., Husain, F.T., 2017. Connectivity of precuneus to the default mode and dorsal attention networks: a possible invariant marker of long-term tinnitus. Neuroimage Clin. 16, 196–204.
Sedley, W., Gander, P.E., Kumar, S., Oya, H., Kovach, C.K., Nourski, K.V., et al., 2015a. Intracranial mapping of a cortical tinnitus system using residual inhibition. Curr. Biol. 25, 1–7.
Sedley, W., Parikh, J., Edden, R.A.E., Tait, V., Blamire, A., Griffiths, T.D., 2015b. Human auditory cortex neurochemistry reflects the presence and severity of tinnitus. J. Neurosci. 35 (44), 14822–14828.

Shore, S.E., Wu, C., 2019. Mechanisms of noise-induced tinnitus: insights from cellular studies. Neuron 103, 8–20.

Song, J.-J., De Ridder, D., Weisz, N., Schlee, W., Van de Heyning, P., Vanneste, S., 2014. Hyperacusis-associated pathological resting-state brain oscillations in the tinnitus brain: a hyperresponsiveness network with paradoxically inactive auditory cortex. Brain Struct. Funct. 219, 1113–1128.

Tyler, R., Cacace, A., Stocking, C., Tarver, B., Engineer, N., Martin, J., et al., 2017. Vagus nerve stimulation paired with tones for the treatment of tinnitus: a prospective randomized double-blind controlled pilot study in humans. Sci. Rep. 7, 11960. https://doi.org/10.1038/s41598-017-12178-w.

Vanneste, S., De Ridder, D., 2016. Deafferentation-based pathophysiological differences in phantom sound: tinnitus with and without hearing loss. NeuroImage 129, 80–94.

Vanneste, S., van de Heyning, P., De Ridder, D., 2011. The neural network of phantom sound changes over time: a comparison between recent-onset and chronic tinnitus patients. Eur. J. Neurosci. 34, 718–731.

Vanneste, S., Martin, J., Rennaker 2nd, R.L., Kilgard, M.P., 2017. Pairing sound with vagus nerve stimulation modulates cortical synchrony and phase coherence in tinnitus: an exploratory retrospective study. Sci. Rep. 7, 17345. https://doi.org/10.1038/s41598-017-17750-y.

Vanneste, S., Alsalman, O., De Ridder, D., 2019. Top-down and bottom-up regulated auditory phantom perception. J. Neurosci. 39 (2), 364–378.

Wang, H., Brozoski, T.J., Caspary, D.M., 2011. Inhibitory neurotransmission in animal models of tinnitus: maladaptive plasticity. Hear. Res. 279 (1–2), 111–117.

Weisz, N., Moratti, S., Meinzer, M., Dohrmann, K., Elbert, T., 2005. Tinnitus perception and distress is related to abnormal spontaneous brain activity as measured by magnetoencephalography. PLoS Med. 2 (6), e153. https://doi.org/10.1371/journal.pmed.0020153.

Xu, J.-J., Cui, J., Feng, Y., Yong, W., Chen, H., Chen, Y.-C., et al., 2019. Chronic tinnitus exhibits bidirectional functional dysconnectivity in frontostriatal circuit. Front. Neurosci. 13, 1299. https://doi.org/10.3389/fnins.2019.01299.

Zimmerman, B.J., Abraham, I., Schmidt, S.A., Baryshnikov, Y., Husain, F.T., 2018. Dissociating tinnitus patients from healthy controls using resting-state cyclicity analysis and clustering. Network Neurosci. 3 (1), 67–89. https://doi.org/10.1162/netn_a_00053.

Zobay, O., Adjamian, P., 2015. Source-space cross-frequency amplitude-amplitude coupling in tinnitus. Biomed. Res. Int. 2015, 489619. https://doi.org/10.1155/2015/489619.

Zobay, O., Palmer, A.R., Hall, D.A., Sereda, M., Adjamian, P., 2015. Source space estimation of oscillatory power and brain connectivity in tinnitus. PLoS One 10 (3), e0120123. https://doi.org/10.1371/journal.pone.0120123.

CHAPTER

11

Developmental disorders: Dyslexia and autism

11.1 Dyslexia

Developmental dyslexia affects around 5% of school children and is a specific learning disability of reading and spelling, which cannot be attributed to low intellectual ability or inadequate schooling (Lehongre et al., 2011). Intact phonological processing is crucial for successful literacy acquisition. Consequently, we will emphasize the auditory aspects of dyslexia. Three phonological symptoms can be distinguished: (1) poor ability to pay attention to, and mentally manipulate individual speech sounds; (2) poor verbal short-term memory, that is, the ability to repeat, for example, pseudowords or digit series; and (3) slow performance in naming a series of pictures, colors, or digits (Lehongre et al., 2011).

Earlier (Chapter 14 in Eggermont, 2015), I discussed the auditory temporal processing deficits in dyslexia. I will proceed here by reviewing the developmental disorder aspects affecting both auditory and visual structures.

11.1.1 Structural changes in dyslexia

11.1.1.1 White matter

Reviewed studies (Vandermosten et al., 2012) generally show that poorer reading ability or dyslexia is correlated with lower fractional anisotropy values (Chapter 6), that is, in nerve tracts in left temporoparietal and frontal areas (Klingberg et al., 2000; Odegard et al., 2009). Most studies indicate that these regions comprise the left arcuate fasciculus and corona radiate (Mamah et al., 2019; Fig. 11.1). Some studies (Mandonnet et al., 2007) suggest a role for the posterior part of the corpus callosum (CC) or for more ventral tracts such as the inferior longitudinal fasciculus or the inferior frontooccipital fasciculus. The arcuate fasciculus is the principal

https://doi.org/10.1016/B978-0-12-819818-6.00001-7

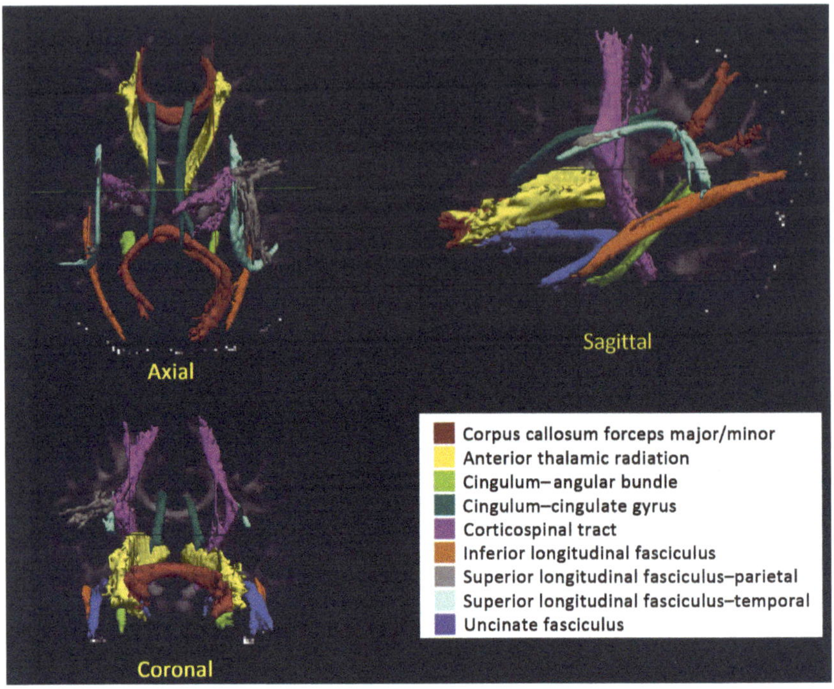

FIG. 11.1 Depiction of white matter tracts identified by a tract-based approach in sample control participant. This method determines the connection that best fits between two selected endpoints unique to each (18) white matter tract of interest using diffusion data. Bilateral tracts are given a single color in this figure, but were treated as independent for all analyses. *From Mamah, D., Ji, A., Rutlin, J., Shimony, J.S., 2019. White matter integrity in schizophrenia and bipolar disorder: tract- and voxel-based analyses of diffusion data from the Connectom scanner. NeuroImage: Clin. 21, 101649. Open access.*

connection between two speech-related areas, Broca's and Wernicke's area, whereas the inferior frontooccipital fasciculus connects the posterior temporal areas with the orbitofrontal region, and the inferior longitudinal fasciculus pathway indirectly projects from Broca's area to Wernicke's area (Steinmann and Mulert, 2012).

Linguistic processing depends mainly on the perisylvian region in the left hemisphere, whereas homeotopic regions in the right hemisphere are specialized in the implementation of prosodic information, such as individual intonation, accent pattern, and rhythm. Thus both hemispheres contribute to speech processes differentially and have to coordinate their information through the CC. The region, which is relevant for speech processing forms the posterior third of the CC, which is crossed by the interhemispheric pathways connecting primary (A1) and secondary (A2) auditory cortices (cf. Fig. 4.1). The regions connecting the homeotopic A1 and A2 involve the largest diameter fibers in the CC (Aboitiz, 1992), which provide fast conduction times. Hofer and Frahm (2006) showed with diffusion-tensor imaging (DTI) striking interindividual differences in the posterior CC. Phonological awareness appears associated with larger diameter temporocallosal fibers (Dougherty et al., 2007).

Leonard and Eckert (2008) performed a quantitative analysis of the DTI brain scans from children and adults with reading and language disorders. They identified two clusters with contrasting anatomical and reading profiles. Individuals with small symmetrical brain structures tended to have deficits in multiple aspects of written and oral language whereas those with larger asymmetrical structures were more likely to have the isolated phonological deficits seen in adults with compensated dyslexia. Individuals with moderate brain size and asymmetry typically had the best overall performance (Fig. 11.2). Johnson et al. (2013) corroborated this symmetry with concurrent electroencephalography (EEG) and magnetoencephalography (MEG) from a group of children with dyslexia aged 8–12 years, and a group of age-matched children with normal reading ability. Auditory responses were elicited using 500-ms duration broadband noise. These responses were strongly lateralized in control children. Children with dyslexia showed significantly less lateralization of auditory

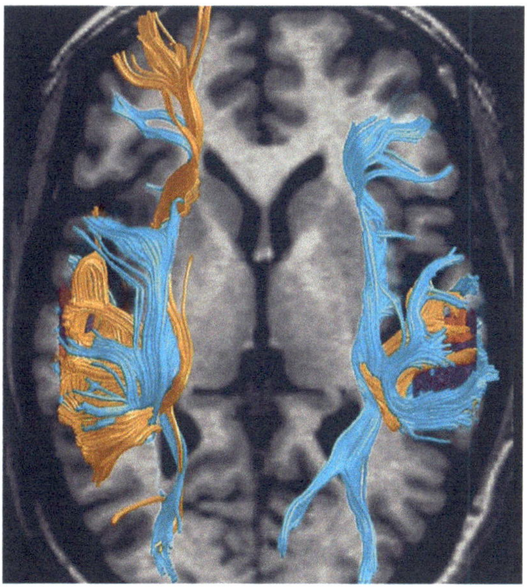

FIG. 11.2 The diffusion tensor imaging of an overlain fiber tract from a man with normal leftward planar asymmetry (*yellow*). Note the absence of connections between the temporal and frontal lobes in the right hemisphere and a dyslexic man (*blue*) with symmetrical plana temporale due to a larger planum temporale in the right hemisphere. *From Leonard, CM., Eckert, M.A., 2008. Asymmetry and dyslexia. Dev. Neuropsychol. 33 (6), 663–681, with permission from Taylor & Francis Ltd., http://www.tandfonline.com.*

cortical functioning, and a different pattern of development of auditory lateralization with age.

Vandermosten et al. (2013) investigated the structure-function relationship between DTI and auditory steady state responses (ASSR) in a group of normal reading adults and a group of dyslexic adults. They found reduced white matter lateralization in the posterior superior temporal gyrus and the arcuate fasciculus in the dyslexic readers. Additionally, white matter lateralization in the posterior superior temporal gyrus and white matter integrity in the splenium of the corpus callosum positively related to interhemispheric coherence to phonemic-rate ASSR modulations (i.e., 20 Hz). This correlation was negative in normal readers. Thus less pronounced left white matter dominance in dyslexic adults might relate to their problems to process phonemic-rate acoustic information and to integrate them into the phonological system.

Abnormalities in the connectivity between the thalamus and cortex, both in the visual and auditory domains have also been reported. By using fMRI, Díaz et al. (2012) found in dyslexic adults that the medial geniculate body (MGB) responded abnormally when the task required attending to phonemes compared with other speech features. No other structure in the auditory pathway showed distinct functional neural patterns. Using ultrahigh-resolution structural MRI, diffusion MRI, and probabilistic tractography, Müller-Axt et al. (2017) showed that individuals with dyslexia had reduced connections in the direct pathway between the left visual thalamus (lateral geniculate nucleus, LGN) and left middle temporal visual cortical area (V5/MT) but not between the left LGN and left primary visual cortex. They also found that the connectivity strength between left V5/MT and LGN correlated with a key deficit in dyslexia; rapid naming ability. By means of DTI, the same group (Tschentscher et al., 2019) found reduced white matter connectivity in a corticothalamic auditory pathway between the left auditory motion-sensitive planum temporale and the left MGB in dyslexic male adults. In neurotypical readers connectivity strength of this pathway is associated with measures of reading fluency.

11.1.1.2 Gray matter

Temple et al. (2001) used functional magnetic resonance imaging (fMRI) in dyslexic and normal-reading children (8–12 years old) during phonological and orthographic tasks of rhyming and matching visually presented letter pairs. During letter rhyming (requires phonological judgment), both normal and dyslexic reading children showed activity in left frontal brain regions, whereas only normal-reading children had activity in left temporoparietal cortex. During letter matching (needs orthographic judgment), normal-reading children showed activity throughout extra-striate visual cortex (ESC), especially in occipitoparietal regions, whereas dyslexic children had little activity in ESC during this task. Thus childhood dyslexia may be characterized by disruptions in the neural bases of both phonological (rhyming) and orthographic (letter matching) processes important for reading.

van Atteveldt et al. (2004) investigated the integration of visually presented letters and speech sounds using fMRI. They found that letters and speech sounds are integrated in heteromodal superior temporal cortex. Simultaneously auditory and visually presented letters modified responses to speech sounds in A1. Because A1 does not respond to visual letters alone, this represents feedback modulation from heteromodal supra-temporal areas regions where integration takes place. The same group (Blau et al., 2010) studied letter-speech sound integration in children with and without developmental dyslexia. Dyslexic children had reduced neural integration of visually presented letters and speech sounds in the planum temporale/Heschl's sulcus (PT/HS) and the superior temporal sulcus (STS). Whole-brain analyses of visual and auditory group differences also showed reduced single modality responses to letters in the fusiform gyrus in dyslexic children, and reduced activity for processing speech sounds in the anterior superior temporal gyrus (STG), PT/HS, and STS (Fig. 11.3). Neural integration of letters and speech sounds in the PT/HS and the neural response to letters in the fusiform gyrus explained almost 40% of the variance in individual reading performance. This suggested to Blau et al. (2010) that: "an interrelated network of visual, auditory and heteromodal brain areas contributes to the skilled use of letter–speech sound associations necessary for learning to read."

Krafnick et al. (2014) contrasted dyslexic children not only with age-matched controls but also with younger controls who read at the same level as the dyslexics. They found that dyslexics showed less gray matter volume (GMV) in multiple left and right hemisphere regions, including left superior temporal sulcus when compared with age-matched controls. However, only some of these differences persisted when dyslexics were compared with controls matched on reading abilities, with only right precentral gyrus GMV being different in this second analysis. When similar analyses were performed for white matter volume, no differences emerged from either comparison. These results indicated to Krafnick et al. (2014) that: "the GMV differences in dyslexia reported here and in prior studies are in large part the outcome of experience (e.g., disordered reading experience) compared with controls, with only a fraction of the differences being driven by dyslexia per se." This was also advocated by Goswami (2015) and corroborated by Moreau et al. (2019) and Torre and Eden (2019). The latter study using voxel-based morphometry in a large sample of typically developing participants aged 6–22 years, demonstrated no general linear relationship between GMV and single real word reading ability.

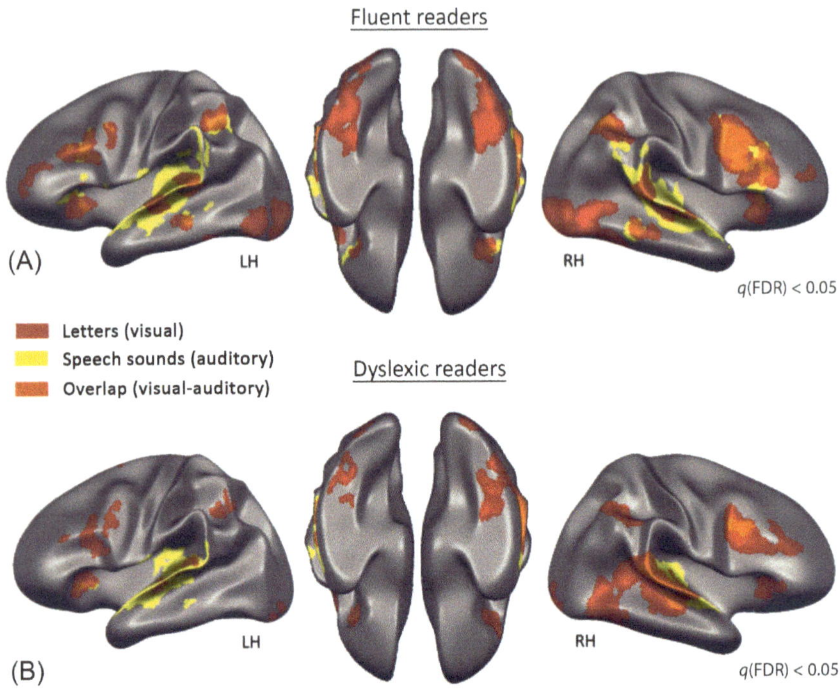

FIG. 11.3 Spatial cortical networks involved in processing letters (*red*), speech sounds (*yellow*), or both unisensory conditions (*orange*) in fluent (A) and dyslexic readers (B). *FDR*, false discovery rate. *From Blau, V., Reithler, J., van Atteveldt, N., Seitz, J., Gerretsen, P., Goebel, R., et al., 2010. Deviant processing of letters and speech sounds as proximate cause of reading failure: a functional magnetic resonance imaging study of dyslexic children. Brain 133 (Pt 3), 868–879, by permission from Oxford University Press on behalf of the Guarantors of Brain.*

11.1.2 Brain rhythm changes in dyslexia

11.1.2.1 Resting-state EEG-band power

Goswami (2011) found impaired neural coding of low-frequency oscillatory networks in the right hemisphere a putative causal factor for dyslexia. She stated that the core deficit in dyslexia is phonological. Temporal sampling of speech by cortical oscillations that encode incoming information at different frequencies could explain the perceptual and phonological difficulties with syllables, rhymes, and phonemes found in individuals with dyslexia. According to her temporal sampling theory of dyslexia, atypical phase-locking of amplitude rise time (stimulus onset to maximum amplitude) prevents the assimilation of the phonological primitives that are needed subsequently for phonemic processing (see also Chapter 14 in Eggermont, 2015).

From eyes-closed EEG, Babiloni et al. (2012) found that resting-state alpha rhythms were abnormal in dyslexic children and related to reading deficits. They split up the alpha range in alpha1 (6–8 Hz), alpha2 (8–10 Hz), and alpha3 (10–12 Hz). When compared with the control children, the dyslexics showed lower amplitude in regions where alpha2 and alpha3 power is strong—parietal, occipital, and temporal—whereas other rhythms, including gamma, were not affected (Fig. 11.4). In dyslexics, they found that shorter reading time correlated with higher alpha power. Babiloni et al. (2012) concluded: "Dyslexic children are characterized by limited abnormalities of resting state EEG rhythms as to topography (posterior regions) and frequency (alpha), which were related to phonological encoding (pseudo-words reading). Consequently, dyslexia may be associated to some functional impairment of cortical neuronal synchronization mechanisms involved in the resting state condition."

Using simultaneous EEG and fMRI recordings during passive viewing, that is, eyes open, of an audiovisual movie, Lehongre et al. (2013) compared the distribution of brain oscillations in the delta, theta, and gamma ranges over left and right auditory cortices. In controls, the results were consistent with the hypothesis that there is a dominance of gamma oscillations in the left hemisphere and a dominance of delta–theta oscillations in the right hemisphere (Giraud et al., 2007). In dyslexics, such an interaction was not found; oscillations were similar in both hemispheres. This confirms that the primary cortical disruption in dyslexia lies in a lack of hemispheric specialization (Fig. 11.2) for gamma oscillations, which might disrupt the representation of or the access to phonemic units. Tierney et al. (2014) examined resting-state gamma activity and language-dependent behavioral performance in adolescents aged 14–15 years. Resting-state gamma power was inversely related to reading fluency, rapid naming, and basic reading proficiency.

11.1.2.2 Entrainment of brain oscillations

Using MEG and behavioral tests in dyslexic subjects, Lehongre et al. (2011) showed a hemisphere bias for phonemic processing, reflected in less entrainment to ~ 30 Hz ASSRs in left auditory cortex. This deficit correlated with measures of phonological processing and rapid naming. There was, however, enhanced cortical entrainment at rates beyond 40 Hz, which are dominated by brainstem activity, suggesting a lack of adaptation (Jaffe-Dax et al., 2018) in dyslexics and this was associated with a verbal memory deficit. Poelmans et al. (2012) recorded multichannel-EEG ASSRs in normal-reading and dyslexic adults in response to speech-spectrum noise stimuli that were amplitude modulated (AM) at 80, 20, and 4 Hz. They found that normal-reading and dyslexic participants showed similar processing at the brainstem (80 Hz) level. At the auditory cortex level, dyslexic subjects demonstrated deviant phonemic-rate (20 Hz) responses compared with normal readers, but no group differences were found for the 4 Hz syllabic rate. Furthermore, there was a relationship between

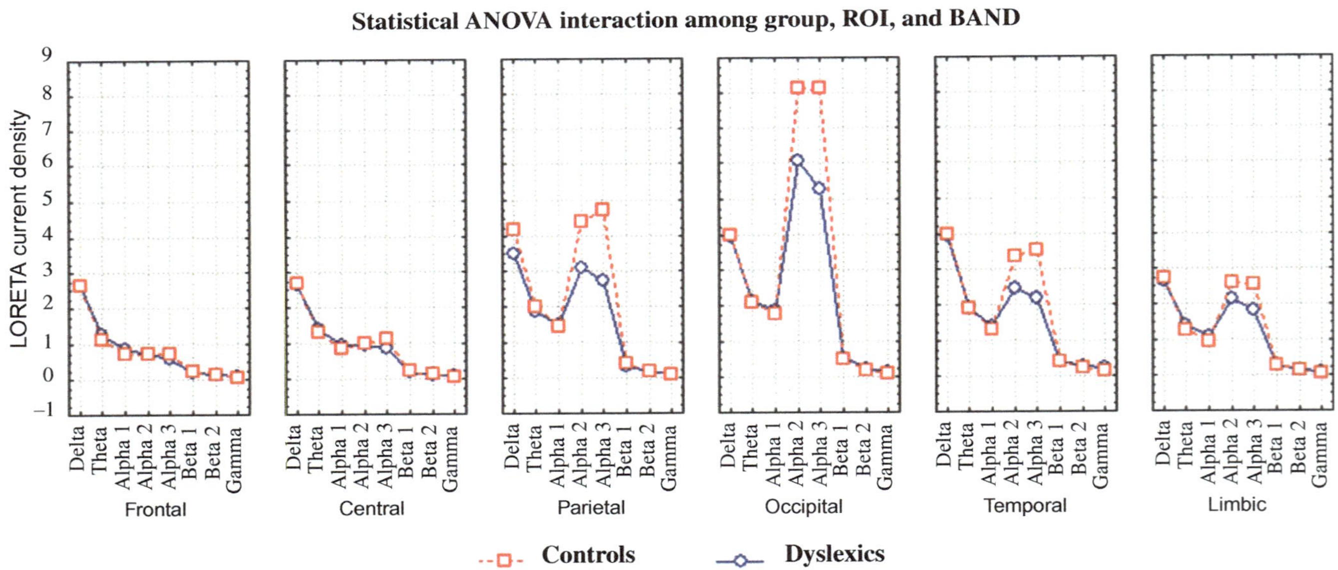

FIG. 11.4 Regional normalized LORETA solutions (mean across subjects) relative to a statistical ANOVA interaction among the factors Group (Controls, Dyslexics), Band (delta, theta, alpha 1, alpha 2, alpha 3, beta 1, beta 2, gamma), and ROI (frontal, central, parietal, occipital, temporal, limbic). Regional normalized LORETA solutions modeled the EEG relative power spectra as revealed by a sort of "virtual" intracranial macroelectrodes located on the macrocortical region of interest. *From Babiloni, C., Stella, G., Buffo, P., Vecchio, F., Onorati, P., Muratori, C., et al., 2012. Cortical sources of resting state EEG rhythms are abnormal in dyslexic children. Clin. Neurophysiol. 123, 2384–2391, with permission from Elsevier on behalf of the International Federation of Clinical Neurophysiology.*

phonemic-rate ASSRs with speech-in-noise perception and phonological awareness. Focusing on the 4 and 20 Hz ASSRs, the same group (De Vos et al., 2017) found that, compared with normal readers, dyslexic readers showed reduced nonsynchronized theta power (syllabic rate), reduced synchronized alpha power and enhanced synchronized beta power (phonemic rate). Positive correlations between alpha synchronization and phonological skills were found in normal readers but were absent in dyslexic readers. In contrast, dyslexic readers exhibited positive correlations between beta synchronization and phonological skills. Thus auditory neural synchronization of alpha and beta oscillations is atypical in dyslexia, indicating deviant neural processing of both syllabic and phonemic rate information. The same group (Van Hirtum et al., 2019) found that such atypical neural synchronization for alpha and beta oscillations was modulated by stimulus rise time. Deficits found at 10 and 20 Hz were only evident when the envelope's rise time was significantly shortened. This impaired tracking of rise time cues may potentially lead to the speech and phonological processing difficulties observed in dyslexia.

Leong and Goswami (2014) reported somewhat different findings from assessing rhythmic entrainment to three dominant AM frequencies in speech—each associated with a different phonological unit (0.9–2.5 Hz, prosodic stress; 2.5–12 Hz, syllables; 12–40 Hz, phonemes). This showed atypical rhythmic entrainment by dyslexic participants to syllable patterns (~ 5 Hz) in speech, suggesting that neural mechanisms for syllable perception and production may also be atypical. The same group (Power et al., 2016) tested children with dyslexia and chronological age- and reading-level matched younger children, who listened to semantically unpredictable sentences in a word report task. The sentences were noise-vocoded to increase reliance on envelope cues. Envelope reconstruction for envelopes $\leq$ 10 Hz showed that the children with dyslexia had significantly poorer speech encoding in the 0–2 Hz band compared with both age- and reading-level matched controls. These data suggest that impaired neural encoding of low frequency speech envelopes, related to speech prosody, may underpin the phonological deficit that causes dyslexia across languages.

11.1.3 Functional connectivity in dyslexia

Intact phonetic representations in bilateral A1 and A2 were found in combined MRI with voxel-based morphometry (VBM) pattern analysis and functional connectivity (FC) analysis (Boets et al., 2013). In contrast, the functional and structural connectivity between the bilateral auditory cortices and the left inferior frontal gyrus (IFG; a region involved in higher-level phonological processing) was significantly impaired in dyslexics, suggesting deficient access to otherwise intact phonetic representations.

Molinaro et al. (2016) recorded MEGs from dyslexic readers and age-matched controls while they were listening to ~10-s long sentences. Compared with controls, dyslexic readers had (1) impaired neural entrainment to speech in the delta band (0.5–1 Hz); (2) reduced delta synchronization in both the right auditory cortex and the left IFG; and (3) impaired feed-forward functional coupling between neural oscillations in the right auditory cortex and the left inferior frontal regions (Fig. 11.5). Molinaro et al. (2016) confirmed the findings by Boets et al. (2013) and also found that the reduced speech-brain synchronization in dyslexic readers compared with normal readers appears preserved through the developmental period from childhood to adulthood. In a recent EEG study with age and reading-level controls, Di Liberto et al. (2018) confirmed atypical tracking of right delta/theta (< 10 Hz) in children with dyslexia and confirmed that it was correlated with poor phonological skills.

Morken et al. (2017) took a longitudinal approach to cortical connectivity based on fMRI data in reading tasks in children with dyslexia and children with typical reading development. The participants were followed with repeated measurements through pre-literacy (6 years old), emergent literacy (8 years old), and literacy (12 years old) stages. They found that the difference between groups centered on connections going to and from the IFG (two connections) and the occipitotemporal cortex (three connections). For these five connections, the typical developing group showed stable or decreasing connectivity measures with age. The dyslexia group, on the other hand, showed a marked up-regulation (occipitotemporal connections) or down-regulation (IFG connections) from testing at 6 years to 8 years, followed by normalization from 8 years to 12 years. Morken et al. (2017) interpreted this as a delay in the dyslexia group to develop into the preliteracy and emergent-literacy stages. By age 12, there was no statistically significant difference in connectivity between the groups but differences in literacy skills were still present, and were in fact larger than when measured at younger ages. Note that this analysis found a different connectivity set than Molinaro et al. (2016) among others an involvement of the visual areas but again identified the role of the IFG.

Using fMRI, Bailey et al. (2018) found a positive relationship between the global modularity of an individual's brain network and reading skill, with the visual, default mode (DMN) and cinguloopercular networks (CON) showing the highest correlations. Bailey et al. (2018) suggested that "an efficient network organization, that is, one in which brain areas form cohesive resting-state networks, is important for skilled reading, and that dyslexia can be characterized by abnormal functioning of hub regions that map information between multiple systems." Lou et al. (2019) reconstructed white matter networks for dyslexic children and age-matched controls. A graph-theoretical analysis identified a network connecting the left-occipital-temporal cortex and temporoparietal cortex that

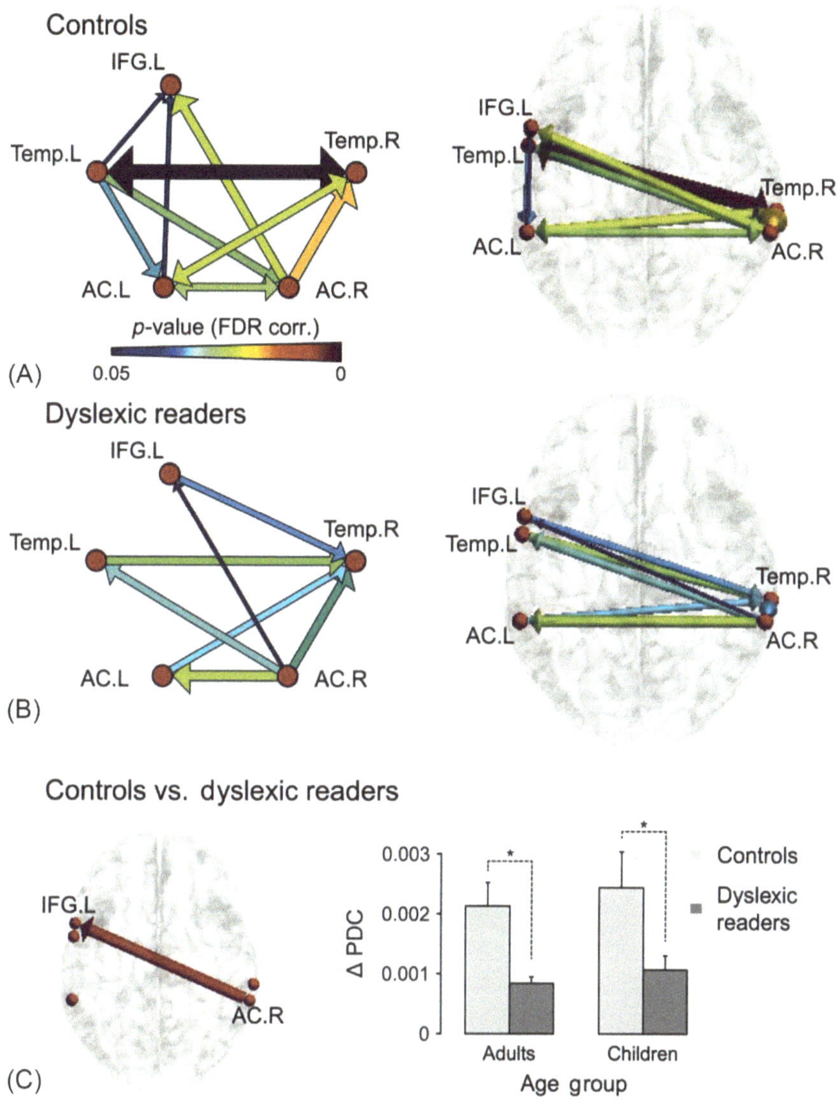

FIG. 11.5 Partial direct coherence analysis. Network dynamics for control (A) and dyslexic participants (B) among the five seeds in the 0.5–1-Hz frequency band (during Speech perception compared with Baseline) plotted on both connectivity graphs and dorsal views of the brain renderings. *Arrow* orientation represents the causal direction of the observed coupling; *arrow color and thickness* represent the statistical strength of the connection (*P* values). (C) Left panel: Differential connection strength between control and dyslexic readers (padre < 0.05, age and IQ corrected). Right panel: Strength of AC.R→IFG.L connection (for dyslexic readers and their control peers) plotted separately for adults and children. *AC*, Auditory cortex; *IFG*, inferior frontal gyrus; *Temp*, temporal gyrus. *From Molinaro, N., Lizarazu, M., Allier, M., Bourguignon, M., Carreiras, M., 2016. Out-of-synchrony speech entrainment in developmental dyslexia. Hum. Brain Mapp. 37 (8), 2767–2783, with permission from Wiley Periodicals, Inc.*

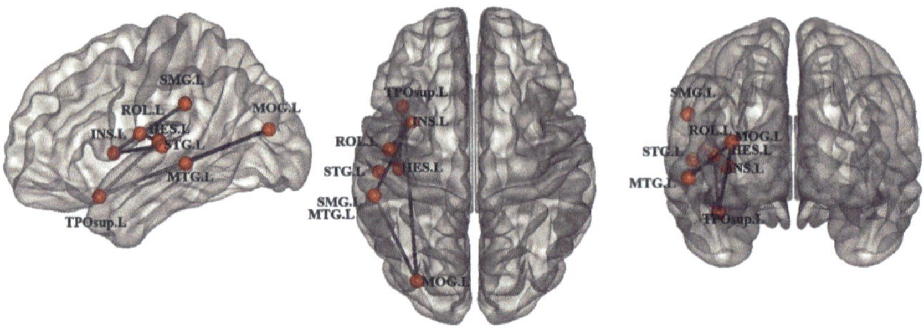

FIG. 11.6 Abnormalities in a number-of-streamline weighted brain subnetwork in developmental dyslexia. The subnetwork with significantly weaker connections compared with controls consists of a central cluster of eight nodes and seven edges located in the left occipitotemporo-parietal lobe. *L*, left; *TPOsup, superior temporal pole; MOG*, middle occipital gyrus; *MTG*, middle temporal gyrus; *STG*, superior temporal gyrus; *INS, insula; HES, Heschl's gyrus; ROL*, rolandic operculum; *SMG*, supramarginal gyrus. *From Lou, C., Duan, X., Altarelli, I., Sweeney, J.A., Ramus, F., Zhao. J., 2019. White matter network connectivity deficits in developmental dyslexia. Hum. Brain Mapp. 40 (2), 505–516, with permission from Wiley Periodicals, Inc.*

had decreased streamlines in dyslexic children (Fig. 11.6). Four network topological parameters (clustering coefficient, local efficiency, transitivity, and global efficiency) were positively correlated with literacy skills of dyslexic children, and explained a substantial proportion of additional variance in literacy skills.

Summarizing, the connectivity between bilateral ACx and IFG is impaired (Boets et al., 2013; Molinaro et al., 2016). IFG shows decreased connections with occipitotemporal cortex (Molinaro et al., 2016; Morken et al., 2017), as do left occipitotemporal cortex and temporal–parietal cortex (Lou et al., 2019). Modularity of DMN and CON correlates with reading skills (Bailey et al., 2018).

11.1.4 Summary

Goswami (2015) noted that the temporal sampling theory (Goswami, 2011) and rise-time theory (Van Hirtum et al., 2019) can explain why deficient phonemic processing and atypical rapid-rate acoustic processing may occur in dyslexia. If extraction of primitive phonological structure is mediated by neuronal oscillatory entrainment at different speech timescales (Poelmans et al., 2012; Leong and Goswami, 2014), and if entrainment to slower amplitude modulation rates is impaired in developmental dyslexia, then the mental lexicon would develop a different "auditory organization" from birth. Potentially, a primary deficit in pan-sensory oscillatory entrainment (Lehongre et al., 2013) could explain both the auditory and visual sensory difficulties.

11.2 Autism spectrum disorder

Autism spectrum disorder (ASD) is a developmental *disorder* that affects communication and behavior and occurs in ~0.7% of children (Amaral et al., 2008). Although *autism* can be diagnosed at any age, it is said to be a "developmental disorder" because symptoms generally appear in the first 2 years of life. Several brain regions have been implicated in social behavior; these include regions of the frontal lobe, the superior temporal cortex, the parietal cortex and the amygdala. ASD shows also abnormal development in the use of language. Expressive language function comprises Broca's area in the inferior frontal gyrus and portions of the supplementary motor cortex. Receptive language function requires Wernicke's area. The superior temporal sulcus plays a role in both language processing and social attention. In the following, we will expand on this based on recent publications.

11.2.1 Structural brain changes in autism spectrum disorder

11.2.1.1 Nerve tracts

Watanabe and Rees (2015) analyzed a publicly shared DTI dataset, and found that, during adolescence, brains of typically developing (TD) individuals showed increases in rich-club organization (Chapter 9) and network functionality, whereas individuals with ASD did not. ASD individuals did not show increases in gray matter volume as the TD, and this relative anatomical immaturity was correlated with the severity of ASD social symptoms. The differences between TD and ASD groups were statistically significant for both structural and functional properties. Moreover, these typical age-related changes in rich-club organization were characterized by a progressive involvement of the right anterior insula.

In a review of DTI studies, Ameis and Catani (2015) reported microstructural differences in ASD that were indicative of developmental alterations in white matter organization and potentially in myelination. Alterations of the uncinate fasciculus and frontal and temporal thalamic projections have been associated with social symptoms in ASD (Fig. 11.7). A graph theoretical analysis of DTI data (Li et al., 2018) subsequently demonstrated a decreased characteristic path length, increased global efficiency, and increased clustering coefficient compared with typical development. These all reflected increased structural connectivity between a large number of areas, in childhood ASD. Notably, a single connected network with 13 nodes and 12 connections in the left hemisphere was altered in the ASD group compared with the TD group. Compared with the TD group, ASD showed only increased connections in those nodal regions, and connectivity differences between

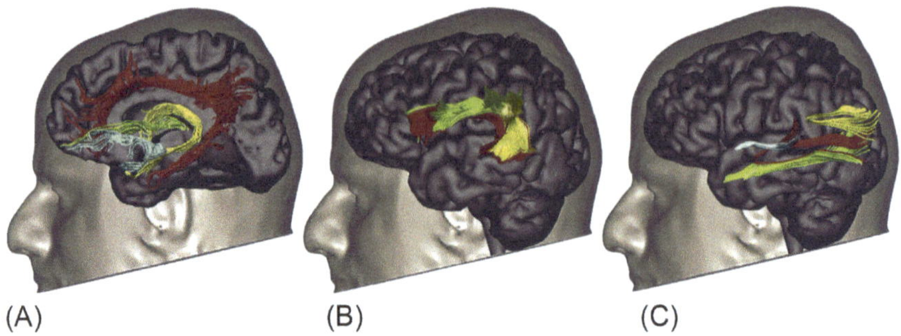

FIG. 11.7 White matter tracts of the socio-emotional processing system. (A) White matter tracts of the limbic system: cingulum bundle (*red*), uncinate fasciculus (*light blue*), fornix (*yellow*), mammillo-thalamic tract (*black*), anterior thalamic projections (*green*). (B) White matter tracts linking the mirror neuron system; segments of the arcuate fasciculus are presented: long segment (*red*), anterior segment (*green*), posterior segment (*yellow*). (C) White matter tracts of the face processing system; the inferior longitudinal fasciculus (*green*) is presented along with white matter tracts of the visual pathway: splenium fibers (*yellow*), optic radiations (*red*), optic tract (*blue*). *From Ameis, S.H., Catani, M., 2015. Altered white matter connectivity as a neural substrate for social impairment in Autism Spectrum Disorder. Cortex 62, 158–181, with permission from Elsevier.*

the ASD and TD groups were only observed in the left hemisphere. These altered patterns of global and local topologic properties may underlie the abnormal brain development process in preschool children with ASD.

Dimond et al. (2019) calculated structural metrics specific to differently oriented fiber bundles within a voxel, termed "fixels." Fixel-based analyses were used to compare diffusion-weighted MRI data from adolescents with ASD and typically developing age-matched controls. Group comparisons of fiber density and bundle morphology were run on a fixel-wise, tract-wise, and global white matter basis. Individuals with ASD showed reduced fiber density, suggesting decreased axonal count, in several major white matter tracts (cf. Fig. 11.7), including the corpus callosum, bilateral inferior frontal-occipital fasciculus, right arcuate fasciculus, and right uncinate fasciculus, as well as a global white matter reduction. Secondary analyses assessed associations with social impairment in participants with ASD, and showed that lower fiber density in the splenium of the CC was associated with greater social impairment. Dimond et al. (2019) suggested that reduced fiber density could be the primary microstructural white matter abnormality in ASD. This is in contrast to the findings of increased connectivity by Li et al. (2018).

11.2.1.2 Gray matter

From a systematic review of post-2007 MRI studies of brain morphology, Pagnozzi et al. (2018) concluded that patients with ASD had increased whole brain volume, particularly those younger than 6 years of age. Compared with typically developing controls, they identified increases in volume of the frontal and temporal lobes, cortical thickness in the frontal lobe, surface area and cortical gyrification, and cerebrospinal fluid volume. This was accompanied by reductions in cerebellum volume and corpus callosum volume. Pereira et al. (2018) used fMRI-based multimodal morphometry to investigate brain changes in adolescents and young adults with high-functioning autism compared with age-, and intelligence quotient-matched healthy controls. ASD patients had gray matter reduction, *decreased* cortical thickness and larger cortical surface areas in several brain regions, including the cingulate, temporal lobes, and amygdala, as well as increased gyrification in regions associated with encoding visual memories and areas of the sensorimotor component of the DMN, more pronounced in the left hemisphere.

Summarizing, some studies found increased whole brain volume and cortical thickness, and reduced volume for cerebellum and CC (Pagnozzi et al., 2018) whereas others reported decreased cortical volume and thickness combined with increased surface area in cingulate, temporal lobes, and amygdala (Pereira et al., 2018).

11.2.2 Brain rhythms in autism spectrum disorder

Orekhova et al. (2007) recorded EEG during sustained visual attention in two independent samples of 3–8 year old boys with autism from Moscow and Gothenburg and in age-matched typically developing boys. In both samples, boys with autism demonstrated a pathological increase of gamma (24.4–44.0 Hz) power, which correlated positively with degree of developmental delay in boys with autism. The same group (Stroganova et al., 2007) found that boys with ASD comprised a non-homogeneous group in relation to theta- and alpha-power. They also demonstrated a typical leftward—stronger—broadband EEG asymmetry with a maximum effect over the mid-temporal regions in ASD. Concurrently the normal leftward asymmetry of the alpha rhythm in mid-temporal regions was absent in ASD (Stroganova et al., 2007). Using child-size MEG equipment, Kikuchi et al. (2013) obtained measurements of the brain activity of young children (3–7 years old) with ASD and age-matched TD children. In contrast to Stroganova et al. (2007), who studied a similar age group, they found significant rightward connectivity between the parietotemporal areas, by increased gamma-band oscillations, in the ASD group.

Clarke et al. (2016) recorded EEG in 20 boys with Asperger's syndrome, aged 7–12 years, and an age and sex matched control group in the eyes closed condition. They found that the Asperger's group had a global increase in absolute delta and an anterior increase in relative delta. Both absolute and relative theta was globally increased and relative alpha was globally decreased. Subjects with Asperger's syndrome exhibited reduced anterior interhemispheric coherence in the alpha and beta bands. These results suggested to Clarke et al. (2016) the existence of frontal lobe abnormalities in children with Asperger's syndrome, and possible abnormalities in normal brain maturational processes. Asperger children may have CNS abnormalities different from children with other forms of ASD.

11.2.3 The role of the thalamus

Alpha circuits (8–12 Hz), necessary for basic and complex brain processes (Chapters 3 and 4), are abnormal in ASD. Matlis et al. (2015) corroborated the general finding that ASD subjects have decreased alpha power gradients and network connectivity compared with controls. Because the thalamus modulates cortical resting state alpha rhythms, Edgar et al. (2015) examined associations between thalamic structure and alpha activity. They found a lack of associations between thalamic structure and alpha in ASD suggesting thalamic involvement in these resting state alpha abnormalities. This confirmed the DTI findings by Cheon et al. (2011) who found hypoconnectivity between the frontal cortex and thalamus in ASD. In an fMRI study, Iidaka et al. (2019) in contrast showed that patients in the ASD group exhibited significant hyperconnectivity between the thalamus and various cortical regions distributed throughout the frontoparietal lobes. In addition, they found significant hypoconnectivity between the amygdala and six cortical regions.

11.2.4 Functional connectivity in autism spectrum disorder

11.2.4.1 MRI based

Keown et al. (2017) used resting-state fMRI data from ASD and typically developing participants and applied several graph theory metrics. They found that while differences in hub architecture were not significant, rich club connectivity (i.e., among the hubs; Chapter 6) was increased in the ASD group. Lynch et al. (2017) assessed 7–13-year-old children with ASD and age- and IQ-matched typically developing control subjects in both a resting-state and a selective attention task fMRI session. They found that: "The participation coefficient of the frontoparietal-insular cortex, including core nodes of the frontoparietal control and salience networks, significantly increased in typically developing children but not in children with ASD during the task relative to rest."

Pereira et al. (2018) examined the fMRI-based FC of the default mode network to investigate superregional, system-level alterations in adolescents and young adults with high-functioning autism compared with age-, and intelligence quotient-matched healthy controls. The main findings were that patients with ASD had decreased connectivity between the posterior cingulate cortex (PCC), and areas of the executive control component of the DMN and increased FC between the anteromedial prefrontal cortex (PFC) and areas of the sensorimotor component of the DMN. Bi et al. (2018) showed abnormal FC of RSNs in ASD patients (aged 8–18 years) compared with healthy controls. The RSNs with the FC in ASD patients included DMN, executive control network (ECN), core network, visual network, and self-referential network. The RSNs with the FC in ASD patients included the auditory network and the somatomotor network. The dorsal attention network (DAN) in ASD patients showed decreased FC. This indicates that findings of abnormal FCs in RSNs are extensive in ASD patients. Using resting-state fMRI data in a large sample of ASD toddlers and other non-ASD comparison groups, Lombardo et al. (2019) found ASD-related functional hypoconnectivity between "social brain" circuitry such as the DMN and visual and attention networks. An ASD subtype with pronounced early social visual engagement difficulties is characterized by marked DMN-occipitotemporal cortex (OTC) hypoconnectivity. Increased DMN-OTC hypoconnectivity was also related to increased severity of social-communication difficulties, but only in ASD with early social visual engagement difficulties.

Oldehinkel et al. (2019) obtained resting-state fMRI data in individuals with ASD (7.5–30.3 years) and typically developing individuals (6.9–29.8 years), all with IQ > 70. They found that higher autism traits were associated with connectivity within the salience, medial motor, and orbitofrontal networks. The between-network analysis showed on average (1) decreased connectivity of the visual association network with somatosensory, medial, and lateral motor networks, and (2) increased connectivity of the cerebellum with these sensory and motor networks in ASD compared with typically developing subjects. These between-network alterations are potentially underlying impairments in multisensory and visual-motor integration frequently observed in ASD. Jung et al. (2019) applied machine-learning techniques to investigate patterns of surface-based brain resting-state connectivity in boys (7–15 years) with ASD and age-matched boys with typical development. Compared with the typical development group, the ASD group showed increased connectivity in brain areas associated with the limbic, visual, default mode, somatomotor, dorsal attention, frontoparietal, and ventral attention networks. These results appear to be at variance with the findings of Bi et al. (2018) and Oldehinkel et al. (2019). Different ASD subtypes may underlie this (e.g., Lombardo et al., 2019).

Using resting-state fMRI, Hong et al. (2019) demonstrated atypical connectivity transitions between sensory and higher-order default mode regions in a large cohort of individuals with autism relative to typically developing controls. The present results suggest a diminished segregation in ASD between sensory systems and unimodal convergence regions such as posterior middle temporal gyrus (pMTG) and occipitotemporal network (OTN) on the one hand, and transmodal hubs such as mPFC and PCC/precuneus on the other hand. In typically developing individuals, an important role of OTN and pMTG in the integration of feedforward and feedback streams has previously been recognized in the visual and auditory system, particularly for face and language processing.

Using rs-fMRI, Guo et al. (2019) observed significantly impaired FC in ASD in the ventral medial prefrontal cortex and PCC, which are two critical hubs of the DMN. States in which ASD showed decreased connectivity between the right anterior insula and these regions were those more relevant to socio-cognitive imaging. Lawrence et al. (2019) examined the FC of three neurocognitive resting-state networks—the DMN, salience network, and ECN—in a longitudinal sample of youth with ASD and without ASD studied during early/mid- and late adolescence. FC between the ECN and the DMN displayed significantly altered developmental trajectories in ASD. TD controls—but not youth with ASD—exhibited an increase in negative FC between these two networks with age. Lawrence et al. (2019) concluded: "This significant interaction was due to the ASD group displaying less negative functional connectivity than the TD group during late adolescence only, with no significant group differences in early/mid-adolescence. These preliminary findings suggest a localized age-dependency of functional connectivity alterations in ASD and underscore the importance of considering age when examining brain connectivity." Compare with Iidaka et al. (2019) with respect to the thalamic-cortical connections.

From an rs-fMRI study, Rolls et al. (2020) found (Fig. 11.8) that directed connectivity from the MTG, implicated in face expression processing and theory of mind, to the precuneus and cuneus implicated in processing about the self was reduced. In addition, connectivity from the MTG to the VMPFC, involved in emotion and emotion-related decision-making, was reduced in autism. Another set of findings was that effective connectivity from memory and emotion-related areas, the hippocampus and amygdala, respectively, to the MTGs was increased in autism. Effective connectivity from the basal ganglia to the precentral and postcentral gyrus and midcingulate cortex was increased in autism.

11.2.4.2 MEG/EEG based

From a critical review of MEG/EEG data, O'Reilly et al. (2017) found relatively strong support for long-range under-connectivity in ASD, whereas the status of local connectivity remained unclear. There was exemplified by lower frequencies, reflecting long-range connections, being often associated with under-connectivity and higher frequencies, underlying local connectivity, being associated with both under- and

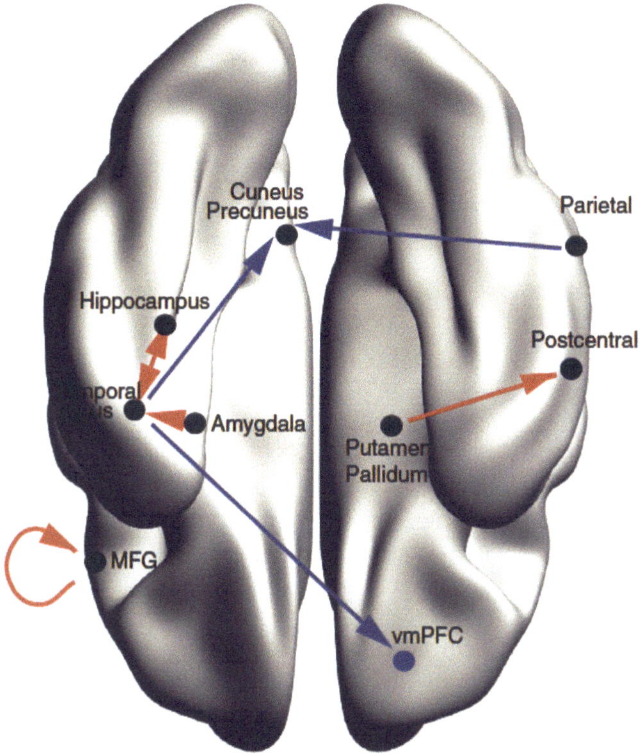

FIG. 11.8 Summary of some of the main changes in effective connectivity in autism. The increased effective connectivities in autism are shown in *red* and decreased in *blue*. A *blue circle* indicates a decrease in Sigma in autism. *From Rolls, E.T., Zhou, Y., Cheng, W., Gilson, M., Deco, G., Feng, J., 2020. Effective connectivity in autism. Autism Res. 13 (1), 32–44, with permission from International Society for Autism Research, Wiley Periodicals, Inc.*

over-connectivity. Abnormal lateralization in ASD, specifically an elevated left-over-right EEG and MEG functional connectivity ratio, has been also reported consistently across studies. They concluded that: "ASD is characterized by a general trend toward an under-expression of lower-band wide-spread integrative processes compensated by more focal, higher-frequency, locally specialized, and segregated processes." Ghuman et al. (2017) recorded spontaneous MEG data and found that: "Individuals with ASD showed local hyper-synchrony in the theta-band (4–7 Hz) in the lateral occipitotemporal cortex. Long-range hyposynchrony was seen in the alpha-band (10–13 Hz), which was most prominent in neural circuitry underpinning social processing. The magnitude of this alpha-band hyposynchrony was correlated with social symptom severity." Vakorin et al. (2017) recorded resting-state MEG from children and adolescents with and without ASD, and calculated resting spectral power and functional connectivity. Although no overall group differences were observed, significant alterations in age-related changes in resting oscillatory power and functional connectivity were found. These findings establish that, rather than overall differences in network synchrony or spectral power, brain function in ASD is characterized by atypical maturational shifts in neurophysiological oscillations and inter-regional phase synchrony. This is consistent with evidence from other modalities indicating that it is the developmental trajectory of brain connectivity that is associated with ASD, rather than over-connectivity or under-connectivity per se (Vakorin et al., 2017).

11.2.4.3 Auditory system involvement

I have outlined the deficits in auditory temporal processing in autism in Chapter 15 of a previous book (Eggermont, 2015). I largely found neural synchrony changes responsible for the auditory problems. Jochaut et al. (2015) concurrently recorded EEG and fMRI in ASD subjects and controls (15–40 years of age) and showed that in autism, gamma, and theta cortical activity do not engage synergistically in response to speech (as seen in TD; Chapter 2). Theta activity in left auditory cortex failed to track speech modulations, and to downregulate gamma oscillations in the group with autism. This deficit predicts the severity of both verbal impairment and autism symptoms in the affected sample. Jochaut et al. (2015) also found that oscillation-based connectivity between auditory and other language cortices is altered in autism (Fig. 11.9). These results suggest that the verbal disorder in autism could be associated with an altered balance of slow and fast auditory oscillations and that this anomaly could compromise the mapping between sensory input and higher-level cognitive representations.

11.2.5 Summary

Some studies suggest increased whole brain volume and cortical thickness, and reduced volume for cerebellum and corpus callosum, whereas others reported decreased cortical volume and thickness combined with increased surface area in cingulate, temporal lobes, and amygdala. Pathological increase of gamma power correlated positively with degree of developmental delay in boys with autism. Asperger's had a global increase in absolute delta and an anterior increase in relative delta. Both absolute and relative theta was globally increased and relative alpha was globally decreased. Verbal disorders in autism were related to an altered balance of slow and fast auditory oscillations.

As the compilation in Fig. 11.10 shows, local connectivity (diagonal) is only consistently down in the ECN, but changes in DMN, auditory, visual and somatomotor networks and FPN depend on the age of testing. Consistent changes are found in interarea connectivity strengths;

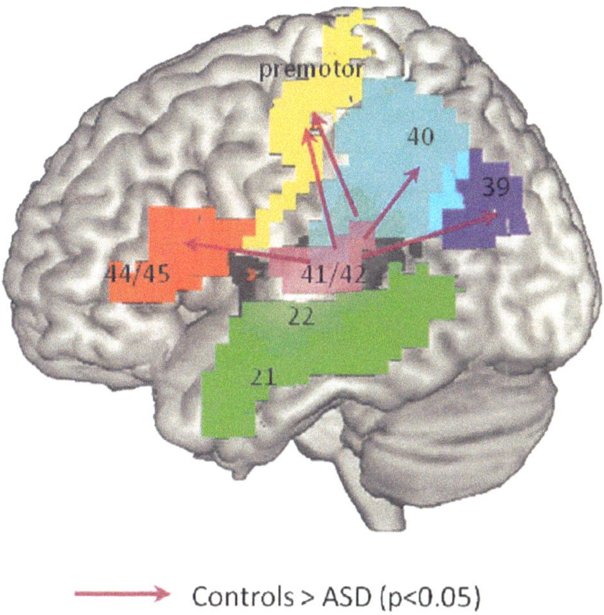

FIG. 11.9 Oscillation-based functional disconnection of auditor cortex in autism. Statistics are reported on a brain template; they show that the primary auditory cortex (BA41/42) has reduced connectivity with Brodmann Areas 44, 45, 40, 39, and the premotor cortex relative to controls. *From Jochaut, D., Lehongre, K., Saitovitch, A., Devauchelle, A.-D., Olasagasti, I., Chabane, N., et al., 2015. Atypical coordination of cortical oscillations in response to speech in autism. Front. Hum. Neurosci. 9, 171; https://doi.org/10.3389/fnhum.2015.00171, Open, access.*

	DMN	PFC	MTG	precun	PCC	AUD	VIS	SoMo	Cereb	FPN	ECN
DMN	⇓⇑	⇑			⇓						⇓
PFC	⇑		⇓								
MTG		⇓		⇓	⇓						
precun			⇓								
PCC	⇓		⇓		⇓						⇓
AUD						⇑⇓		⇑⇓	⇑		
VIS							⇓⇑	⇓	⇑		
SoMo						⇑⇓	⇓	⇓⇑			
Cereb						⇑	⇑				
FPN										⇓⇑	
ECN	⇓				⇓						⇓

FIG. 11.10 Area-wise changes in activity (diagonal) and functional connectivity (off-diagonal) in ASD. *AUD*, Auditory network; *cereb*, cerebellum; *ECN*, executive control network; *DMX, default mode network*; *FPN*, frontoparietal network; *MTG*, medial temporal gyrus; *PCC*, posterior cingulate cortex; *PFC*, prefrontal cortex; *precun*, precuneus; *SoMo*, somato-motor network; *VIS*, visual network.

these were enhanced between PFC and DMN, and between cerebellum and auditory and visual areas. Decreased connectivity is more prevalent and found between MTG and PFS; between precuneus and PCC; between DMN and ECN; between PCC and DMN; and between ECN and PCC.

References

Aboitiz, F., 1992. Brain connections: interhemispheric fiber systems and anatomical brain asymmetries in humans. Biol. Res. 25 (2), 51–61.

Amaral, D.G., Mills Schumann, C., Wu Nordahl, C., 2008. Neuroanatomy of autism. Trends Neurosci. 31 (3), 137–145.

Ameis, S.H., Catani, M., 2015. Altered white matter connectivity as a neural substrate for social impairment in autism spectrum disorder. Cortex 62, 158–181.

Babiloni, C., Stella, G., Buffo, P., Vecchio, F., Onorati, P., Muratori, C., et al., 2012. Cortical sources of resting state EEG rhythms are abnormal in dyslexic children. Clin. Neurophysiol. 123, 2384–2391.

Bailey, S.K., Aboud, K.S., Nguyen, T.Q., Cutting, L.E., 2018. Applying a network framework to the neurobiology of reading and dyslexia. J. Neurodev. Disord. 10 (1), 37. https://doi.org/10.1186/s11689-018-9251-z.

Bi, X., Zhao, J., Xu, Q., Sun, Q., Wang, Z., 2018. Abnormal functional connectivity of resting state network detection based on linear ICA analysis in autism spectrum disorder. Front. Physiol. 9, 475. https://doi.org/10.3389/fphys.2018.00475.

Blau, V., Reithler, J., van Atteveldt, N., Seitz, J., Gerretsen, P., Goebel, R., et al., 2010. Deviant processing of letters and speech sounds as proximate cause of reading failure: a functional magnetic resonance imaging study of dyslexic children. Brain 133 (Pt 3), 868–879.

Boets, B., Op de Beeck, H.P., Vandermosten, M., Scott, S.K., Gillebert, C.R., Mantini, D., et al., 2013. Intact but less accessible phonetic representations in adults with dyslexia. Science 342, 1251–1254.

Cheon, K.-A., Kim, Y.-S., Oh, S.-H., Park, S.-Y., Yoon, H.-W., Herrington, J., et al., 2011. Involvement of the anterior thalamic radiation in boys with high functioning autism spectrum disorders: a diffusion tensor imaging study. Brain Res. 1417, 77–86.

Clarke, A.C., Barry, R.J., Indraratna, A., Dupuy, F.E., McCarthy, R., Selikowitz, M., 2016. EEG activity in children with Asperger's syndrome. Clin. Neurophysiol. 127, 442–451.

De Vos, A., Vanvooren, S., Vanderauwera, J., Ghesquière, P., Wouters, J., 2017. Atypical neural synchronization to speech envelope modulations in dyslexia. Brain Lang. 164, 106–117.

Di Liberto, G.M., Peter, V., Kalashnikova, M., Usha Goswami, U., Burnham, D., Lalor, E.C., 2018. Atypical cortical entrainment to speech in the right hemisphere underpins phonemic deficits in dyslexia. NeuroImage 175, 70–79.

Díaz, B., Hintz, F., Kiebel, S.J., von Kriegstein, K., 2012. Dysfunction of the auditory thalamus in developmental dyslexia. Proc. Natl. Acad. Sci. U. S. A. 109, 13841–13846.

Dimond, D., Schuetze, M., Smith, R.E., Dhollander, T., Cho, I., Vinette, S., et al., 2019. Reduced white matter fiber density in autism spectrum disorder. Cereb. Cortex 29 (4), 1778–1788.

Dougherty, R.F., Ben-Shachar, M., Deutsch, D.K., Hernandez, A., Fox, G.R., Wandell, B.A., 2007. Temporal-callosal pathway diffusivity predicts phonological skills in children. Proc. Natl. Acad. Sci. U. S. A. 104 (20), 8556–8561.

Edgar, J.C., Heiken, K., Chen, Y.-H., Herrington, J.D., Chow, V., Liu, S., et al., 2015. Resting-state alpha in autism spectrum disorder and alpha associations with thalamic volume. J. Autism Dev. Disord. 45, 795–804. https://doi.org/10.1007/s10803-014-2236-1.

Eggermont, J.J., 2015. Auditory Temporal Processing and Its Disorders. Oxford University Press, Oxford, UK.

Ghuman, A.S., van den Honert, R.N., Huppert, T.J., Wallace, G.L., Martin, A., 2017. Aberrant oscillatory synchrony is biased toward specific frequencies and processing domains in the autistic brain. Biol. Psychiatry: Cogn. Neurosci. Neuroim. 2, 245–252.

Giraud, A.L., Kleinschmidt, A., Poeppel, D., Lund, T.E., Frackowiak, R.S.J., Laufs, H., 2007. Endogenous cortical rhythms determine cerebral specialization for speech perception and production. Neuron 56, 1127–1134.

Goswami, U., 2011. A temporal sampling framework for developmental dyslexia. Trends Cogn. Sci. 15 (1), 3–10.

Goswami, U., 2015. Sensory theories of developmental dyslexia: three challenges for research. Nat. Rev. Neurosci. 16, 43–54.

Guo, X., Duan, X., Suckling, J., Chen, H., Liao, W., Cui, Q., et al., 2019. Partially impaired functional connectivity states between right anterior insula and default mode network in autism spectrum disorder. Hum. Brain Mapp. 40, 1264–1275.

Hofer, S., Frahm, J., 2006. Topography of the human corpus callosum revisited – comprehensive fiber tractography using tensor magnetic resonance imaging. NeuroImage 32, 989–994.

Hong, S.-J., Vos de Wael, R., Bethlehem, R.A.I., Lariviere, S., Paquola, C., Valk, S.L., et al., 2019. Atypical functional connectome hierarchy in autism. Nat. Commun. 10, 1022. https://doi.org/10.1038/s41467-019-08944-1.

Iidaka, T., Kogata, T., Mano, Y., Komeda, H., 2019. Thalamocortical hyperconnectivity and amygdala-cortical hypoconnectivity in male patients with autism spectrum disorder. Front. Psychiatry 10, 252. https://doi.org/10.3389/fpsyt.2019.00252.

Jaffe-Dax, S., Kimel, E., Ahissar, M., 2018. Shorter cortical adaptation in dyslexia is broadly distributed in the superior temporal lobe and includes the primary auditory cortex. elife 7. https://doi.org/10.7554/eLife.30018.

Jochaut, D., Lehongre, K., Saitovitch, A., Devauchelle, A.-D., Olasagasti, I., Chabane, N., et al., 2015. Atypical coordination of cortical oscillations in response to speech in autism. Front. Hum. Neurosci. 9, 171. https://doi.org/10.3389/fnhum.2015.00171.

Johnson, B.W., McArthur, G., Hautus, M., Reid, M., Brock, J., Castles, A., et al., 2013. Lateralized auditory brain function in children with normal reading ability and in children with dyslexia. Neuropsychologia 51, 633–641.

Jung, M., Tu, Y., Park, J., Jorgenson, K., Lang, C., Song, W., et al., 2019. Surface-based shared and distinct resting functional connectivity in attention-deficit hyperactivity disorder and autism spectrum disorder. Br. J. Psychiatry 214, 339–344. https://doi.org/10.1192/bjp.2018.248.

Keown, C.L., Datko, M.C., Chen, C.P., Maximo, J.O., Jahedi, A., Müller, R.-A., 2017. Network organization is globally atypical in autism: a graph theory study of intrinsic functional connectivity. Biol. Psychiatry: Cogn. Neurosci. Neuroim. 2, 66–75.

Kikuchi, M., Shitamichi, K., Yoshimura, Y., Ueno, S., Hiraishi, H., Hirosawa, T., et al., 2013. Altered brain connectivity in 3-to 7-year-old children with autism spectrum disorder. NeuroImage: Clin. 2, 394–401.

Klingberg, T., Hedehus, M., Temple, E., Salz, T., Gabrieli, J.D.E., Moseley, M.E., et al., 2000. Microstructure of temporo-parietal white matter as a basis for reading ability. Evidence from diffusion tensor magnetic resonance imaging. Neuron 25, 493–500.

Krafnick, A.J., Flowers, D.L., Luetje, M.M., Napoliello, E.M., Eden, G.F., 2014. An investigation into the origin of anatomical differences in dyslexia. J. Neurosci. 34 (3), 901–908.

Lawrence, K.E., Hernandez, L.M., Bookheimer, S.Y., Dapretto, M., 2019. Atypical longitudinal development of functional connectivity in adolescents with autism spectrum disorder. Autism Res. 12, 53–65.

Lehongre, K., Ramus, F., Villiermet, N., Schwartz, D., Giraud, A.L., 2011. Altered low-gamma sampling in auditory cortex accounts for the three main facets of dyslexia. Neuron 72, 1080–1090.

Lehongre, K., Morillon, B., Giraud, A.L., Ramus, F., 2013. Impaired auditory sampling in dyslexia: further evidence from combined fMRI and EEG. Front. Hum. Neurosci. 7, 454. https://doi.org/10.3389/fnhum.2013.00454.

Leonard, C.M., Eckert, M.A., 2008. Asymmetry and dyslexia. Dev. Neuropsychol. 33 (6), 663–681.

Leong, V., Goswami, U., 2014. Assessment of rhythmic entrainment at multiple timescales in dyslexia: evidence for disruption to syllable timing. Hear. Res. 308, 141–161.

Li, S.-J., Wang, Y., Qian, L., Liu, G., Liu, S.-F., Zou, L.-P., et al., 2018. Alterations of white matter connectivity in preschool children with autism spectrum disorder. Radiology 288, 209–217.

Lombardo, M.V., Eyler, L., Moore, A., Datko, M., Barnes, C.C., Cha, D., et al., 2019. Default mode-visual network hypoconnectivity in an autism subtype with pronounced social visual engagement difficulties. elife 8. https://doi.org/10.7554/eLife.47427.

Lou, C., Duan, X., Altarelli, I., Sweeney, J.A., Ramus, F., Zhao, J., 2019. White matter network connectivity deficits in developmental dyslexia. Hum. Brain Mapp. 40 (2), 505–516.

Lynch, C.J., Breeden, A.L., You, X., Ludlum, R., Gaillard, W.D., Kenworthy, L., et al., 2017. Executive dysfunction in autism spectrum disorder is associated with a failure to modulate frontoparietal-insular hub architecture. Biol. Psychiatry: Cogn. Neurosci. Neuroim. 2, 537–545.

Mamah, D., Ji, A., Rutlin, J., Shimony, J.S., 2019. White matter integrity in schizophrenia and bipolar disorder: tract- and voxel-based analyses of diffusion data from the Connectom scanner. NeuroImage Clin. 21, 101649.

Mandonnet, E., Nouet, A., Gatignol, P., Capelle, L., Duffau, H., 2007. Does the left inferior longitudinal fasciculus play a role in language? A brain stimulation study. Brain 130, 626–629.

Matlis, S., Boric, K., Chu, C.J., Kramer, M.A., 2015. Robust disruptions in electroencephalogram cortical oscillations and large-scale functional networks in autism. BMC Neurol. https://doi.org/10.1186/s12883-015-0355-8.

Molinaro, N., Lizarazu, M., Allier, M., Bourguignon, M., Carreiras, M., 2016. Out-of-synchrony speech entrainment in developmental dyslexia. Hum. Brain Mapp. 37, 2767–2783.

Moreau, D., Wiebels, K., Wilson, A.J., Waldie, K.E., 2019. Volumetric and surface characteristics of gray matter in adult dyslexia and dyscalculia. Neuropsychologia 127, 204–210.

Morken, F., Helland, T., Hugdahl, K., Specht, K., 2017. Reading in dyslexia across literacy development: a longitudinal study of effective connectivity. NeuroImage 144, 92–100.

Müller-Axt, C., Anwander, A., von Kriegstein, K., 2017. Altered structural connectivity of the left visual thalamus in developmental dyslexia. Curr. Biol. 27, 3692–3698.

Odegard, T.N., Farris, E.A., Ring, J., McColl, R., Black, J., 2009. Brain connectivity in non-reading impaired children and children diagnosed with developmental dyslexia. Neuropsychologia 47, 1972–1977.

Oldehinkel, M., Mennes, M., Marquand, A., Charman, T., Tillmann, J., Ecker, C., et al., 2019. Altered connectivity between cerebellum, visual, and sensory-motor networks in autism spectrum disorder: results from the EU-AIMS longitudinal European autism project. Biol. Psychiatry: Cogn. Neurosci. Neuroim. 4, 260–270.

O'Reilly, C., Lewis, J.D., Elsabbagh, M., 2017. Is functional brain connectivity atypical in autism? A systematic review of EEG and MEG studies. PLoS One 12 (5). https://doi.org/10.1371/journal.pone.0175870.

Orekhova, E.V., Stroganova, T.A., Nygren, G., Tsetlin, M.M., Posikera, I.N., Gillberg, C., et al., 2007. Excess of high frequency electroencephalogram oscillations in boys with autism. Biol. Psychiatry 62, 1022–1029.

Pagnozzi, A.M., Conti, E., Calderoni, S., Fripp, J., Rose, S.E., 2018. A systematic review of structural MRI biomarkers in autism spectrum disorder: a machine learning perspective. Int. J. Dev. Neurosci. 71, 68–82.

Pereira, A.M., Campos, B.M., Coan, A.C., Pegoraro, L.F., de Rezende, T.J.R., Obeso, I., et al., 2018. Differences in cortical structure and functional MRI connectivity in high functioning autism. Front. Neurol. 9, 539. https://doi.org/10.3389/fneur.2018.00539.

Poelmans, H., Luts, H., Vandermosten, M., Boets, B., Ghesquière, P., Wouters, J., 2012. Auditory steady state cortical responses indicate deviant phonemic-rate processing in adults with dyslexia. Ear Hear. 33, 134–143.

Power, A.J., Colling, L.J., Mead, N., Barnes, L., Goswami, U., 2016. Neural encoding of the speech envelope by children with developmental dyslexia. Brain Lang. 160, 1–10.

Rolls, E.T., Zhou, Y., Cheng, W., Gilson, M., Deco, G., Feng, J., 2020. Effective connectivity in autism. Autism Res. 13 (1), 32–44.

Steinmann, S., Mulert, C., 2012. Functional relevance of interhemispheric fiber tracts in speech processing. J. Neurolinguistics 25, 1–12.

Stroganova, T.A., Nygren, G., Tsetlin, M.M., Posikera, I.N., Gillberg, C., Elam, M., et al., 2007. Abnormal EEG lateralization in boys with autism. Clin. Neurophysiol. 118, 1842–1854.

Temple, E., Poldrack, R.A., Salidis, J., Deutsch, G.K., Tallal, P., Merzenich, M.M., et al., 2001. Disrupted neural responses to phonological and orthographic processing in dyslexic children: an fMRI study. NeuroReport 12, 299–307.

Tierney, A., Strait, D.L., Kraus, N., 2014. Resting gamma power is linked to reading ability in adolescents. Dev. Sci. 17 (1), 86–93.

Torre, G.-A.A., Eden, G.F., 2019. Relationships between gray matter volume and reading ability in typically developing children, adolescents, and young adults. Dev. Cogn. Neurosci. 36, 100636. https://doi.org/10.1016/j.dcn.2019.100636.

Tschentscher, N., Ruisinger, A., Blank, H., Díaz, B., von Kriegstein, K., 2019. Reduced structural connectivity between left auditory thalamus and the motion-sensitive planum temporale in developmental dyslexia. J. Neurosci. 39 (9), 1720–1732.

Vakorin, V.A., Doesburg, S.M., Leung, R.C., Vogan, V.M., Anagnostou, E., Taylor, M.J., 2017. Developmental changes in neuromagnetic rhythms and network synchrony in autism. Ann. Neurol. 81, 199–211.

van Atteveldt, N., Formisano, E., Goebel, R., Blomert, L., 2004. Integration of letters and speech sounds in the human brain. Neuron 43, 271–282.

Van Hirtum, T., Ghesquière, P., Wouters, J., 2019. Atypical neural processing of rise time by adults with dyslexia. Cortex 113, 128–140.

Vandermosten, M., Boets, B., Wouters, J., Ghesquière, P., 2012. A qualitative and quantitative review of diffusion tensor imaging studies in reading and dyslexia. Neurosci. Biobehav. Rev. 36, 1532–1552.

Vandermosten, M., Poelmans, H., Sunaert, S., Ghesquière, P., Wouters, J., 2013. White matter lateralization and interhemispheric coherence to auditory modulations in normal reading and dyslexic adults. Neuropsychologia 51, 2087–2099.

Watanabe, T., Rees, G., 2015. Age-associated changes in rich-club organisation in autistic and neurotypical human brains. Sci. Rep. 5, 16152. https://doi.org/10.1038/srep16152.

CHAPTER

12

Schizophrenia and Alzheimer's disease

12.1 Introduction

Here, we pay attention to two dissynchrony disorders that have an auditory component; schizophrenia (SZ) and Alzheimer's disease (AD). I covered the auditory temporal processing aspects in SZ in a previous book (Eggermont, 2015a). Synchronization in different electroencephalogram (EEG) frequency bands has been associated with various cognitive functions and the integration of information in the healthy brain (Uhlhaas and Singer, 2010). In neurological disorders, this process of functional integration can become disrupted and give rise to various symptoms of cognitive dysfunction (Stam, 2010). Synchronization here refers to the existence of a consistent relationship between activity patterns of two or more spatially separated neuronal groups. It often implies that there is a consistent relation between the phases of the oscillatory activity of two brain regions (Chapter 6). Many neurological and psychiatric disorders can be described as disconnection syndromes because their symptoms and clinical manifestations can be related to disrupted integration of spatially distributed regions of the brain that are part of large-scale networks underlying specific functions (Filippi et al., 2013).

Brain networks are more-and-more analyzed with graph theoretical methods (Chapter 6). A brief recapitulation of the methods and terminology of graph-theory is provided by Fig. 12.1. Crossley et al. (2016) mapped the nodes of a normative connectome defined by a prior meta-analysis of 1641 functional neuroimaging studies of task-related activation in healthy volunteers. Under-activations and over-activations in neuropsychiatric patients, for example, SZ, relative to controls were found in a wide diversity of brain regions. Both under- and over-activations in neurological disorders were significantly more likely to be located in hub nodes that constitute the "rich club" or core of the normative connectome.

https://doi.org/10.1016/B978-0-12-819818-6.00006-6

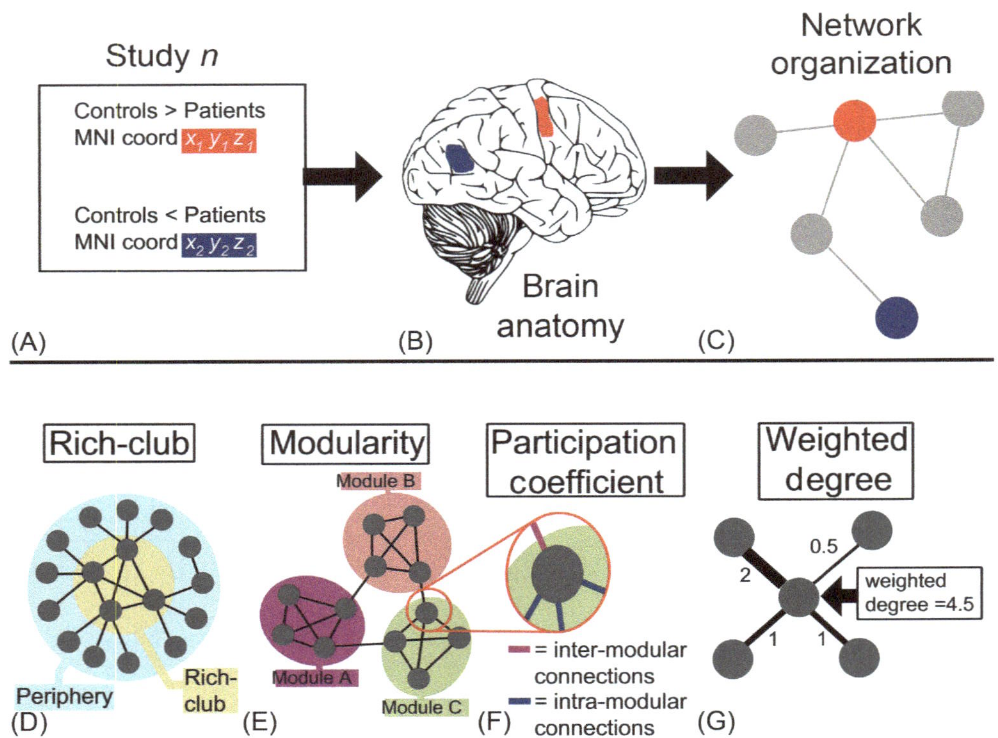

FIG. 12.1 Overview of the graph-theoretical methods and terminology. (A) The coordinates of under-activations (*red*) and over-activations (*blue*) during a task in schizophrenia were retrieved. (B) These were mapped onto regions of interest (ROIs) defined in standard neuroanatomical space. (C) The coordinates were also mapped onto a normative connections (brain network) based on the same ROIs. This network "mapping" exercise provided information about the position in the network where each differential activation was located. Interest was focused on the relationship between abnormal activations and specific network configurations such as the rich club (highly connected high degree nodes) (D) or network modules (groups of densely connected nodes that are sparsely connected to nodes m other modules) (E). These were associated with the participation coefficient (F), which refers to the relative proportions of intramodular and intermodular connections a node mediates, and the weighted degree (G), which is the sum of connection strengths to a node. *From Crossley, N.A., Mechelli, A., Ginestet, C., Rubinov, M., Bullmore, E.T., McGuire, P., 2016. Altered hub functioning and compensatory activations in the connectome: a meta-analysis of functional neuroimaging studies in schizophrenia. Schizophr. Bull. 42 (2), 434–442, Open access.*

12.2 Schizophrenia

Schizophrenia is a neuropsychiatric illness with a prevalence of 1%–2%. Core impairments include positive symptoms (hallucinations and delusions), negative symptoms (e.g., impoverished speech, social deficits, inability to feel pleasure, and apathy), and cognitive deficits (attention, working memory, and executive functions). Additionally, there are significant abnormalities in sensory and perceptual processing (Gandal et al., 2012). Specifically, SZ patients display deficits in both basic nonverbal auditory processing—pitch, loudness, amplitude modulation, and duration (Eggermont, 2015a,b)—and source monitoring of speech (Dondé et al., 2019).

12.2.1 MRI-based network changes

Hu et al. (2017) noted that changes in the default mode network (DMN) have become central in neuropsychiatric disorders, including SZ. They (cf. Fig. 1.1) outlined the DMN and described its functional and anatomical connectivity, and anti-correlation of its activity with that of the task positive network (TPN), often referred to as executive control network (ECN). From a review on the rationale for the TPN and DMN and preliminary findings of the two networks in schizophrenia, Hu et al. (2017) described that these two anti-correlated networks (Chapter 1) may be important for task performance and self-referencing (Williamson, 2007). Strong support for the involvement of anti-correlated networks in schizophrenia patients came from Whitfield-Gabrieli et al. (2009), who detected significantly reduced anti-correlations between the two networks in both patients and their relatives, suggesting a genetic risk of schizophrenia. Other studies (Chai et al., 2011; Wolf et al., 2009) have also noted these significantly reduced anti-correlations between the DMN and TPN in schizophrenia patients (cited in Hu et al., 2017).

Using diffusion tensor imaging (DTI) and graph theoretical analysis, Van den Heuvel et al. (2010) examined brain networks of schizophrenia patients and healthy controls. SZ patients had a preserved overall small-world network organization (Chapter 6) but significantly longer node-specific path lengths of frontal and temporal regions, especially of bilateral inferior/superior frontal cortex and temporal pole regions. Furthermore, frontal hubs of patients showed a less central hub role of these regions in the overall network structure (cf. Fig. 1.5). Van den Heuvel et al. (2010) suggested that this results in a reduced capacity to integrate information a cross brain regions. The same group (Collin et al., 2014) then examined the "rich club" organization in networks derived from DTI in schizophrenia patients, unaffected siblings of patients, and healthy control subjects. Connectivity between "rich club" hubs (Fig. 12.1) was highest in controls, intermediate in siblings of SZ (~ 8% reduced relative to controls) and lowest in SZ patients (~ 20% reduced compared to controls). Furthermore, in SZ patients, lower levels of "rich club" connectivity were related to longer duration of illness and worse overall functioning. Together, these findings again suggest that impaired rich-club connectivity is related to familial, possibly reflecting genetic, vulnerability for schizophrenia, and that abnormal rich club organization plays a central role in the etiology of schizophrenia (Collin et al., 2014).

In a meta-analysis of 121 studies that reported both under- and over-activations in the same SZ patients, Crossley et al. (2016) found that these abnormalities were located in close topological proximity to each other. Under-activation in a peripheral node (Fig. 12.1) was more frequently associated with over-activation of "rich-club" nodes than with over-activation of another peripheral node. Although SZ is associated with altered brain functional activation in a wide variety of regions, abnormal responses are concentrated in hubs of the normative connectome. Task-specific under-activation in SZ appears accompanied by over-activation of topologically central, less functionally specialized network nodes, which may represent a compensatory response. Mamah et al. (2019) investigated white matter integrity in SZ using voxel-based and tract-based approaches (cf. Fig. 11.1). Specifically, they found decreased fractional anisotropy (FA) in multiple, primarily left-sided, white matter regions in SZ using a voxel-based approach.

Klauser et al. (2017) compared FA and mean diffusivity between patients with schizophrenia or schizoaffective disorder and age-matched healthy controls. The corpus callosum, cingulum, and thalamic radiations exhibited the most extensive pathology according to effect size. More than 50% of corticocortical and subcortical–cortical white matter fiber bundles were disrupted in SZ. Connections between hub nodes comprising the rich club (Fig. 12.1) including the precuneus, cingulate cortex, superior frontal cortex, and superior parietal cortex were disproportionately affected (Zhao et al., 2017b). Thus although connectivity between rich-club hubs is most extensively disturbed in SZ, white matter pathology is widespread, affecting all cerebral lobes and the cerebellum, leading to disruptions in the majority of the brain's fiber bundles (Gao et al., 2019). Zhao et al. (2018a) demonstrated callosal and subcortical white-matter alterations in SZ patients. These disruptions were evident in both voxel and connectivity levels and further supported by associations between FA values and illness duration. The structural network showed weaker global integration in SZ patients than in healthy subjects. Some of the hub regions, especially the thalamus and precuneus, had impaired connections, suggesting that they not only play a crucial role in the white-matter structural network, but also are vulnerable to damage in patients. The same group, Zhao et al. (2019) then used functional magnetic resonance imaging (fMRI), DTI, and structural MRI data to construct the functional, anatomical, and morphological networks of each participant, respectively. The anatomical network obtained by tractography (cf. Fig. 11.1), was extensively distributed throughout the brain. Connections between thalamus and cortex were enhanced in the patient group. Most connections, mainly in the thalamus and from the right globus pallidus to the inferior frontal gyrus were weakened in SZ. These results highlight the critical role of the basal ganglia–thalamus–cortex network in the pathophysiology of schizophrenia.

12.2.2 Functional connectivity

Menon (2011) proposed a triple network model of aberrant saliency mapping and cognitive dysfunction in psychopathology (Fig. 12.2; Krajcovic et al., 2019), emphasizing the parallels that are emerging across psychiatric and neurological disorders. The salience network (SAL), interacting with both the DMN and ECN (also known as the TPN; Mulders et al., 2016), modulates the strength of the anti-correlation between the DMN and ECN. The model proposes that weak salience detection and mapping of goal-relevant external stimuli and internal mental events from, and into, the SAL plays a major role in psychopathology. Weak mapping from the insular-cingulate SAL gives rise to aberrant engagement of the frontoparietal ECN, compromising cognition and goal-relevant adaptive behavior. Aberrant DMN organization and weak engagement or disengagement of the DMN by salient events is associated with altered self-referential mental activity (Menon, 2011).

Fox et al. (2017) used resting-state neuroimaging data to evaluate connectivity between the main DMN hubs (i.e., the medial prefrontal cortex, mPFC and the posterior cingulate cortex–anterior precuneus) in individuals with schizophrenia and controls. They observed that greater connectivity between structures of the DMN was associated with better social attainment and social competence in individuals with schizophrenia, despite the lack of group differences in average DMN connectivity. Across two SZ samples, Spreng et al. (2019) found reliable and robust reductions in structural integrity of the frontoparietal control and salience networks, but not DMN, dorsal attention (DAN), motor and sensory networks.

Fan et al. (2019) examined abnormalities of different DMN subsystems in first episode SZ and associations between these abnormalities and individual psychopathology. They reported markedly reduced strength of DMN connectivity in SZ, with the degree centrality of the medial temporal lobe (MTL) subsystem significantly reduced, including the posterior inferior parietal lobule (pIPL), parahippocampal cortex (PHC) and lateral temporal cortex (LTC). Patients also exhibited reduced connectivity within the MTL subsystem and between the MTL and dorsomedial prefrontal cortex (dmPFC) subsystems (Fig. 12.3). Clinical symptoms were negatively correlated with degree centrality of LTC, pIPL, and PHC.

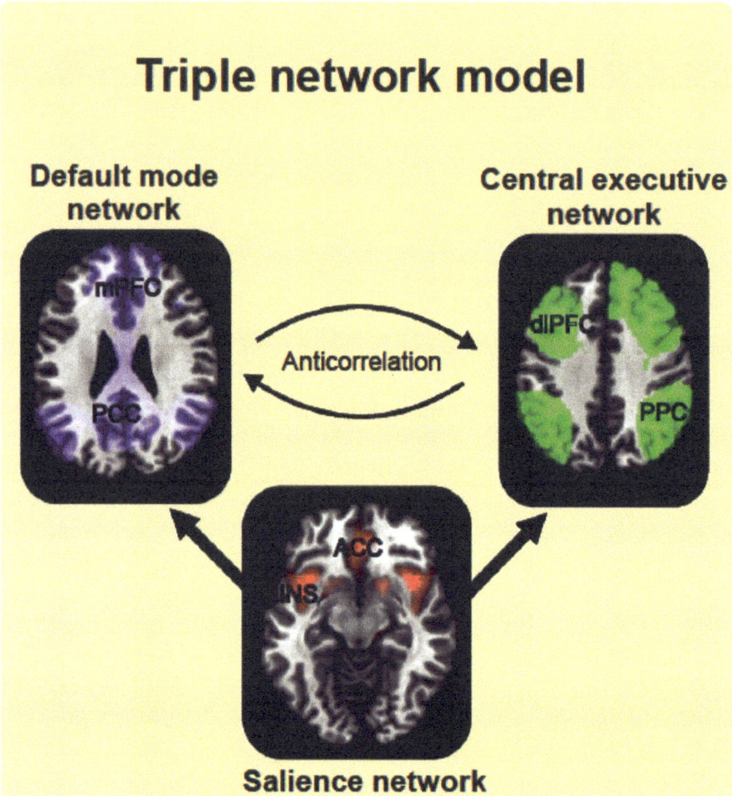

FIG. 12.2 Schematic illustration of the triple network model consisting of the salience network (SAL), default mode network (DMN), and central executive network (CEN). According to this model, the SAL coordinates the CEN/DMN activity, that is, activates the CEN and deactivates the DMN in response to salient stimuli. *With permission from Krajcovic, B., Fajnerova, I., Horacek, J., Kelemen, E., Kubik, S., Svoboda, J., et al., 2019. Neural and neuronal discoordination in schizophrenia: from ensembles through networks to symptoms. Acta Physiol. 226 (4), e13282. https://doi.org/10.1111/apha.13282. Copyright 2019, Scandinavian Physiological Society, Published by John Wiley & Sons Ltd.*

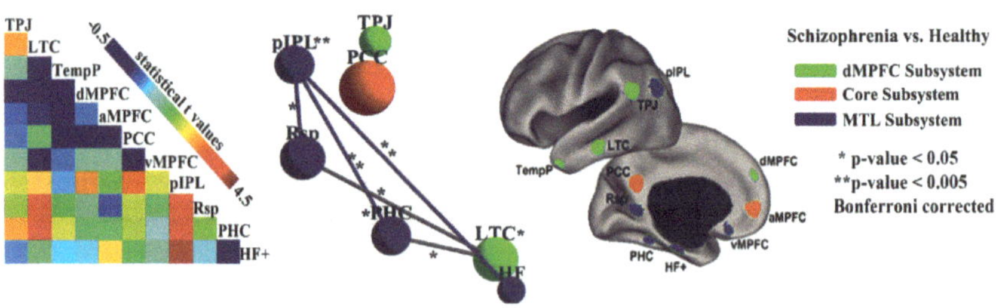

FIG. 12.3 Group differences in nodal centrality and edge strength, and its relationship with clinical symptoms in schizophrenia. *With permission from Fan, F., Tan, Y., Wang, Z., Yang. F., Fan, H., Xiang. H., et al., 2019. Functional fractionation of default mode network in first episode schizophrenia. Schizophrenia Res. 210, 115–121, Elsevier.*

Hyperfractionation of different DMN components implies that communication and coordination throughout the DMN are functionally over-segregated in schizophrenia. Huang et al. (2019) found both increased- and reduced-connectivity of the salience network in first episode SZ. Increased functional connectivity (FC) mainly existed between the SAL and cortico-cerebellar sub-circuit and PFC, whereas reduced FC was found within the cortico-striatal-thalamic-cortical sub-circuit. Abnormalities in insular functional connectivity have been implicated in many clinical features of SZ. Li et al. (2019) found that compared with healthy controls, SZ patients showed reduced connectivity between posterior insula (PI) and somatosensory areas, and between dorsal anterior insula (AI) and putamen. SZ patients also showed dorsal- and ventral-AI increased connectivity with visual areas relative to controls. Ventral AI connectivity with dorsolateral PFC was negatively correlated with the severity of the symptoms of first episode SZ.

12.2.3 Synchrony and resting-state brain rhythms

A wide range of problems in schizophrenia may be caused by synchronization failures between different brain regions (Ford et al., 2007). Gamma oscillations are induced in PFC during working memory tasks and schizophrenic patients perform poorly on working memory tasks and have reduced gamma-phase locking (Gordon et al., 2001). Using magnetoencephalography (MEG), Kissler et al. (2000) examined SZ patients and healthy comparison subjects at rest and performing a complex mental arithmetic task. Both groups showed a frontal and frontotemporal increase in high frequency (45–71 Hz) gamma activity. However, this was left lateralized in controls but right lateralized in SZ. Abnormal oscillations have since then been reported in schizophrenia for both power and phase in the gamma- (Ford and Mathalon, 2005; Spencer et al., 2004) and theta- (decreased), alpha- and beta- (increased) range (Koukkou et al., 2000). Downregulation of glutamatergic transmission at NMDA receptors may underlie this (Benes et al., 1996; Javitt et al., 1996; Lewis, 2000), and a net reduction of GABA-mediated inhibitory modulation (Benes et al., 1996; Lewis, 2000; Rotaru et al., 2011).

Using EEG, Hirvonen et al., (2017) tested alterations in large-scale phase synchronization in chronic SZ patients, and healthy participants, during a perceptual task. For this purpose, they presented a set of Mooney faces (Mooney, 1957) in the upright orientation, and inverted-scrambled orientation. SZ patients had a reduction in gamma-band (30–120 Hz) oscillation amplitudes accompanied by a pronounced deficit in large-scale synchronization at gamma-band frequencies. In beta- and gamma-bands synchronization was reduced within visual regions and between visual and frontal cortex and the reduction of synchronization correlated with increased clinical disorganization. Using MEG, Zeev-Wolf et al. (2018) found that patients high in positive symptoms showed widespread low-alpha power, which was negatively correlated with positive symptoms. In contrast, patients with high negative symptoms showed greater beta power in left hemisphere regions than those low in negative symptoms.

12.2.4 Sound entrainment of oscillations

Kwon et al. (1999) were the first to examine stimulus-driven phase locking to 40 Hz click trains in the auditory steady-state response (ASSR) in SZ. Compared with matched healthy controls, ASSRs in SZ were slower to phase lock with the stimulus trains and never reached the same degree of maximal phase locking. This supports the hypothesis of abnormal neural synchrony in schizophrenia. Light et al. (2006) found that compared with healthy controls, SZ patients had reductions in both evoked power and phase synchronization in ASSRs to 30- and 40-Hz click stimulation but normal responses to 20-Hz stimulation. Maximal deficits were detected in response to 40-Hz stimulation. Brenner et al. (2009) noted that SZ patients exhibit pronounced ASSR deficits within the gamma-frequency stimulation range (35–50 Hz) for click trains and amplitude-modulated tones (Krishnan et al., 2009). Visual SSR deficits were also observed, most prominently in the alpha- and beta-frequency stimulation ranges (7–30 Hz) in response to high-contrast, high-luminance stimuli. Roach et al. (2018) recorded ASSRs from SZ patients and healthy controls listening to repeated 40-Hz 500-ms-duration click trains. Evoked power, total power, phase-locking factor (PLF), and phase-locking angle (PLA) were calculated for single trial data from electrode Fz. They found that: "in patients with schizophrenia, 40-Hz PLA was significantly reduced and phase delayed, and was unrelated to reductions in their 40-Hz power or PLF" (Fig. 12.4). Thereby they confirmed the findings by Kwon et al. (1999).

12.2.5 Neurochemistry

Using proton magnetic resonance spectroscopy (^{1}H-MRS), Théberge et al. (2002) found the level of glutamine significantly higher in the left ACC and thalamus of the patients with schizophrenia than in the healthy subjects. Glutamine is formed by glutamine synthetase from glutamate. A high level of glutamine suggests greater than normal glutamatergic activity. Fuente-Sandoval et al. (2011) compared, using ^{1}H-MRS, glutamate levels in the precommissural dorsal-caudate (a dopamine-rich region) and the cerebellar cortex (negligible dopamine) in antipsychotic-naive subjects with prodromal symptoms and considered to be at ultrahigh-risk for schizophrenia (UHR), antipsychotic-naive first-episode psychosis patients, and age- and sex-matched healthy controls. In the precommissural dorsal-caudate, the ultrahigh risk for schizophrenia and first-episode

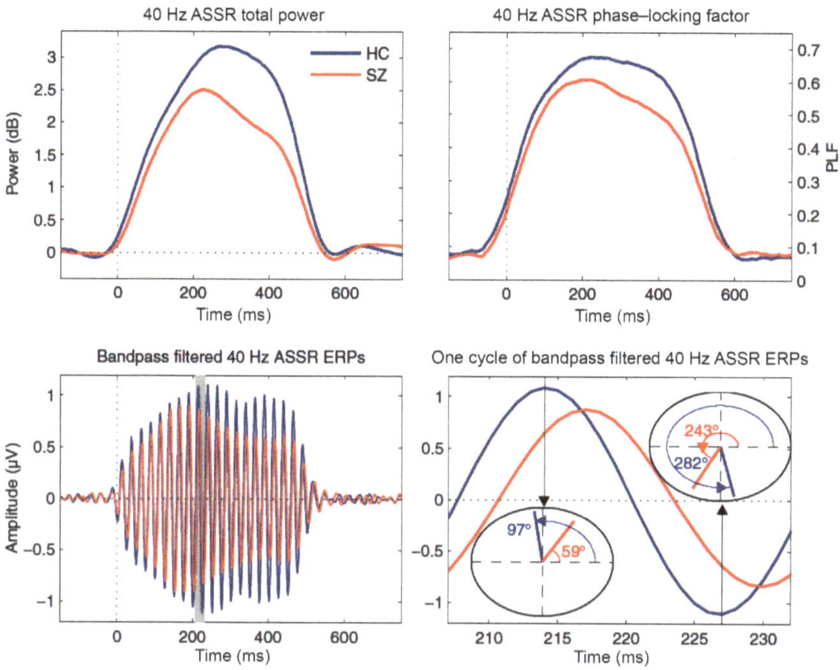

FIG. 12.4 Group differences in total power (top left panel) and phase-locking factor (PLF) (top right panel) group averages for the 40-Hz auditory steady-state response (ASSR) are plotted. The 30- to 50-Hz band-pass filtered 40-Hz ASSR electroencephalography data are overlaid for the healthy control (HC; *blue group*) and the schizophrenia (SZ; *red*) group on the bottom left. Data in the *gray* highlighted area (bottom left panel) are magnified (bottom right panel) to show group differences in the mean phase angles on the unit circle at 214ms (inset lower left *circle*) and 227ms (inset upper right *circle*), approximately corresponding to the HC group event-related potential (EBP) peak and trough. These points differ by approximately 13ms, or half a 40-Hz sine wave cycle. The mean phase locking angles (PLA) of the SZ group at these time points indicate that the phase of their 40-Hz click train–driven gamma band oscillation lags behind the HC group by approximately 38 degrees (3ms; phase lags indicated by *red*, relative to *blue*, mean phase angle vectors). *From Roach, B.J., Ford, J.M., Mathalon, D.H., 2019. Gamma band phase delay in schizophrenia. Biol. Psychiatry: Cogn. Neurosci. Neuroim. 94 (2), 131–139, With permission from Elsevier on behalf of the Society of Biological Psychiatry.*

psychosis patients showed higher levels of glutamate than controls. In the cerebellum, no differences were seen between the three groups. This suggests that a high glutamate level precedes the onset of schizophrenia and is present in a dopamine-rich region previously implicated in the pathophysiology of schizophrenia (Fuente-Sandoval et al., 2011). By conducting meta-analyses of glutamate ^{1}H-MRS studies, Merritt et al. (2016) and Cadena et al. (2018) found that schizophrenia is associated with elevations in glutamatergic metabolites across several brain regions. This finding supports the hypothesis that schizophrenia is associated with excess glutamatergic neurotransmission in several limbic areas. Note that this may contradict one of the leading hypotheses of schizophrenia implying the downregulation of glutamatergic transmission at NMDA receptors (Benes et al., 1996; Javitt et al., 1996; Lewis, 2000).

12.2.6 Auditory verbal hallucinations

Auditory verbal hallucinations (AVH) are a common symptom of schizophrenia. When auditory stimuli are absent in SZ, primary auditory cortex (A1) becomes hyperactive, whereas A1 responses to auditory stimuli are reduced. Such activation pattern of A1 responses during auditory hallucination is consistent with aberrant gamma rhythms in schizophrenia observed during auditory tasks, raising the possibility that the pathology underlying abnormal gamma rhythms can account for auditory hallucinations (Kompus et al., 2013; Lee, 2017).

Zhang et al. (2018) found decreased FA and increased radial diffusivity in AVH compared with the healthy control group in multiple white matter tracts the language, auditory and memory/limbic networks. These include the corpus callosum, superior longitudinal fasciculus, inferior frontooccipital fasciculus, uncinate fasciculus, cingulum, external capsule, and anterior limb of the internal capsule (Xie et al., 2019). The chronic-AVH group showed more widespread white matter impairment than the first-episode-AVH group. Looijestijn et al. (2018) reported that areas with the most influential roles in AVH were those within the salience network (comprising the anterior cingulate gyrus, right insula, Broca's homologue, premotor cortex, and supramarginal gyrus) (Mallikarjun et al., 2018). From rs-fMRI, Hare et al. (2018) found that hippocampal amplitude of low-frequency fluctuations was negatively associated with FC between A1 and the salience network in healthy subjects but was positively associated with FC between these networks in patients reporting hallucinations.

Using Granger causality analysis, Zhao et al. (2018b) reported that, compared with the non-AVHs group, FC decreased from anteromedial PFC (amPFC) to left inferior temporal gyrus (ITG) and from PCC to left cerebellum posterior lobe, ITG, and right middle frontal gyrus in SZ patients with AVHs. They also found significant correlations between clinical assessments and mean strengths of connectivity from amPFC to left ITG and from PCC to left ITG (Scheinost et al., 2019).

12.2.7 Summary

Schizophrenia patients display deficits in basic non-verbal auditory processing pitch, loudness, amplitude modulation, and duration. Abnormal rich-club organization plays a central role in the etiology of schizophrenia. Some of the hub regions, especially the thalamus and precuneus, had impaired connections, and are thus vulnerable to damage in patients. Robust reductions are found in structural integrity of the frontoparietal control and salience networks but not default mode, dorsal attention, motor and sensory networks. Schizophrenia patients have abnormal oscillation in the gamma- and theta- (decreased), alpha- and beta- (increased) range. Downregulation of glutamatergic transmission at NMDA receptors may underlie this, as well as a net reduction of GABA-mediated inhibitory modulation. MEG of patients high in positive symptoms showed widespread low-alpha power, whereas patients with high negative symptoms showed greater beta power in left hemisphere regions. During silence, primary auditory cortex in SZ with audio verbal hallucinations becomes hyperactive, whereas responses to auditory stimuli are reduced. This is consistent with aberrant gamma rhythms in schizophrenia.

12.3 Mild cognitive impairment and Alzheimer's disease

Alzheimer's disease is the most common form of dementia. It is a progressive and neurodegenerative disease, and causes cognitive and memory deficits. Mild cognitive impairment (MCI) is considered a precursor of AD. The two hallmark pathologies required for a diagnosis of Alzheimer's disease (AD) are the extracellular plaque deposits of the β-amyloid peptide (Aβ) and the neurofibrillary tangles of the tau protein. Tau is a protein contained within the axons of the nerve cells. More specifically, tau helps form microtubules that transport nutrients within nerve cells. In Alzheimer's, tau forms tangles (Jones et al., 2016).

12.3.1 Genotype risk, amyloid-deposits, and tau tangles

12.3.1.1 ApoE ε4 genotype risk

Alzheimer's disease is characterized by the earlier-mentioned molecular pathologies and the ApoE ε4 genotype risk factor. Machulda et al. (2011) measured alterations in DMN and SAL connectivity in APoE ε4 carriers compared with APoE ε4 noncarriers aged from 63 to 91 years. In the carriers relative to non-carriers, there was decreased in-phase connectivity between a PCC seed (in the DMN) and regions of the posterior DMN that included the left inferior parietal lobe, left middle temporal gyrus, and bilateral anterior temporal lobes. In contrast, an ACC seed (in the DAN) showed greater in-phase connectivity with regions in the SAL including the cingulate gyrus, medial prefrontal cortex, bilateral insular cortex, striatum, and thalamus in APoE ε4 carriers vs non-carriers. There were no group-wise differences in brain anatomy.

Using resting-state fMRI and graph-theoretic approaches, Wang et al. (2015) found that APoE 4 significantly disrupted whole-brain topological organization. In addition there was decreased intramodular connectivity

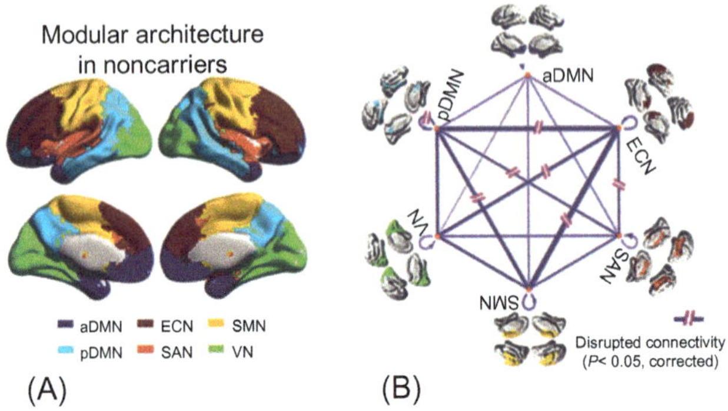

FIG. 12.5 The effects of APOE ε4 on intermediate modular structure. Six modules were identified for the group-level mean network of APOE ε4 noncarriers (A) including the aDMN, pDMN, VN, SMN, SAN, and ECN. Further statistical analysis revealed significantly ($P<0.05$, False Discovery Rate corrected) disrupted intramodule and intermodule functional connectivity in the APOE ε4 carriers versus noncarriers (B). Connectivity network changes. The line width indicates between-group differences (all carriers < noncarriers). *aDMN*, Anterior default mode network; *pDMN*, posterior default mode network; *ECN*, executive control network; *SMN*, sensorimotor network; *SAN*, salience network; *VN*, visual network. *From Wang, J., Wang, X., He, Y., Yu, X., Wang, H., He, Y., 2015. Apolipoprotein E e4 modulates functional brain connectome in Alzheimer's Disease. Hum. Brain Mapp. 36, 1828–1846, by permission copyright 2015, Wiley Periodicals Ltd.*

within the posterior DMN and intermodular connectivity of the posterior DMN and ECN with other neuroanatomical systems. They also found impaired functional hubs in their rich-club connectivity that primarily involved the posterior DMN, ECN, and sensorimotor systems (Fig. 12.5).

Koelewijn et al. (2019) studied resting-state oscillatory connectivity using MEG in healthy young humans genotyped for APOE 4. Connectivity in a set of mostly right-hemisphere connections, including lateral parietal and precuneus regions of the DMN was increased in APOE ε4 across frequencies, but most prevalent in alpha/beta. Similar regions also demonstrated hyperactivity, but only in the gamma band (40–160 Hz). In a separate study of AD patients, they found reduced connectivity in an extended bilateral network that partially overlapped with the hyperconnected regions seen in young APOE 4 carriers.

12.3.1.2 Amyloid-β plaques and tau tangles

Drzezga et al. (2011) had found significant disruptions of whole-brain connectivity in amyloid-positive patients with MCI in typical cortical hubs (PCC/precuneus). Whole-brain connectivity and metabolism were positively correlated whereas there was a negative correlation with amyloid burden. The spatial overlap between hypometabolism and disruption of connectivity in cortical hubs suggests a susceptibility of these regions to early Alzheimer's-type neurodegeneration and may reflect a link between synaptic dysfunction and functional disconnection (Buckner et al., 2009). This indicates that disruption of functional connectivity and hypometabolism may represent early functional consequences of emerging molecular Alzheimer's disease pathology, evolving prior to clinical onset of dementia. Mutlu et al. (2017) found that atrophy at baseline was predicted by specific connectivity Aβ deposition and positively influenced by both global and specific connectivity. Atrophy spread in regions with high specific connectivity while hypometabolism propagated in areas showing high global connectivity. Thus the high level of functional connectivity found in hubs exerts a differential influence on these Alzheimer's disease lesions. Schultz et al. (2017) described fMRI connectivity across four major cortical association networks—DMN, SAL, DAN, and frontoparietal network (cf. Fig. 4.4)—as a function of global A and regional tau in entorhinal, inferior temporal cortex (ITC), and inferior parietal cortex (IPC). They found that Aβ-positive individuals had increased connectivity in the DMN and SAL when neocortical tau levels were low, and decreased connectivity in these networks when tau levels were high. This pattern suggests a hyperconnectivity phase followed by a hypoconnectivity phase in the course of preclinical AD as reflected in increasing tau levels. FMRI of humans with early symptomatic AD has demonstrated excessive neuronal activity in the hippocampus and neocortex, where A accumulates in abundance (Zott et al., 2018). Zott et al. (2019) showed that soluble A dimers and oligomers, which are formed during A aggregation, are neurotoxic. They preferentially affect repeatedly activated neurons by blocking glutamate reuptake by neurons and astrocytes. This leads to hyperactivation and eventually neurodegeneration.

Using various positron emission tomography (PET) tracers, Ossenkoppele et al. (2019) found convergence between filamentous tau covariance patterns and intrinsic connectivity maps, which corresponded with expected functional networks. The topography of Aβ covariance patterns was more diffuse and less network-specific, while glucose hypometabolic patterns were more spatially restricted than tau but overlapped with functional networks. Ossenkoppele et al. (2019) suggested that the spatial patterns of tau and glucose hypometabolism observed in AD resemble the functional organization of the healthy brain, supporting the notion that tau pathology spreads through circumscribed brain networks and drives neurodegeneration. Pereira et al. (2019) applied a joint independent component analysis to ^{18}F-Flutemetamol (Aβ) and ^{18}F-Flortaucipir (tau) PET images to identify Aβ and tau networks across different stages of AD. They found nine patterns that were linked across tau and Aβ data, and showed a fair to moderate overlap with distinct functional networks. Only tau was associated with white matter integrity loss and multiple cognitive functions. Pereira et al. (2019) concluded that Aβ and tau "not only have distinct spatial patterns but their accumulation might have a different impact on functional and anatomical brain connectivity." Maass et al. (2019) quantified uptake of ^{11}CPiB PET (amyloid-) and ^{18}F-flortaucipir (tau) in a sample of Aβ positive cognitively normal adults aged 20–93 years and Aβ-positive 65–95 year-old patients with MCI or AD dementia. Tau burden was relatively higher in anterior-temporal regions in normal aging and this difference was further pronounced in the presence of amyloid- and cognitive impairment. In contrast, Aβ deposition dominated in posterior-medial regions. With aging, Maass et al. (2019) found tau deposits mainly in the anterior-temporal system, which results in deficits in mnemonic object discrimination. As Alzheimer's disease develops, Aβ deposits preferentially in posterior-medial regions additionally compromising scene discrimination and anterior-temporal tau deposition worsens further. In a group of patients with early symptomatic AD stages, La Joie et al. (2020) found that the global intensity of tau-PET, but not -amyloid–PET, signal predicted the rate of subsequent atrophy. Additional investigations demonstrated that the specific distribution of tau-PET signal was a strong indicator of the topography of future atrophy at the single patient level and that the relationship between baseline tau-PET and subsequent atrophy was particularly strong in younger patients.

Summarizing, higher amyloid-load in cognitively normal individuals with subjective cognitive decline is associated with lower concurrent global cognition and with faster rate of decline in a variety of cognitive domains (Timmers et al., 2019). As Alzheimer's disease develops, Aβ deposits preferentially in posterior-medial regions additionally compromising scene discrimination and anterior-temporal tau deposition worsens further (Maass et al., 2019). The vulnerability of hubs is illustrated by the sites of Aβ deposition in patients with AD, which correlate with the location of major hubs, as defined by graph theoretical analysis of fMRI in healthy adults (Buckner et al., 2009). These regions include the PCC and precuneus, the IPL, and the medial pre-frontal cortex (MPFC), suggesting that hubs are preferentially affected in the progression of AD. Furthermore, convergent evidence from MRI, EEG and MEG suggests that AD is associated with perturbations of the organization of small-world networks in the brain (Chapter 6).

12.3.2 Connectome changes: The importance of hub regions

Historically, AD is viewed as a disease of abnormally aggregated proteins by pathologists and molecular biologists and as a disease of clinical symptoms by neurologists and psychologists, which are not synonymous (Jack et al., 2019). Bridging the divide between these viewpoints may be based on the connectivity in brain networks.

12.3.2.1 MEG and EEG studies

Stam et al. (2009) recorded MEGs in AD patients and nondemented controls in a no-task, eyes-closed condition. For the main frequency bands, synchronization between all pairs of MEG channels was assessed using a phase lag index (PLI; Chapter 6). AD patients showed a decrease of mean PLI in the lower alpha- and beta-band. In the lower alpha-band, the clustering coefficient and path length were both decreased in Alzheimer's disease patients. Consequently, AD patients display a loss of resting-state functional connectivity in lower alpha- and beta-bands. Moreover, the large-scale structure of lower alpha-band functional networks in AD is more random. Thus highly connected neural network "hubs" may be especially at risk in AD, and supports the concept of AD as a disconnection syndrome. Engels et al. (2017) confirmed their previously reported (Engels et al., 2015) posterior-to-anterior information flow in the higher frequency bands in the healthy controls, and found it to be disturbed in the beta-band in AD. Most prominently, the information flow from the precuneus and the visual cortex, towards frontal and subcortical structures, was decreased in AD. These disruptions did not correlate with cognitive impairment or cerebrospinal fluid biomarkers. The same group (Yu et al., 2017) performed eyes-closed resting-state MEG recordings from patients with AD and age-matched controls. Nodal centrality metrics for source space showed consistently that several hub regions, particularly left hippocampus, posterior parts of the default mode network and occipital regions were vulnerable in patients with AD compared with control subjects. Of note, these detected vulnerable hubs in Alzheimer's disease were absent in each individual frequency-specific network (delta, theta, alpha 1, alpha 2, and beta band), thus showing the value of integrating the networks.

Using whole-brain measures of local neural oscillatory activity and connectivity in the resting state, Koelewijn et al. (2017) found that patients with AD present a distinct pattern of dysfunction across the brain compared with that in healthy aging. Oscillatory envelope amplitude for alpha and beta bands increased with healthy aging, whereas both local amplitude and node-to-global connectivity decreased with AD. In the beta band, a significant proportion of the variance in both oscillatory envelope variability and connectivity in areas of peak group difference was explained by cognitive integrity. This again supports the disconnection syndrome hypothesis and strongly suggests that AD shows distinct and unique patterns of disrupted neural functioning, rather than patterns of exaggerated healthy aging.

12.3.2.2 MRI studies

From resting-state fMRI data of young subjects, old normal controls and AD patients, Miao et al. (2011) inferred the causal influences between any pair of nodes within the DMN using Granger causality analysis and

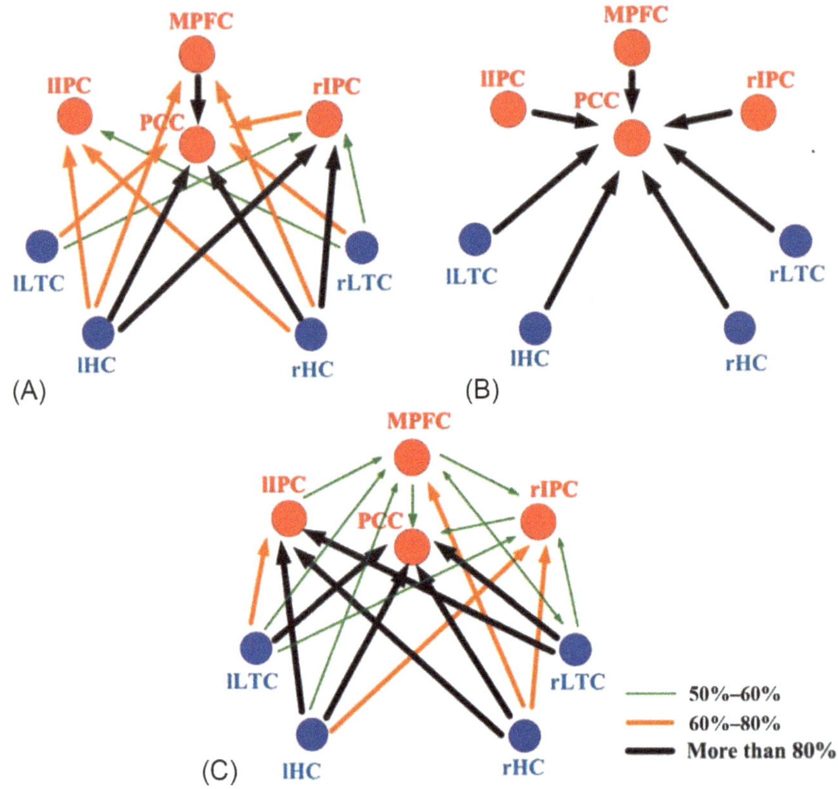

FIG. 12.6 Granger causality pattern of the DMN hubs in the young subjects (A) and in the normal aging (B) and AD subjects (C). DMN hubs are colored in *red* and the non-hub regions in *blue*. Line width and color indicate the proportion of subjects showing significant causal relationship in this direction ($P=0.05$). The *thickest black lines* represent more than 80% subjects showing significant causal influence in this direction, the *saffron line* represent 60%–80%, and the *green dashed lines* represent 50%–60%. *MPFC*, medial prefrontal cortex; *PCC*, posterior cingulate cortex; *IPC*, inferior parietal cortex; *ITC*, inferior temporal cortex (labeled LTC in the figure); *HC*, (para) hippocampal formation. *From Miao, X., Wu, X., Li, R., Chen, K., Yao, L., 2011. Altered connectivity pattern of hubs in default-mode network with Alzheimer's disease: an Granger causality modeling approach. PLoS ONE 6 (10), e25546, https://doi.org/10.1371/journal.pone.0025546; Open access.*

graph-theoretic methods. They found that hubs in the DMN (PCC, MPFC, and IPC) but especially the PCC showed the widest and distinctive causal effects on the DMN dynamics in the young group. In AD, but not in normal aging, the MPFC and IPC nodes showed causal disruption of interaction with other nodes, and the PCC was the only region having significant causal relations with all other nodes (Fig. 12.6).

In concordance herewith, Filippi et al. (2013) noted that "compared with controls, patients with AD or MCI retained their hub regions in the frontal lobe but not in the temporal lobe. Increased inter-regional correlations within the local brain lobes and disrupted long-distance inter-regional correlations were also detected in these patients." Specifically, disruptive organization in the whole-brain networks in AD appears involved in the loss of small-world characteristics and the re-organization of hub distributions (Xie and He, 2012). Using rs-fMRI and voxel-based graph-theoretic analysis, Dai et al. (2015) found that AD selectively targeted highly connected hub regions (in terms of nodal functional connectivity strength) of brain networks, involving the medial and lateral prefrontal and parietal cortices, insula, and thalamus. This impairment was most severe in the long-range connections (e.g., 100–130 mm). Moreover, AD also disrupted functional connections within the DMN, SAL and ECN, and connections between the SAL and ECN. Dai et al. (2015) concluded that the "disrupted brain hubs patterns in AD are connection-distance-dependent, being characterized by the disruptions of longer distance connections, which tend to consume more energy." See also Dai et al. (2019) for similar effects on structural and functional connections.

Gray matter alterations in AD do not occur randomly across the brain but follow identifiable patterns of distribution (Manuello et al., 2018). This alteration pattern exhibits a network-like structure composed of co-altered areas that can be defined as a co-atrophy network (Fig. 12.7 *left*). Within the co-atrophy network of AD, they further identified a core sub-network of co-altered areas that includes the left hippocampus, left and right amygdalae, right parahippocampal gyrus, and right temporal inferior gyrus (Fig. 12.7 *right*).

12.3.3 From mild cognitive impairment to Alzheimer's disease

Based on source-based EEG maps of resting-state activity, Hsiao et al. (2013) found alterations of cortical spectral power in mild AD when compared with MCI. These included attenuated alpha or beta activities in the DMN, and enhanced delta or theta activities in the MTL, IPC, PCC, and precuneus. Connection hubs in the precuneus, PCC, ACC, and MTL showed altered functional interconnections. Posterior theta and alpha power and altered connectivity in the MTL correlated significantly with scores obtained on the mini-mental state examination.

There is significantly decreased global and local network efficiency in patients with amnestic MCI compared with controls, and a significant decline in network efficiency with aging in the amnestic MCI patients compared with healthy controls (Zhao et al., 2017a). The age-related decreases of nodal efficiency in amnestic MCI patients were mainly distributed in the key regions of the DMN, such as precuneus, ACC, and parahippocampal gyrus. In addition, age-related decreases in the connection strength between peripheral nodes were observed in amnestic MCI patients. Moreover, the

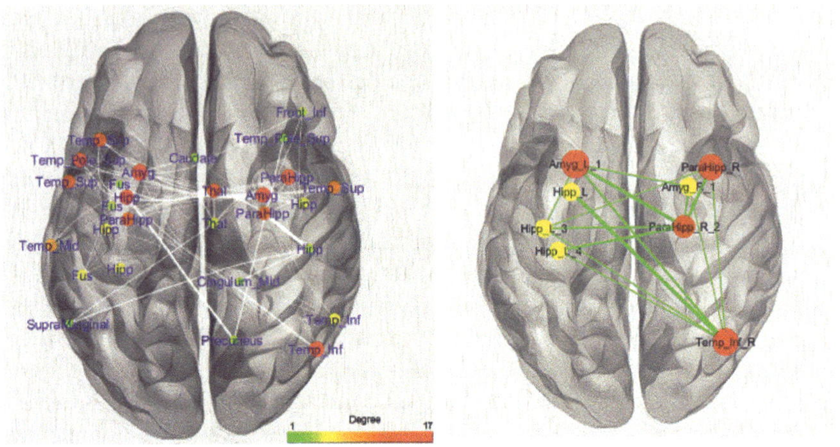

FIG. 12.7 Topological analysis of the co-atrophy network of Alzheimer's disease. Left nodes referring to the same brain areas or strictly close one to the other have been collapsed in a single node. Right network clustering with k-core decomposition algorithm. Colors and dimensions of nodes indicate their topological degree (smaller = node = lower degree; from *green* to *red* = from lower to higher values). Thickness of edges indicate the degree of edge betweenness (smaller edge = lower degree). Both node degree and edge betweenness values refer to the original co-atrophy network. *From Manuello, J., Nani, A., Premi, E., Borroni, B., Costa, T., Tatu, K., et al., 2018. The pathoconnectivity profile of Alzheimer's disease: a morphometric coalteration network analysis. Front. Neurol. 8, 739, https://doi.org/10.3389/fneur.2017.00739, Open access.*

decreased regional efficiency of the parahippocampal gyrus was correlated with impaired memory performances in patients (Zhao et al., 2017a).

In healthy subjects, DMN and TPN (cf. Fig. 1.1) interaction is modulated by the thalamus suggesting that abnormal task-based DMN deactivation in amnestic MCI may be a consequence of impaired thalamocortical white matter circuitry. Alderson et al. (2017) identified significant structural deficits in thalamic white matter projection fibers to posterior DMN components, PCC and lateral IPL together with significantly reduced effective connectivity from left thalamus to left IPL. Crucially, impaired thalamocortical white matter circuitry correlated with memory performance. Disrupted thalamocortical structure was accompanied by significant reductions in IPL and PCC corticocortical effective connectivity. No structural deficits were found between DMN nodes.

To evaluate the structural differences between forms of MCI and AD, Yan et al. (2018) recruited normal controls, subjective cognitive decline patients, amnestic MCI and AD patients and collected DTI data to analyze rich-club organization. They demonstrated that rich-club organization was disrupted, with reduced structural connectivity among rich-club nodes, in amnestic MCI and AD patients but remained stable in patients with subjective cognitive decline. In addition, patients with subjective cognitive decline, amnestic MCI and AD patients showed similar patterns of disrupted peripheral regions and reduced connectivity involving these regions, suggesting that peripheral regions might contribute to cognitive decline. Thus rich-club organization may provide a new perspective on how AD affects brain topology and function—as previously suggested by previous structural and functional studies (Dai et al., 2015; Daianu et al., 2015, 2016).

12.3.4 Summary

AD is the most common form of dementia. It is a progressive and neurodegenerative disease, and causes cognitive and memory deficits. MCI is considered a precursor of AD. The two hallmark pathologies required for a diagnosis of AD are the extracellular plaque deposits of the β-amyloid peptide (Aβ) and the neurofibrillary tangles of the tau protein.

The sites of Aβ deposition in patients with AD correlate with the location of major hubs. These regions include the posterior cingulate cortex and precuneus, the inferior parietal lobule, and the medial prefrontal cortex, suggesting that hubs are preferentially affected in the progression of AD. Furthermore, convergent evidence from MRI, EEG, and MEG suggests that AD is associated with perturbations of the organization of small-world networks in the brain.

AD selectively targeted highly connected hub regions of brain networks, involving the medial and lateral prefrontal and parietal cortices, insula, and thalamus. This impairment is most severe in the long-range connections (e.g., 100–130 mm). Moreover, AD also disrupts functional connections within the default mode, salience and ECNs, and connections between the salience and ECNs.

References

Alderson, T., Kehoe, E., Maguire, L., Farrell, D., Lawlor, B., Kenny, R.A., et al., 2017. Disrupted thalamus white matter anatomy and posterior default mode network effective connectivity in amnestic mild cognitive impairment. Front. Aging Neurosci. 9, 370. https://doi.org/10.3389/fnagi.2017.00370.

Benes, F.M., Vincent, S.L., Marie, A., Khan, Y., 1996. Up-regulation of $GABA_A$ receptor binding on neurons of the prefrontal cortex in schizophrenic subjects. Neuroscience 75, 1021–1031.

Brenner, C.A., Krishnan, G.P., Vohs, J.L., Ahn, W.-Y., Hetrick, W.P., Morzorati, S.L., et al., 2009. Steady state responses: electrophysiological assessment of sensory function in schizophrenia. Schizophr. Bull. 35 (6), 1065–1077.

Buckner, R.L., Sepulcre, J., Talukdar, T., Krienen, F.M., Liu, H., Hedden, T., et al., 2009. Cortical hubs revealed by intrinsic functional connectivity: mapping, assessment of stability, and relation to Alzheimer's disease. J. Neurosci. 29 (6), 1860–1873.

Cadena, E.J., White, D.M., Kraguljac, N.V., Reid, M.A., Maximo, J.O., Nelson, E.A., et al., 2018. A longitudinal multimodal neuroimaging study to examine relationships between resting state glutamate and task related bold response in schizophrenia. Front. Psychiatry 9, 632. https://doi.org/10.3389/fpsyt.2018.00632.

Chai, X.J., Whitfield-Gabrieli, S., Shinn, A.K., Gabrieli, J.D.E., Nieto, A., Castañón, A., et al., 2011. Abnormal medial prefrontal cortex resting-state connectivity in bipolar disorder and schizophrenia. Neuropsychopharmacology 36, 2009–2017.

Collin, G., Kahn, R.S., de Reus, M.A., Cahn, W., van den Heuvel, M.P., 2014. Impaired rich club connectivity in unaffected siblings of schizophrenia patients. Schizophr. Bull. 40 (2), 438–448.

Crossley, N.A., Mechelli, A., Ginestet, C., Rubinov, M., Bullmore, E.T., McGuire, P., 2016. Altered hub functioning and compensatory activations in the connectome: a meta-analysis of functional neuroimaging studies in schizophrenia. Schizophr. Bull. 42 (2), 434–442.

Dai, Z., Yan, C., Li, K., Wang, Z., Wang, J., Cao, M., et al., 2015. Identifying and mapping connectivity patterns of brain network hubs in Alzheimer's disease. Cereb. Cortex 25, 3723–3742.

Dai, Z., Lin, Q., Li, T., Wang, X., Yuan, H., Yu, X., et al., 2019. Disrupted structural and functional brain networks in Alzheimer's disease. Neurobiol. Aging 75, 71–82.

Daianu, M., Jahanshad, N., Nir, T.M., Jack Jr., C.R., Weiner, M.W., Bernstein, M.A., et al., 2015. Rich club analysis in the Alzheimer's disease connectome reveals a relatively undisturbed structural core network. Hum. Brain Mapp. 36, 3087–3103.

Daianu, M., Mezher, A., Mendez, M.F., Jahanshad, N., Jimenez, E.E., Thompson, P.M., 2016. Disrupted rich club network in behavioral variant frontotemporal dementia and early-onset Alzheimer's disease. Hum. Brain Mapp. 37, 868–883.

Dondé, C., Mondino, M., Leitman, D.I., Javitt, D.C., Suaud-Chagny, M.-F., D'Amato, T., et al., 2019. Are basic auditory processes involved in source-monitoring deficits in patients with schizophrenia? Schizophr. Res. 210, 135–142. Epub.

Drzezga, A., Becker, J.A., Van Dijk, K.R.A., Sreenivasan, A., Talukdar, T., Sullivan, C., et al., 2011. Neuronal dysfunction and disconnection of cortical hubs in non-demented subjects with elevated amyloid burden. Brain 134, 1635–1646.

Eggermont, J.J., 2015a. Auditory Temporal Processing and Its Disorders. Oxford University Press, Oxford, UK.

Eggermont, J.J., 2015b. Animal models of auditory temporal processing. Int. J. Psychophysiol. 95, 201–215.

Engels, M.M.A., Stam, C.J., van der Flier, W.M., Scheltens, P., de Waal, H., van Straaten, E.C.W., 2015. Declining functional connectivity and changing hub locations in Alzheimer's disease: an EEG study. BMC Neurol. 15, 145. https://doi.org/10.1186/s12883-015-0400-7.

Engels, M.M.A., Yu, M., Stam, C.J., Gouw, A.A., van der Flier, W.M., Scheltens, P., et al., 2017. Directional information flow in patients with Alzheimer's disease. A source-space resting-state MEG study. NeuroImage: Clin. 15, 673–681.

Fan, F., Tan, Y., Wang, Z., Yang, F., Fan, H., Xiang, H., et al., 2019. Functional fractionation of default mode network in first episode schizophrenia. Schizophr. Res. 210, 115–121.

Filippi, M., van den Heuvel, M.P., Fornito, A., He, Y., Hulshoff Pol, H.E., Agosta, F., et al., 2013. Assessment of system dysfunction in the brain through MRI-based connectomics. Lancet Neurol. 12, 1189–1199.

Ford, J.M., Mathalon, D.H., 2005. Corollary discharge dysfunction in schizophrenia: can it explain auditory hallucinations? Int. J. Psychophysiol. 58, 179–189.

Ford, J.M., Krystal, J.H., Mathalon, D.H., 2007. Neural synchrony in schizophrenia: from networks to new treatments. Schizophr. Bull. 33 (4), 848–852.

Fox, J.M., Abram, S.V., Reilly, J.L., Eack, S., Goldman, M.B., Csernansky, J.G., et al., 2017. Default mode functional connectivity is associated with social-functioning in schizophrenia. J. Abnorm. Psychol. 126 (4), 392–405.

Fuente-Sandoval, C., León-Ortiz, P., Favila, R., Stephano, S., Mamo, D., Ramírez-Bermúdez, J., et al., 2011. Higher levels of glutamate in the associative-striatum of subjects with prodromal symptoms of schizophrenia and patients with first-episode psychosis. Neuropsychopharmacology 36, 1781–1791.

Gandal, M.J., Edgar, J.C., Klook, K., Siegel, S.J., 2012. Gamma synchrony: towards a translational biomarker for the treatment-resistant symptoms of schizophrenia. Neuropharmacology 62, 1504–1518.

Gao, J., Tang, X., Wang, C., Yu, M., Sha, W., Wang, X., et al., 2019. Aberrant cerebellar neural activity and cerebro-cerebellar functional connectivity involving executive dysfunction in schizophrenia with primary negative symptoms. Brain Imag. Behav. https://doi.org/10.1007/s11682-018-0032-9. Epub.

Gordon, E., Williams, L., Haig, A.R., Wright, J., Meares, R.A., 2001. Symptom profile and "gamma" processing in schizophrenia. Cogn. Neuropsychiatry 6, 7–19.

Hare, S.M., Law, A.S., Ford, J.M., Mathalon, D.H., Ahmadi, A., Damaraju, E., et al., 2018. Disrupted network cross talk, hippocampal dysfunction and hallucinations in schizophrenia. Schizophr. Res. 199, 226–234.

Hirvonen, J., Wibra, W., Palva, J.M., Singer, W., Uhlhaas, P., Satu Palva, S., 2017. Whole-brain source-reconstructed MEG-data reveal reduced long-range synchronization in chronic schizophrenia. eNeuro 4 (5), e0338. https://doi.org/10.1523/ENEURO.0338-17.2017.

Hsiao, F.-J., Wang, Y.-J., Yan, S.-H., Chen, W.-T., Lin, Y.-Y., 2013. Altered oscillation and synchronization of default-mode network activity in mild Alzheimer's disease compared to mild cognitive impairment: an electrophysiological study. PLoS One 8 (7). https://doi.org/10.1371/journal.pone.0068792.

Hu, M.-L., Zong, X.-F., Mann, J.J., Zheng, J.-J., Liao, Y.-H., Li, Z.-C., et al., 2017. A review of the functional and anatomical default mode network in schizophrenia. Neurosci. Bull. 33 (1), 73–84.

Huang, H., Botao, Z., Jiang, Y., Tang, Y., Zhang, T., Tang, X., et al., 2019. Aberrant resting-state functional connectivity of salience network in first-episode schizophrenia. Brain Imag. Behav. https://doi.org/10.1007/s11682-019-00040-8. Epub.

Jack Jr., C.R., Therneau, T.M., Weigand, S.D., Wiste, H.J., Knopman, D.S., Vemuri, P., et al., 2019. Prevalence of biologically vs clinically defined Alzheimer spectrum entities using the National Institute on Aging–Alzheimer's Association Research Framework. JAMA Neurol. 76 (10), 1174–1183. https://doi.org/10.1001/jamaneurol.2019.1971.

Javitt, D.C., Steinschneider, M., Schroeder, C.E., Arezzo, J.C., 1996. Role of cortical N-methyl-D-aspartate receptors in auditory sensory memory and mismatch negativity generation: implications for schizophrenia. Proc. Natl. Acad. Sci. U. S. A. 93, 11962–11967.

Jones, D.T., Knopman, D.S., Gunter, J.L., Graff-Radford, J., Vemuri, P., Boeve, B.F., et al., 2016. Cascading network failure across the Alzheimer's disease spectrum. Brain 139, 547–562.

Kissler, J., Muller, M.M., Fehr, T., Rockstroh, B., Elbert, T., 2000. MEG gamma band activity in schizophrenia patients and healthy subjects in a mental arithmetic task and at rest. Clin. Neurophysiol. 111, 2079–2087.

Klauser, P., Baker, S.T., Cropley, V.L., Bousman, C., Fornito, A., Cocchi, L., et al., 2017. White matter disruptions in schizophrenia are spatially widespread and topologically converge on brain network hubs. Schizophr. Bull. 43 (2), 425–435.

Koelewijn, L., Bompas, A., Tales, A., Brookes, M.J., Muthukumaraswamy, S.D., Bayer, A., et al., 2017. Alzheimer's disease disrupts alpha and beta-band resting-state oscillatory network connectivity. Clin. Neurophysiol. 128, 2347–2357.

Koelewijn, L., Lancaster, T.M., Linden, D., Dima, D.C., Routley, B.C., Magazzini, L., et al., 2019. Oscillatory hyperactivity and hyperconnectivity in young APOE-4 carriers and hypoconnectivity in Alzheimer's disease. eLife 8, e36011. https://doi.org/10.7554/eLife.36011.

Kompus, K., Falkenberg, L.E., Bless, J.J., Johnsen, E., Kroken, R.A., Kråkvik, B., et al., 2013. The role of the primary auditory cortex in the neural mechanism of auditory verbal hallucinations. Front. Hum. Neurosci. 7, 144. https://doi.org/10.3389/fnhum.2013.00144.

Koukkou, M., Federspiel, A., Braker, E., Hug, C., Kleinlogel, H., Merlo, M.C.G., et al., 2000. An EEG approach to the neurodevelopmental hypothesis of schizophrenia studying schizophrenics, normal controls and adolescents. J. Psychiatr. Res. 34, 57–73.

Krajcovic, B., Fajnerova, I., Horacek, J., Kelemen, E., Kubik, S., Svoboda, J., et al., 2019. Neural and neuronal discoordination in schizophrenia: from ensembles through networks to symptoms. Acta Physiol. 226 (4). https://doi.org/10.1111/apha.13282.

Krishnan, G.P., Hetrick, W.P., Brenner, C.A., Shekhar, A., Steffen, A.N., O'Donnell, B.F., 2009. Steady state and induced auditory gamma deficits in schizophrenia. NeuroImage 47, 1711–1719.

Kwon, J.S., O'Donnell, B.F., Wallenstein, G.V., Greene, R.W., Hirayasu, Y., Nestor, P.G., et al., 1999. Gamma frequency-range abnormalities to auditory stimulation in schizophrenia. Arch. Gen. Psychiatry 56, 1001–1005.

La Joie, R., Visani, A.V., Baker, S.L., Brown, J.A., Bourakova, V., Cha, J., et al., 2020. Prospective longitudinal atrophy in Alzheimer's disease correlates with the intensity and topography of baseline tau-PET. Sci. Transl. Med. 12, eaau5732.

Lee, J.H., 2017. Disrupted cholinergic modulation can underlie abnormal gamma rhythms in schizophrenia and auditory hallucination. J. Comput. Neurosci. 43, 173–187.

Lewis, D.A., 2000. GABAergic local circuit neurons and prefrontal cortical dysfunction in schizophrenia. Brain Res. Rev. 31, 270–276.

Li, X.-B., Wang, L.-B., Xiong, Y.-B., Bo, Q.-J., He, F., Li, F., et al., 2019. Altered resting-state functional connectivity of the insula in individuals with clinical high-risk and patients with first-episode schizophrenia. Psychiatry Res. 282, 112608.

Light, G.A., Hsu, J.L., Hsieh, M.H., Meyer-Gomes, K., Sprock, J., Swerdlow, N.R., et al., 2006. Gamma band EEG oscillations reveal neural network cortical coherence dysfunction in schizophrenia patients. Biol. Psychiatry 60, 1231–1240.

Looijestijn, J., Blom, J.D., Hoek, H.W., Renken, R., Liemburg, E., Sommer, I.E.C., et al., 2018. Draining the pond and catching the fish: uncovering the ecosystem of auditory verbal hallucinations. NeuroImage: Clin. 20, 830–843.

Maass, A., Berron, D., Harrison, T.M., Adams, J.N., La Joie, R., Baker, S., et al., 2019. Alzheimer's pathology targets distinct memory networks in the ageing brain. Brain 142, 2492–2509.

Machulda, M.M., Jones, D.T., Vemuri, P., McDade, E., Avula, R., Przybelski, S., et al., 2011. Effect of APOE 4 status on intrinsic network connectivity in cognitively normal elderly subjects. Arch. Neurol. 68 (9), 1131–1136.

Mallikarjun, P.K., Lalousis, P.A., Dunne, T.F., Heinze, K., Reniers, R.L.E.P., Broome, M.R., et al., 2018. Aberrant salience network functional connectivity in auditory verbal hallucinations: a first episode psychosis sample. Transl. Psychiatry 8, 69. https://doi.org/10.1038/s41398-018-0118-6.

Mamah, D., Ji, A., Rutlin, J., Shimony, J.S., 2019. White matter integrity in schizophrenia and bipolar disorder: tract- and voxel-based analyses of diffusion data from the Connectom scanner. NeuroImage: Clin. 21, 101649.

Manuello, J., Nani, A., Premi, E., Borroni, B., Costa, T., Tatu, K., et al., 2018. The pathoconnectivity profile of Alzheimer's disease: a morphometric coalteration network analysis. Front. Neurol. 8, 739. https://doi.org/10.3389/fneur.2017.00739.

Menon, V., 2011. Large-scale brain networks and psychopathology: a unifying triple network model. Trends Cogn. Sci. 15 (10), 483–506.

Merritt, K., Egerton, A., Kempton, M.J., Taylor, M.J., McGuire, P.K., 2016. Nature of glutamate alterations in schizophrenia. A meta-analysis of proton magnetic resonance spectroscopy studies. JAMA Psychiat. 73 (7), 665–674.

Miao, X., Wu, X., Li, R., Chen, K., Yao, L., 2011. Altered connectivity pattern of hubs in default-mode network with Alzheimer's disease: an granger causality modeling approach. PLoS One 6 (10). https://doi.org/10.1371/journal.pone.0025546.

Mooney, C.M., 1957. Age in the development of closure ability in children. Can. J. Psychol. 11 (4), 219–226.

Mulders, P.C., van Eijndhoven, P.F., Beckman, C.F., 2016. Identifying large-scale neural networks using fMRI. In: Frodl, T. (Ed.), Systems Neuroscience in Depression. Elsevier, pp. 209–237.

Mutlu, J., Landeau, B., Gaubert, M., de La Sayette, V., Desgranges, B., Chételat, G., 2017. Distinct influence of specific versus global connectivity on the different Alzheimer's disease biomarkers. Brain 140, 3317–3328.

Ossenkoppele, R., Iaccarino, L., Schonhaut, D.R., Brown, J.A., La Joie, R., O'Neil, J.P., et al., 2019. Tau covariance patterns in Alzheimer's disease patients match intrinsic connectivity networks in the healthy brain. NeuroImage: Clin. 23, 101848. https://doi.org/10.1016/j.nicl.2019.101848.

Pereira, J.B., Ossenkoppele, R., Palmqvist, S., Strandberg, T.O., Smith, R., Westman, E., Oskar Hansson, O., 2019. Amyloid and tau accumulate across distinct spatial networks and are differentially associated with brain connectivity. eLife 8, e50830. https://doi.org/10.7554/eLife.50830.

Roach, B.J., Ford, J.M., Mathalon, D.H., 2018. Gamma band phase delay in schizophrenia. Biol. Psychiatry: Cogn. Neurosci. Neuroim. 94 (2), 131–139.

Rotaru, D.C., Lewis, D.A., Gonzalez-Burgos, G., 2011. The role of glutamatergic inputs onto parvalbumin-positive interneurons: relevance for schizophrenia. Rev. Neurosci. 23 (1), 97–109. https://doi.org/10.1515/revneuro-2011-0059.

Scheinost, D., Tokoglu, F., Hampson, M., Hoffman, R., Constable, R.T., 2019. Data-driven analysis of functional connectivity reveals a potential auditory verbal hallucination network. Schizophr. Bull. 45 (2), 415–424.

Schultz, A.P., Chhatwal, J.P., Hedden, T., Mormino, E.C., Hanseeuw, B.J., Sepulcre, J., et al., 2017. Phases of hyperconnectivity and hypoconnectivity in the default mode and salience networks track with amyloid and tau in clinically normal individuals. J. Neurosci. 37 (16), 4323–4331.

Spencer, K.M., Nestor, P.G., Perlmutter, R., Niznikiewicz, M.A., Klump, M.C., Frumin, M., et al., 2004. Neural synchrony indexes disordered perception and cognition in schizophrenia. Proc. Natl. Acad. Sci. U. S. A. 101, 17288–17293.

Spreng, R.N., DuPre, E., Ji, J.L., Yang, G., Diehl, C., Murray, J.D., et al., 2019. Structural covariance reveals alterations in control and salience network integrity in chronic schizophrenia. Cereb. Cortex 29 (12), 5269–5284.

Stam, C.J., 2010. Use of magnetoencephalography (MEG) to study functional brain networks in neurodegenerative disorders. J. Neurol. Sci. 289, 128–134.

Stam, C.J., de Haan, W., Daffertshofer, A., Jones, B.F., Manshanden, I., van Cappellen van Walsum, A.M., et al., 2009. Graph theoretical analysis of magnetoencephalographic functional connectivity in Alzheimer's disease. Brain 132, 213–224.

Théberge, J., Bartha, R., Drost, D.J., Menon, R.S., Malla, A., Takhar, J., et al., 2002. Glutamate and glutamine measured with 4.0 T proton MRS in never-treated patients with schizophrenia and healthy volunteers. Am. J. Psychiatry 159, 1944–1946.

Timmers, T., Ossenkoppele, R., Verfaillie, S.C.J., van der Weijden, C.W.J., Slot, R.E.R., Wesselman, L.M.P., et al., 2019. Amyloid PET and cognitive decline in cognitively normal individuals: the SCIENCe project. Neurobiol. Aging 79, 50–58.

Uhlhaas, P.J., Singer, W., 2010. Abnormal neural oscillations and synchrony in schizophrenia. Nat. Rev. Neurosci. 11, 100–113.

Van den Heuvel, M.P., Mandl, R.C.W., Stam, C.J., Kahn, R.S., Hulshoff Pol, H.E., 2010. Aberrant frontal and temporal complex network structure in schizophrenia: a graph theoretical analysis. J. Neurosci. 30 (47), 15915–15926.

Wang, J., Wang, X., He, Y., Yu, X., Wang, H., He, Y., 2015. Apolipoprotein E4 modulates functional brain connectome in Alzheimer's disease. Hum. Brain Mapp. 36, 1828–1846.

Whitfield-Gabrieli, S., Thermenos, H.W., Milanovic, S., Tsuang, M.T., Faraone, S.V., McCarley, R.W., et al., 2009. Hyperactivity and hyperconnectivity of the default network in schizophrenia and in first-degree relatives of persons with schizophrenia. Proc. Natl. Acad. Sci. U. S. A. 106, 1279–1284.

Williamson, P., 2007. Are anticorrelated networks in the brain relevant to schizophrenia? Schizophr. Bull. 33 (4), 994–1003.

Wolf, R.C., Vasic, N., Sambataro, F., Höse, A., Frasch, K., Schmid, M., et al., 2009. Temporally anticorrelated brain networks during working memory performance reveal aberrant prefrontal and hippocampal connectivity in patients with schizophrenia. Progr. Neuro-Psychopharmacol. Biol. Psychiatry 33, 1464–1473.

Xie, T., He, Y., 2012. Mapping the Alzheimer's brain with connectomics. Front. Psychiatry 2, 77. https://doi.org/10.3389/fpsyt.2011.00077.

Xie, S., Liu, B., Wang, J., Zhou, Y., Cui, Y., Song, M., et al., 2019. Hyperconnectivity in perisylvian language pathways in schizophrenia with auditory verbal hallucinations: a multi-site diffusion MRI study. Schizophr. Res. 210, 262–269.

Yan, T., Wang, W., Yang, L., Chen, K., Chen, R., Han, Y., 2018. Rich club disturbances of the human connectome from subjective cognitive decline to Alzheimer's disease. Theranostics 8 (12), 3237–3255.

Yu, M., Engels, M.M.A., Hillebrand, A., van Straaten, E.C.W., Gouw, A.A., Teunissen, C., et al., 2017. Selective impairment of hippocampus and posterior hub areas in Alzheimer's disease: an MEG-based multiplex network study. Brain 140, 1466–1485.

Zeev-Wolf, M., Levy, J., Jahshan, C., Peled, A., Levkovitz, Y., Grinshpoon, A., et al., 2018. MEG resting-state oscillations and their relationship to clinical symptoms inschizophrenia. NeuroImage: Clin. 20, 753–761.

Zhang, X., Gao, J., Zhu, F., Wang, W., Fan, Y., Ma, Q., et al., 2018. Reduced white matter connectivity associated with auditory verbal hallucinations in first-episode and chronic schizophrenia: a diffusion tensor imaging study. Psychiatry Res. Neuroimaging 273, 63–70.

Zhao, T., Sheng, C., Bi, Q., Niu, W., Shu, N., Han, Y., 2017a. Age-related differences in the topological efficiency of the brain structural connectome in amnestic mild cognitive impairment. Neurobiol. Aging 59, 144–155.

Zhao, X., Tian, L., Yan, J., Yue, W., Yan, H., Zhang, D., 2017b. Abnormal rich-club organization associated with compromised cognitive function in patients with schizophrenia and their unaffected parents. Neurosci. Bull. 33 (4), 445–454.

Zhao, W., Guo, S., He, N., Yang, A.C., Lin, C.-P., Tsai, S.-J., 2018a. Callosal and subcortical white matter alterations in schizophrenia: a diffusion tensor imaging study at multiple levels. NeuroImage: Clin. 20, 594–602.

Zhao, Z., Li, X., Feng, G., Shen, Z., Li, S., Xu, Y., et al., 2018b. Altered effective connectivity in the default network of the brains of first-episode, drug-naïve schizophrenia patients with auditory verbal hallucinations. Front. Hum. Neurosci. 12, 456. https://doi.org/10.3389/fnhum.2018.00456.

Zhao, W., Guo, S., Linli, Z., Yang, A.C., Lin, C.-P., Tsai, S.-J., 2019. Functional, anatomical, and morphological networks highlight the role of basal ganglia–thalamus–cortex circuits in schizophrenia. Schizophr. Bull. https://doi.org/10.1093/schbul/sbz062.

Zott, B., Busche, M.A., Sperling, R.A., Konnerth, A., 2018. What happens with the circuit in Alzheimer's disease in mice and humans? Annu. Rev. Neurosci. 41, 277–297.

Zott, B., Simon, M.M., Hong, W., Unger, F., Chen-Engerer, H.-J., Frosch, M.P., et al., 2019. A vicious cycle of -amyloid–dependent neuronal hyperactivation. Science 365, 559–565.

List of abbreviations

A1	primary auditory cortex
A2	secondary auditory area
AAC	amplitude-amplitude coupling
Aβ	amyloid-β
ACC	anterior cingulate cortex
ACx	auditory cortex (general)
AD	Alzheimer's disease
AI	anterior insula
AM	amplitude modulation
ANF	auditory nerve fiber
ASD	autism spectrum disorder
ASSR	auditory steady state responses
ATN	anterior thalamic nucleus
BA	Brodman area
BOLD	blood oxygen level dependent
CAEP	cortical auditory evoked potentials
CC	corpus callosum
CF	characteristic frequency
CFC	cross-frequency coupling
CFS	cross-frequency synchrony
CON	cingulo-opercular network
CSD	current source density
CTC	communication through coherence
DAN	dorsal attention network DC dorsal caudate
DLPFC	dorsolateral prefrontal cortex
DMN	default mode network
DMPFC	dorsomedial profrontal cortex
DMTN	dorso-medial thalamic nucleus
DTI	diffusion tensor imaging
ECN	executive control network, see also TPN, TCN
ECoG	electrocorticography
EEG	electroencephalogram, electroencephalography
ERD	event-related decrease
ERP	event-related potential
ESC	extra-striate visual cortex
FA	fractional anisotropy
FC	functional connectivity
FCN	functional connectivity networks
fMRI	functional magnetic resonance imaging
FPN	fronto-parietal network
FTC	frequency-tuning curve
GA	gestational age
GABA	gamma-aminobutyric acid
GM	gray matter

GMV gray matter volume
HF hippocampal formation
HS Heschl's sulcus
Hz Hertz
IFG inferior frontal gyrus
IPC inferior parietal cortex
IPG inferior parietal gyrus
IPL inferior parietal lobule
ISI interstimulus interval
ITC inferior temporal cortex
LFP local field potential
LPS lagged phase synchronization
M1 primary motor cortex
MCI mild cognitive impairment
MDN mediodorsal nucleus of the thalamus
MEG magnetoencephalogram, magnetoencephalography
MFG middle frontal gyrus
MGB medial geniculate body
mGluR metabotropic glutamate receptor
mOFC medial orbitofrontal cortex
MOG middle occipital gyrus
mPFC medial prefrontal cortex
MRI magnetic resonance imaging
MRS magnetic resonance spectroscopy
MTL medial temporal lobe
MUA multiunit activity
NAc nucleus accumbens
NICU neonatal intensive care unit
NIH national institutes of health
NMDA *N*-methyl-D-aspartate
OFC orbitofrontal cortex
PAC phase-amplitude coupling
PC phase coding
PCC posterior cingulate cortex
PCUN precuneus
PD Parkinson's disease
PET positron emission tomography
PFC prefrontal cortex
PI posterior insula
PING principal interneuron gamma
PLA phase-locking angle
PLF phase-locking factor
PLI phase lag index
PLV phase-locking values
PMA postmenstrual age
PMC premotor cortex
PnC caudal pontine reticular nucleus
PoCG postcentral gyrus
PPC posterior parietal cortex
PSD power spectral density
PT planum temporale
PV parvalbumin

REM	rapid eye movement
RI	residual inhibition
ROI	region of interest
rs-FC	resting state functional connectivity
rs-fMRI	resting-state fMRI
REM	rapid eye movement
RSN	resting state network
rTMS	repetitive TMS
S1	primary somatosensory cortex
SAL	salience network
SCN	structural connectivity network
sgACC	subgenual ACC
SGC	subgenual cingulate cortex
SM	sensory memory
SMA	supplementary motor area
SMN	sensorimotor network
SNHL	sensorineural hearing loss
SNR	signal to noise ratio
SPG	superior parietal gyrus
SPL	sound pressure level
SR	stochastic resonance
STA	spike-triggered average
STG	superior temporal gyrus
STM	short-term memory
STRF	spectro-temporal receptive field
STS	superior temporal sulcus
SZ	schizophrenia
tACS	transcranial alternating current stimulation
TBS	theta-burst stimulation
TCD	thalamocortical dysrhythmia
TCN	task control network, see also TPN, ECN
TD	typically developing
tDCS	transcranial direct current stimulation
tFUS	transcranial focused ultrasound
TMS	transcranial magnetic stimulation
TPN	task-positive network, see also TCN, ECN
TRN	thalamic reticular nucleus
tRNS	transcranial random noise stimulation
tVNS	transcutaneous vagus nerve stimulation
UHL	unilateral hearing loss
V1	primary visual cortex
V2	secondary visual cortex
VAN	ventral attention network
VBM	voxel-based morphometry
VMPFC	ventromedial prefrontal cortex
WM	white matter

Index

Note: Page numbers followed by *f* indicate figures.

FSC
www.fsc.org
MIX
Papier | Fördert gute Waldnutzung
FSC® C083411

Kolibri 360 GmbH
Ferdinand-Jühlke-Straße 7
99095 Erfurt, Deutschland
produktsicherheit@kolibri360.de